Nonautonomous Dynamical Systems

Mathematical
Surveys
and
Monographs

Volume 176

Nonautonomous Dynamical Systems

Peter E. Kloeden

Martin Rasmussen

American Mathematical Society
Providence, Rhode Island

2010 *Mathematics Subject Classification.* Primary 37B55, 37C60, 37H05, 37B25, 37C75, 37D10, 37G35.

For additional information and updates on this book, visit
www.ams.org/bookpages/surv-176

Library of Congress Cataloging-in-Publication Data

Kloeden, Peter E.
Nonautonomous dynamical systems / Peter E. Kloeden, Martin Rasmussen.
p. cm. — (Mathematical surveys and monographs ; v. 176)
Includes bibliographical references and index.
ISBN 978-0-8218-6871-3 (alk. paper)
1. Dynamics. 2. Ergodic theory. 3. Stability. I. Rasmussen, Martin, 1975– II. Title.

QA845.K56 2011
515′.392—dc23

2011020550

∞ The paper used in this book is acid-free and falls within the guidelines
established to ensure permanence and durability.
Visit the AMS home page at http://www.ams.org/

10 9 8 7 6 5 4 3 2 1 16 15 14 13 12 11

Contents

Preface

Dynamical systems theory has been a very active area of research in mathematics and cognate fields for many years, but most of the results that have been obtained focus almost exclusively on autonomous dynamical systems. There have, of course, been investigations of nonautonomous differential equations, that is with time-dependent vector fields, during this time, but it is only in the recent decade that a theory of nonautonomous dynamical systems has emerged synergizing parallel developments on time-dependent differential equations, control systems and random dynamical systems. There are now abstract formulations of nonautonomous dynamical systems as two-parameter semi-groups or processes and as skew product flows as well as new concepts of nonautonomous attractors, in particular, pullback attractors.

This development is presented in this book for graduate students and others with a general background in dynamical systems and differential equations. The choice of topics and applications covered, especially in the later chapters, reflects the interests of the authors, but nevertheless provides a broad overview of important developments on the subject.

There are fifteen chapters and an appendix. The first chapter briefly reviews the theory of autonomous dynamical systems from the perspective of what is needed later rather than attempting to be comprehensive in itself. The process and skew product flow formalism of nonautonomous dynamical systems are introduced in the second chapter and the various concepts of nonautonomous attractors are presented and compared in the third chapter. These two chapters are essential reading for everything that is to follow. The remaining chapters can be read more or less independently of each other, except the fourth, fifth and sixth chapters on Morse decompositions, linear systems and invariant manifolds, respectively, which are best read as a sequential block. Lyapunov functions are considered in chapter seven and bifurcations in nonautonomous systems in chapter eight. Generalizations to set-valued nonautonomous dynamical systems and nonautonomous semi-dynamical systems are treated in chapters nine and ten, while the effects of perturbations and discretization are discussed in chapter eleven. Up to here the state space is either $\mathbb{R}^d$ or a general complete metric space, but in chapter twelve issues of explicit relevance to infinite-dimensional state spaces are considered. Chapter thirteen applies previous results to switching and affine control systems interpreted as nonautonomous dynamical systems, while chapter fourteen introduces readers to some of the differences arising in random dynamical systems due to their measure-theoretic rather than topological characteristics. The previous deterministic and random results are then applied to the synchronization of dissipative systems in chapter fifteen.

Finally, various background definitions and results needed within the text are given in the appendix.

Readers who are interested in the dynamical behavior of nonautonomous partial differential equations and evolution equations are advised to refer to the monographs of Carvalho, Langa & Robinson **[35]** and Chepyzhov & Vishik **[43]** in conjunction with this book.

Acknowledgements. We are indebted to numerous colleagues for their helpful discussions during the preparation of this book. In particular, we thank Tomás Caraballo, Alexandre Carvalho, David Cheban, Hans Crauel, Jinquiao Duan, Messoud Effendiev, Barnabas Garay, Peter Giesl, Arnulf Jentzen, Victor Kozyakin, Jeroen Lamb, José Langa, Li Desheng, Thomas Lorenz, Pedro Marín-Rubio, Christian Pötzsche, José Real, Janosch Rieger, James Robinson, Björn Schmalfuß, Stefanie Sonner, Aneta Stefanovska, Meihua Yang for carefully reading parts of the book and for their suggestions for improvements, as well as Fritz Colonius and Christoph Kawan for advice on control systems as skew product flows. In addition, we thank Alexandre Carvalho, José Langa and James Robinson for keeping us informed about developments with preparation of their book **[35]** entitled *Attractors of Infinite Dimensional Nonautonomous Dynamical Systems* and their useful comments about our manuscript. We also thank Sofie van Geene for providing Figure 2.1 and Philipp Storck for providing Figures 3.1 and 3.2 in the book. Finally, we would like to thank both Karin and Eva-Maria for their encouragement during the time we wrote this book.

Peter Kloeden thanks the *Departamento de Ecuaciones Diferenciales y Análisis Numérico* at the Universidad de Sevilla for its hospitality over many years, especially during the summer semester of 2009 when much of this book was written. In particular, the financial support of the following grants is gratefully acknowledged: Programa de Movilidad del Profesorado universitario español y extranjero grant SAB2004-0146, the Ministerio de Ciencia e Innovación (Spain) grant MTM2008-00088 and the Junta de Andalucía grant P07-FQM-02468. The final parts of this book were written during an extended stay by the first coauthor at the *Isaac Newton Institute for Mathematical Sciences* at the University of Cambridge during the first half of 2010. Its financial support and congenial working atmosphere are gratefully acknowledged.

Martin Rasmussen thanks both the *Marie Curie* and *EPSRC Career Acceleration Fellowship* for its financial support and *Imperial College London* for providing a stimulating environment.

Peter Kloeden, Frankfurt am Main Martin Rasmussen, London

CHAPTER 1

Autonomous dynamical systems

Autonomous dynamical systems have been investigated intensively for over a century, leading to a well-established dynamical systems theory. Key ideas and results are briefly reviewed in this chapter as background for the development of the theory of nonautonomous dynamical systems that will be presented in this book.

1. Introduction

An autonomous ordinary differential equation is given by

$$\dot{x} := \frac{\mathrm{d}x}{\mathrm{d}t} = f(x)\,, \tag{1.1}$$

where $f : \mathbb{R}^d \to \mathbb{R}^d$ is assumed to be continuous. A continuously differentiable function $x : \mathbb{I} \to \mathbb{R}^d$, where $\mathbb{I} \subset \mathbb{R}$ is an interval, is said to solve (1.1) if $\dot{x}(t) := \frac{\mathrm{d}x}{\mathrm{d}t}(t) = f(x(t))$ for all $t \in \mathbb{I}$. The combination of (1.1) with an initial condition $x(t_0) = x_0$ is called an *initial value problem*, and a solution x of (1.1) is said to solve this initial value problem if $x(t_0) = x_0$.

Assuming global existence and uniqueness of solutions, the solutions form a mapping

$$(t, t_0, x_0) \mapsto x(t, t_0, x_0) \in \mathbb{R}^d \quad \text{for all } (t, t_0, x_0) \in \mathbb{R} \times \mathbb{R} \times \mathbb{R}^d\,.$$

This mapping is called the *general solution* of the differential equation (1.1).

The situation is actually much simpler. For example, the scalar initial value problem

$$\dot{x} = -x\,, \qquad x(t_0) = x_0\,,$$

has the general solution

$$t \mapsto x(t, t_0, x_0) = x_0\, e^{-(t-t_0)}\,,$$

which depends only on the *elapsed time* $t - t_0$ since starting and not separately on the actual time t and the starting time t_0. Another way of saying this is that the solutions are *translation invariant in time*, i.e., the solution mapping satisfies

$$x(t - t_0, 0, x_0) = x(t, t_0, x_0) \quad \text{for all } (t_0, x_0) \in \mathbb{R} \times \mathbb{R}^d\,. \tag{1.2}$$

This is, in fact, a general property of the solutions of autonomous ordinary differential equations.

THEOREM 1.1 (Translation invariance). *Assume that the solutions of the autonomous ordinary differential equation* (1.1) *exist and are unique forward in time. Then they are translation invariant in time, i.e.,* (1.2) *holds.*

PROOF. Write $\mu_1(t) = x(t, t_0, x_0)$ and $\mu_2(t) = x(t - t_0, 0, x_0)$. Then $\mu_1(t_0) = \mu_2(t_0) = x_0$ and

$$\begin{aligned}\frac{\mathrm{d}}{\mathrm{d}t}\mu_2(t) &= \frac{\mathrm{d}}{\mathrm{d}t}x(t - t_0, 0, x_0) = \frac{\partial x}{\partial t}(t - t_0, 0, x_0) \cdot \frac{\mathrm{d}}{\mathrm{d}t}(t - t_0) \\ &= f\big(x(t - t_0, 0, x_0)\big) = f(\mu_2(t)),\end{aligned}$$

so $\mu_2(t) = x(t - t_0, 0, x_0)$ satisfies the same initial value problem as $\mu_1(t) = x(t, t_0, x_0)$. By the uniqueness assumption, $\mu_2(t) \equiv \mu_1(t)$. $\square$

Thus, in discussing autonomous ordinary differential equations, one can restrict attention to the initial time $t_0 = 0$ and write the solutions as $x(t, x_0)$. Then the solution mapping $\phi : \mathbb{R} \times \mathbb{R}^d \to \mathbb{R}^d$, defined by $\phi(t, x_0) := x(t, x_0)$, is (at least) continuous in both variables and satisfies the *initial value condition*

$$\phi(0, x_0) = x_0 \quad \text{for all } x_0 \in \mathbb{R}^d$$

and the *group property*

$$\phi(s + t, x_0) = \phi(s, \phi(t, x_0)) \quad \text{for all } s, t \in \mathbb{R} \text{ and } x_0 \in \mathbb{R}^d,$$

which is a consequence of the uniqueness of solutions.

1.1. Dynamical systems. The mapping ϕ from above is a special case of a dynamical system, which will be introduced as follows.

DEFINITION 1.2 (Dynamical system). Let $\mathbb{T} = \mathbb{R}$ or $\mathbb{T} = \mathbb{Z}$, respectively, and let X be a metric space. A *dynamical system* is a continuous function $\phi : \mathbb{T} \times X \to X$ with the following two properties:

(i) *Initial value condition.* $\phi(0, x_0) = x_0$ for all $x_0 \in X$.
(ii) *Group property.* $\phi(s + t, x_0) = \phi(s, \phi(t, x_0))$ for all $s, t \in \mathbb{T}$ and $x_0 \in X$.

If $\mathbb{T} = \mathbb{R}$, then the dynamical system is called *continuous*, and in case $\mathbb{T} = \mathbb{Z}$, one has a *discrete* dynamical system.

The second property of a dynamical system is called *group property* because the family of mappings $\{\phi(t, \cdot) : t \in \mathbb{T}\}$ of X into itself forms a group under composition.

1.2. Semi-dynamical systems. Let X be a metric space and $f : X \to X$ be a continuous function. Successive iteration of the autonomous difference equation

$$x_{n+1} = f(x_n) \tag{1.3}$$

in $\mathbb{R}^d$ then generates a solution mapping $\phi : \mathbb{N}_0 \times \mathbb{R}^d \to \mathbb{R}^d$ defined by

$$\phi(n, x_0) := x_n = f^n(x_0) = \underbrace{f \circ f \circ \cdots \circ f}_{n \text{ times}}(x_0).$$

One easily verifies the initial value condition $\phi(0, x_0) = x_0$ and the group property

$$\phi(n, \phi(m, x_0)) = f^n(\phi(m, x_0)) = f^n \circ f^m\,(x_0) = f^{n+m}\,(x_0) = \phi(n + m, x_0)$$

for all $x_0 \in \mathbb{R}^d$ and $n, m \in \mathbb{N}_0$. Although these two conditions are fulfilled, the mapping ϕ is not a discrete time dynamical system as described above, since the elapsed time n is supposed to be a natural number, rather than being an integer. If the function f is a homeomorphism, however, ϕ can be extended to be a dynamical system, which will be described below. Note first that ϕ is a semi-dynamical system whose definition is given as follows.

DEFINITION 1.3 (Semi-dynamical system). Let $\mathbb{T} = \mathbb{R}$ or $\mathbb{T} = \mathbb{Z}$, respectively, and let X be a metric space. A *semi-dynamical system* is a continuous function $\phi : \mathbb{T}_0^+ \times X \to X$ with the following two properties:

(i) *Initial value condition.* $\phi(0, x_0) = x_0$ for all $x_0 \in X$.
(ii) *Semi-group property.* $\phi(s + t, x_0) = \phi(s, \phi(t, x_0))$ for all $s, t \in \mathbb{T}_0^+$ and $x_0 \in X$.

The family of mappings $\{\phi(t, \cdot) : t \in \mathbb{T}_0^+\}$ of X into itself forms a semi-group under composition rather than a group. When the function f of (1.3) is a homeomorphism, i.e., it is continuous and invertible with continuous inverse mapping f^{-1}, then ϕ can be extended to $n \in \mathbb{Z}^-$ by

$$\phi(n, x) = f^{-n}(x) := (f^{-1})^{-n}(x) = \underbrace{f^{-1} \circ f^{-1} \circ \cdots \circ f^{-1}}_{-n \text{ times}}(x) \quad \text{for all } x \in \mathbb{R}^d .$$

The mapping ϕ then defines a discrete dynamical system, and the family of mappings $\{\phi(t, \cdot) : t \in \mathbb{Z}\}$ is a group under composition.

1.3. Examples. The examples of dynamical systems and semi-dynamical systems in this subsection will be required later in various contexts. They are mainly infinite-dimensional.

EXAMPLE 1.4. Consider the Banach space ℓ_∞ of bounded real-valued sequences $x = (x_1, x_2, \dots)$ with the norm

$$\|x\|_\infty = \sup_{i \in \mathbb{N}} |x_i| \quad \text{for all } x = (x_1, x_2, \dots) \in \ell_\infty .$$

The shift operator θ on ℓ_∞ defined componentwise by

$$(\theta(x))_i = x_{i+1} \quad \text{for all } i \in \mathbb{N}$$

generates a discrete semi-dynamical system ϕ on ℓ_∞, where

$$\phi(0, x) = x \quad \text{and} \quad \phi(n, x) = \theta^n(x) \quad \text{for all } n \in \mathbb{N} \text{ and } x \in \ell_\infty ,$$

which means that $(\phi(n, x))_j = x_{j+n}$ for all $j \in \mathbb{N}$. The corresponding shift operator on the Banach space $\ell_\infty^\pm$ of two-sided bounded real-valued sequences $x = (\dots, x_{-2}, x_{-1}, x_0, x_1, x_2, \dots)$ with the norm defined by $\|x\|_\infty = \sup_{i \in \mathbb{Z}} |x_i|$ has an inverse and forms a group or a (two-sided) discrete time dynamical system on this space (which can be interpreted as, i.e., identified with, a subspace of ℓ_∞).

An analogous construction gives a continuous time dynamical system.

EXAMPLE 1.5. The Banach space $C(\mathbb{R}, \mathbb{R}^d)$ of bounded continuous functions $x : \mathbb{R} \to \mathbb{R}^d$ with the norm

$$\|x\|_\infty = \sup_{t \in \mathbb{R}} \|x(t)\| \quad \text{for all } x \in C(\mathbb{R}, \mathbb{R}^d)$$

is also an infinite-dimensional space. The family of shift operators $(\theta_t)_{t \in \mathbb{R}}$ on $C(\mathbb{R}, \mathbb{R}^d)$, defined by

$$\theta_t(x(\cdot)) = x(t + \cdot) \quad \text{for all } t \in \mathbb{R} ,$$

generates a continuous time dynamical system ϕ on $C(\mathbb{R}, \mathbb{R}^d)$, where

$$\phi(0,x) = x \quad \text{and} \quad \phi(t,x) = \theta_t(x) \quad \text{for all } t \in \mathbb{R}.$$

The next example shows that autonomous delay differential equations generate infinite-dimensional semi-dynamical systems.

EXAMPLE 1.6. Assuming existence and uniqueness of solutions forward in time, the solutions $\phi(t) = x(t, \mu(\cdot))$ with values in $\mathbb{R}^d$ for an autonomous delay differential equation

$$\dot{x} = f(x(t-1)), \qquad x(s) = \mu(s), \quad \text{for all } -1 \leq s \leq 0,$$

in $\mathbb{R}^d$ form a semi-dynamical system ϕ on the function space $C([-1,0], \mathbb{R}^d)$ of bounded continuous functions $\mu : [-1,0] \to \mathbb{R}^d$, which is a Banach space under the norm

$$\|\mu\|_\infty = \sup_{t\in[-1,0]} \|\mu(t)\|.$$

Here $\phi(t, \mu(\cdot)) \in C([-1,0], \mathbb{R}^d)$ is defined as $\phi(t, \mu(\cdot))(s) := \phi(t+s, \mu(\cdot))$ for $t \geq 0$ and $-1 \leq s \leq 0$.

Finally, discretizations of autonomous differential equations generate discrete time semi-dynamical systems.

EXAMPLE 1.7. The solution mapping $\phi(t,x)$ of the autonomous ordinary differential equation (1.1) in $\mathbb{R}^d$ with existence and uniqueness of all solutions forward in time forms a continuous time semi-dynamical system on $\mathbb{R}^d$. This equation is discretized by considering the *Euler scheme*

$$x_{n+1} = f_h(x_n) := x_n + h\, f(x_n)$$

with constant stepsize $h > 0$. Successive iteration of f_h gives a discrete time semi-dynamical system on $\mathbb{R}^d$.

2. Local asymptotic behavior

2.1. Invariance. The local long-term dynamical behavior of a semi-dynamical system is characterized by its invariant sets.

DEFINITION 1.8 (Invariance). Let $\phi : \mathbb{T}_0^+ \times X \to X$ be a semi-dynamical system on a metric space X. A subset M of X is called *invariant under* ϕ, or *ϕ-invariant*, if

$$\phi(t, M) = M \quad \text{for all } t \in \mathbb{T}_0^+. \tag{1.4}$$

It is called *positively invariant* if $\phi(t,M) \subset M$ for all $t \in \mathbb{T}_0^+$ and *negatively invariant* if $M \subset \phi(t,M)$ for all $t \in \mathbb{T}_0^+$.

If $\phi : \mathbb{T} \times X \to X$ is a dynamical system, then invariance of a set M implies that $\phi(t,M) = M$ for all $t \in \mathbb{T}$. For a discrete time semi-dynamical system generated by the autonomous difference equation (1.3), invariance is equivalent to $f(M) = M$, and positive (resp. negative) invariance is equivalent to $f(M) \subset M$ (resp. $M \subset f(M)$).

Simple examples of invariant sets are given by equilibria and periodic solutions. In the first case, the invariant set consists of a single point, which, for a discrete time system, must be a fixed point of the mapping f, while for a solution with primitive period $r \in \mathbb{N}$, it consists of a finite set of r distinct points $\{p_1, \dots, p_r\}$, which are fixed points of the composite mapping f^r (but not for an f^j with j smaller than r).

Invariant sets can also be much more complicated, for instance, they can be fractal sets, and the dynamics on them may be chaotic. Important classes of invariant sets are given by the ω- or α-limit sets. In particular, such sets characterize the asymptotic behavior.

DEFINITION 1.9 (ω- and α-limit sets). Let $\phi : \mathbb{T}_0^+ \times X \to X$ be a semi-dynamical system on a metric space X. Then the *ω-limit set* of a point $\xi \in X$ is defined by

$$\omega(\xi) := \big\{x \in X \ : \text{there exists a sequence } (t_j)_{j\in\mathbb{N}} \text{ in } \mathbb{T}_0^+ \\ \text{such that } t_j \to \infty \text{ and } \phi(t_j, \xi) \to x\big\}\,.$$

If $\phi : \mathbb{T} \times X \to X$ is even a dynamical system, then the α-limit set of a point $\xi \in X$ is defined by

$$\alpha(\xi) := \big\{x \in X \ : \text{there exists a sequence } (t_j)_{j\in\mathbb{N}} \text{ in } \mathbb{T}_0^- \\ \text{such that } t_j \to -\infty \text{ and } \phi(t_j, \xi) \to x\big\}\,.$$

The proof of the following basic properties of ω-limit sets is left as an exercise.

EXERCISE 1.10. Suppose that the forwards trajectory $\{\phi(t, \xi) : t \in \mathbb{T}_0^+\}$ is pre-compact. Show that ω-limit sets are nonempty, compact and ϕ-invariant. Furthermore, prove that ω-limit sets of continuous time semi-dynamical systems are connected sets, and demonstrate that they need not be connected for a discrete time semi-dynamical system.

It is easy to check that the relations

$$\omega(\xi) = \bigcap_{\tau\geq 0} \overline{\bigcup_{t\geq\tau} \{\phi(t, \xi)\}} \quad \text{and} \quad \alpha(\xi) = \bigcap_{\tau\leq 0} \overline{\bigcup_{t\leq\tau} \{\phi(t, \xi)\}}$$

hold. This motivates the definition of ω- and α-limit sets of subsets $M \subset X$, which are given by

$$\omega(M) := \bigcap_{\tau\geq 0} \overline{\bigcup_{t\geq\tau} \phi(t, M)} \quad \text{and} \quad \alpha(M) := \bigcap_{\tau\leq 0} \overline{\bigcup_{t\leq\tau} \phi(t, M)}\,.$$

EXERCISE 1.11. Find an example of a semi-dynamical system $\phi : \mathbb{T}_0^+ \times X \to X$ and a set $M \subset X$ such that $\omega(M) \neq \bigcup_{x\in M} \omega(x)$.

2.2. Lyapunov stability and attractivity. What can happen in the vicinity of invariant sets is an important indication of long-term dynamical behavior, for example, if nearby solutions stay close to the invariant set or if they go away. In the first case, the invariant set is called *stable*. Stability was extensively investigated by the Russian mathematician and engineer LYAPUNOV in his thesis [**160**] in the 1890s, see [**161, 162**] for French and English translations. The basic definitions of this theory will be briefly reviewed here. More details can be found in the classical books of CESARI [**37**] and HAHN [**89**] for differential equations and in AGARWAL [**3**] for difference equations.

DEFINITION 1.12 (Lyapunov stability). An invariant subset M of a metric space (X, d) is called *Lyapunov stable* for a semi-dynamical system ϕ on X if for every $\varepsilon > 0$, there exists a $\delta = \delta(\varepsilon) > 0$ such that

$$\operatorname{dist}(\phi(t,\xi), M) < \varepsilon \quad \text{for all } t \in \mathbb{T}^+ \text{ whenever } \operatorname{dist}(\xi, M) < \delta\,.$$

Here dist denotes the Hausdorff semi-distance as introduced in the Appendix.

It follows immediately from this definition that if a set M is Lyapunov stable, then so is its closure $\overline{M}$.

It is interesting to compare the Lyapunov stability of an equilibrium with the concept of continuity in initial conditions: the former holds uniformly over the whole future time set $\mathbb{T}_0^+$, whereas the latter holds only uniformly on compact time sets. This is clearly seen in the scalar ordinary differential equation

$$\dot{x} = x\,,$$

which generates the dynamical system $\phi(t, x) = xe^t$ with the steady state solution $\phi(t, 0) \equiv 0$. The mapping $x \mapsto \phi(t, x)$ is continuous at $x = 0$ uniformly on each bounded time interval $[0, T]$, i.e., $|x - 0| = |x| < \delta(\varepsilon, T)$ implies that

$$|\phi(t,x) - \phi(t,0)| = |\phi(t,x)| < \varepsilon \quad \text{for all } t \in [0, T]$$

with $\delta(\varepsilon, T) := \varepsilon e^{-T}$. However, $\lim_{T\to\infty} \delta(\varepsilon, T) = 0$ and the above estimates cannot be extended to all $t \in \mathbb{R}_0^+$. The zero solution is obviously not Lyapunov stable.

DEFINITION 1.13 (Lyapunov attractivity). An invariant subset M of a metric space (X, d) is called *Lyapunov attractive* for a semi-dynamical system ϕ on X if there exists a $\delta > 0$ such that

$$\lim_{t\to\infty} \operatorname{dist}(\phi(t,x), M) = 0 \quad \text{for all } x \in X \text{ with } \operatorname{dist}(x, M) < \delta\,.$$

If the invariant subset M is both Lyapunov stable and Lyapunov attractive, then M is called *Lyapunov asymptotically stable.*

The following exercise shows that Lyapunov stability and Lyapunov attractivity are generally independent concepts. An except is for scalar systems and linear systems in higher dimensions, where Lyapunov stable is a consequence of attractivity.

EXERCISE 1.14. Show that the trivial equilibrium of the system

$$\dot{r} = r(1 - r)\,, \qquad \dot{\varphi} = \sin^2\left(\frac{\varphi}{2}\right)\,.$$

is Lyapunov attractive but not Lyapunov stable.

This motivates a stronger definition of a local attractor which combines both stability and attractivity. It will be given after Lyapunov functions have been introduced.

2.3. Lyapunov functions. The Lyapunov stability of invariant sets can be characterized by Lyapunov functions. There are many theorems which show that the existence of a suitable Lyapunov function is necessary or sufficient, or both, for the asymptotic stability of an invariant set. The following is typical for a sufficient condition. Other theorems with proofs are given in Chapter 7.

THEOREM 1.15 (Lyapunov function implies stability). *Suppose that M is a nonempty compact invariant set for a semi-dynamical system ϕ on a complete metric space (X,d). Suppose also that there exists a continuous mapping $V : X \to \mathbb{R}_0^+$ with the properties:*

(i) *There exist continuous strictly increasing functions α, $\beta : \mathbb{R}^+ \to \mathbb{R}^+$ such that*

$$\alpha(\operatorname{dist}(x,M)) \le V(x) \le \beta(\operatorname{dist}(x,M)) \quad \textit{for all } x \in X\,,$$

(ii) *$V(\phi(t,x)) \le V(x)$ for all $x \notin M$ and $t > 0$.*

Then the set M is Lyapunov stable under ϕ.

Lyapunov functions are like a nonlinear measure of distance from the invariant set M. In mechanics, the graph of a Lyapunov function can be interpreted as an energy basin (in fact, it often is exactly this) with the local attractor at its base at the zero energy level and with the energy not increasing along trajectories of the system outside the local attractor.

2.4. Local attractors and repellers. A local attractor of a semi-dynamical system is an invariant compact set which attracts all trajectories starting in some neighborhood of the attractor in forward time, while a local repeller is an invariant compact set which attracts trajectories in backward time.

DEFINITION 1.16 (Local attractor and repeller). Let $\phi : \mathbb{T}_0^+ \times X \to X$ be a semi-dynamical system on a metric space (X,d). An invariant compact set $A \subset X$ is called a *local attractor* of ϕ if there exists an $\eta > 0$ such that

$$\lim_{t\to\infty} \operatorname{dist}\big(\phi(t, B_\eta(A)), A\big) = 0\,.$$

If $\phi : \mathbb{T} \times X \to X$ is a dynamical system, then an invariant compact set $R \subset X$ is called a *local repeller* if there exists an $\eta > 0$ such that

$$\lim_{t\to\infty} \operatorname{dist}\big(\phi(-t, B_\eta(R)), R\big) = 0\,.$$

A dynamical system can have infinitely many local attractors and repellers.

EXAMPLE 1.17. Consider the differential equation

$$\dot{x} = \sin(\pi x)\,,$$

and let $b \in \mathbb{N}_0$ be an even number. Then, for all even numbers $a \in \mathbb{Z}$, the set $[a,b]$ is a local attractor and, for all odd numbers a, the set $[a, a+b]$ is a local repeller.

Since repulsion means attraction backwards in time, it makes sense to change the direction of time in a dynamical system.

DEFINITION 1.18 (Dynamical system under time reversal). Let $\phi : \mathbb{T} \times X \to X$ be a dynamical system on a metric space (X,d). Then the corresponding *dynamical system under time reversal* $\phi^{-1} : \mathbb{T} \times X \to X$ is defined by

$$\phi^{-1}(t,x) := \phi(-t,x) \quad \text{for all } (t,x) \in \mathbb{T} \times X\,.$$

The dynamical system under time reversal is indeed a dynamical system in the sense of Definition 1.2, and there is thus a direct duality between local attractors and repellers.

PROPOSITION 1.19 (Duality of attractor and repeller). *Let $\phi : \mathbb{T} \times X \to X$ be a dynamical system on a metric space (X, d). Then an invariant compact set $M \subset X$ is a local attractor of ϕ if and only if M is a local repeller of ϕ^{-1}. Similarly, an invariant compact set $M \subset X$ is a local repeller of ϕ if and only if M is a local attractor of ϕ^{-1}.*

When the metric space X is compact, local attractors and repellers can be defined as ω- and α-limit sets of their neighborhoods.

PROPOSITION 1.20 (Characterization of local attractors). *Let $\phi : \mathbb{T}_0^+ \times X \to X$ be a semi-dynamical system on a compact metric space (X, d), $\eta > 0$ and $A \subset X$ be compact and invariant. Then the following statements are equivalent:*

(i) *A is a local attractor.*
(ii) *There exists an $\eta > 0$ such that $\omega(B_\eta(A)) = A$.*

A similar assertion holds for local repellers of the system and α-limit set of neighborhoods of the repellers.

PROOF. (i) $\Rightarrow$ (ii): Since A is a local attractor, there exists an $\eta > 0$ such that $\lim_{t\to\infty} \operatorname{dist}\big(\phi(t, B_\eta(A)), A\big) = 0$. Assume to the contrary that $\omega(B_\eta(A)) \neq A$, which implies that $\omega(B_\eta(A)) \setminus A \neq \emptyset$, since A is invariant. Hence, there exist sequences $(x_n)_{n\in\mathbb{N}}$ in $B_\eta(A)$ and $(t_n)_{n\in\mathbb{N}}$ in $\mathbb{T}_0^+$ such that $\lim_{n\to\infty} \phi(t_n, x_n) = x \notin A$. Since $\operatorname{dist}(x, A) > 0$, this contradicts $\lim_{t\to\infty} \operatorname{dist}\big(\phi(t, B_\eta(A)), A\big) = 0$.
(ii) $\Rightarrow$ (i): Assume to the contrary that there exists an $\varepsilon > 0$ and sequences $(x_n)_{n\in\mathbb{N}}$ in $B_\eta(A)$ and $(t_n)_{n\in\mathbb{N}}$ in $\mathbb{T}_0^+$ such that $\lim_{n\to\infty} t_n = \infty$ and $\operatorname{dist}(\phi(t_n, x_n), A) \geq \varepsilon$ for all $n \in \mathbb{N}$. Since X is compact, a subsequence of $(\phi(t_n, x_n))_{n\in\mathbb{N}}$ is convergent, and the limit lies in $\omega(B_\eta(A)) = A$. This is a contradiction and finishes the proof of this proposition. □

EXERCISE 1.21. Construct an example of a dynamical system on a noncompact phase space such that the equivalence in Proposition 1.20 does not hold.

2.5. Existence of local attractors and repellers. The existence and approximate location of a maximal local attractor follows from that of more easily found locally absorbing sets, which typically have a convenient simpler shape such as a ball or ellipsoid.

DEFINITION 1.22 (Locally absorbing set). Let $\phi : \mathbb{T}_0^+ \times X \to X$ be a semi-dynamical system on a metric space (X, d). A nonempty compact subset B of X is called a *locally absorbing set* of ϕ if there exists an $\eta > 0$ and a $T \in \mathbb{T}^+$ such that $\phi(t, B_\eta(B)) \subset B$ for all $t \geq T$.

Absorbing sets are often called *attracting sets* when they are also *positively invariant*, i.e., with $\phi(t, B) \subseteq B$ for all $t \in \mathbb{T}^+$. For example, this holds for a difference equation (1.3) if $f(B) \subset B$.

Attractors differ from attracting sets in that they consist entirely of limit points of the system and are thus strictly invariant in the sense of (1.4). However, every local attractor has an absorbing set as superset. This can be seen by looking at a local attractor A which attracts an η-neighborhood of itself, i.e., one has $\lim_{t\to\infty} \operatorname{dist}\big(\phi(t, B_\eta(A)), A\big) = 0$. Then the set $B_{\eta/2}(A)$ is an absorbing set.

The converse also holds, as the following fundamental theorem shows.

THEOREM 1.23 (Existence of local attractors). *Suppose that a semi-dynamical system ϕ on X has a locally absorbing set B. Then ϕ has a local attractor $A \subset B$ given by the ω-limit set*

$$A = \omega(B) = \bigcap_{\tau\geq 0} \overline{\bigcup_{t\geq\tau} \phi(t, B)},$$

or simply by $A = \bigcap_{t\geq 0} \phi(t, B)$ when B is positively invariant.

PROOF. For simplicity, the proof will be given for the case of a positively locally invariant absorbing set. Then $\phi(t, B) \subset \phi(s, B)$ for all $t \geq s$ and $\overline{\bigcup_{t\geq s} \phi(t, B)} = \phi(s, B)$. Moreover, $\phi(t, B)$ is compact for all $t \geq 0$, since B is compact and $x \mapsto \phi(t, x)$ is continuous. Thus, as the intersection of a nested family of nonempty compact subsets, $A := \bigcap_{t\geq 0} \phi(t, B)$ is a nonempty compact subset.
Note that $a \in A$ if and only if there are sequences $b_n \in B$ and $t_n \to \infty$ such that $\phi(t_n, b_n) \to a$ as $n \to \infty$. Consider any point $a \in A$ and $T > 0$. Then $\phi(T + t_n, b_n) = \phi(T, \phi(t_n, b_n)) \to \phi(T, a)$ as $n \to \infty$ by continuity. But $b_n \in B$ and $T + t_n \to \infty$, so $\phi(T, a) \in A$ and any $T > 0$. Since $a \in A$ was otherwise arbitrary, this says that $\phi(T, A) \in A$, i.e., A is positively invariant. Consider again any point $a \in A$ and sequences $b_n \in B$ and $t_n \to \infty$ such that $\phi(t_n, b_n) \to a$ as $n \to \infty$. Let $T > 0$ be fixed and define $b_n^* := \phi(t_n - T, b_n)$ for all n for which $t_n \geq T$. Note that $\phi(t_n - T, b_n) \in B$ by the positive invariance of B. Since B is compact, there exists a convergent subsequence $b_{n'}^* \to b^* \in B$ as $n' \to \infty$. By continuity

$$\phi(t_{n'}, b_{n'}) = \phi(T, \phi(t_{n'} - T, b_{n'})) = \phi(T, b_{n'}^*) \to \phi(T, b^*) \quad \text{as } n' \to \infty.$$

However, $\phi(t_{n'}, b_{n'}) \to a$ as $n' \to \infty$, so $\phi(T, b^*) = a$. But $b^* \in A$ since there are (sub)sequences $b_n' \in B$ and $t_n' - T \to \infty$ such that $\phi(t_{n'} - T, b_{n'}) \to b^*$. Thus, $a \in \phi(T, A)$, and since $a \in A$ was otherwise arbitrary, it follows that $A \subset \phi(T, A)$, i.e., A is negatively invariant. Combining the results gives the invariance of A.
To show that A attracts a neighborhood of itself, note that $A \subset B$ which follows from the definition of a locally absorbing set. In particular, this means that it suffices to show that A attracts $B_\eta(B)$. Moreover, the definition of a locally absorbing set also means that showing that A attracts the absorbing set B finishes the proof. Suppose that this is not true. Then there exist sequences $b_n \in B$ and $t_n \to \infty$ and a constant $\varepsilon_0 > 0$ such that

$$\operatorname{dist}(\phi(t_n, b_n), A) \geq \varepsilon_0 > 0 \quad \text{for all } n \in \mathbb{N}. \tag{1.5}$$

Define $c_n^* := \phi(t_n, b_n) \in B$. Since B is compact there is a convergent subsequence $c_{n'}^* = \phi(t_{n'}, b_{n'}) \to c^* \in B$ as $n' \to \infty$. Clearly, $c^* \in A$ and

$$0 = \operatorname{dist}(c^*, A) = \lim_{n'\to\infty} \operatorname{dist}(\phi(t_{n'}, b_{n'}), A),$$

which contradicts (1.5). Hence, A attracts B. □

The existence of a local repeller follows from Theorem 1.23 via time reversal.

THEOREM 1.24 (Existence of local repellers). *Suppose that a dynamical system ϕ on X has a* locally repelling set B, *i.e., there exists an $\eta > 0$ and a $T \in \mathbb{T}^+$ such that $\phi(-t, B_\eta(B)) \subset B$ for all $t \geq T$ in $\mathbb{T}^+$. Then ϕ has a local repeller $R \subset B$ given by the α-limit set*

$$R = \alpha(B) = \bigcap_{\tau \leq 0} \overline{\bigcup_{t \leq \tau} \phi(t, B)},$$

or simply by $A = \bigcap_{t \leq 0} \phi(t, B)$ when B is negatively invariant.

2.6. Invariant sets and entire solutions. Invariant sets provide useful information about the behavior of a dynamical system even when they are not local attractors. This also holds for positive and negative invariant sets, see Definition 1.8. In particular, such sets contain entire solutions of the system.

DEFINITION 1.25 (Entire solution). A mapping $\xi : \mathbb{T} \to X$ is called an *entire solution* of a semi-dynamical system $\phi : \mathbb{T}_0^+ \times X \to X$ on a metric space (X, d) if

$$\xi(t) = \phi(t - s, \xi(s)) \quad \text{for all } s, t \in \mathbb{T} \text{ with } s \leq t. \tag{1.6}$$

A semi-dynamical system cannot always be extended uniquely (or even at all) backwards in time, so entire solutions need not exist. However, this is possible for invariant sets.

LEMMA 1.26. *Let $B \subset X$ be a nonempty ϕ-invariant subset of a semi-dynamical system $\phi : \mathbb{T}_0^+ \times X \to X$, i.e., $\phi(t, B) = B$ for all $t \in \mathbb{T}_0^+$. Then for any $a_0 \in B$, there exists an entire solution ξ through a_0 which is contained in B, i.e., with $\xi(t) \in B$ for all $t \in \mathbb{T}$.*

PROOF. This is easy to see in the forward direction, by defining $\xi(t) := \phi(t, b_0)$ for $t \in \mathbb{T}_0^+$, so in particular $\xi(0) = b_0$. For negative times, note that, since $\phi(1, B) = B$, there is a point $b_{-1} \in B$ such that $\phi(1, b_{-1}) = b_0$ and define $\xi(t) := \phi(t+1, b_{-1})$ for $t \in [-1, 0]$, which is in B by the invariance of B. Then repeat this construction on each subinterval $[-n-1, -n]$ for $n \geq 1$ defining $\xi(t) := \phi(t + n + 1, b_{-n+1})$ for $t \in [-n-1, -n]$ where $b_{-n-1} \in B$ is chosen so that $\phi(1, b_{-n-1}) = b_{-n}$. Finally, use the semi-group property to show that ξ satisfies (1.6). □

A similar result holds also for nonempty positive or negative invariant compact subsets, since by the following results they contain a nonempty invariant subset. Their proofs require the next lemma, the proof of which is left to the reader.

LEMMA 1.27. *Let $(B_n)_{n \in \mathbb{N}}$ be a decreasing nested family of nonempty compact subsets of a complete metric space (X, d). Then*

$$B_\infty = \bigcap_{n \in \mathbb{N}} B_n$$

is a nonempty compact set, and

$$\lim_{n \to \infty} \operatorname{dist}(B_n, B_\infty) = 0.$$

THEOREM 1.28. *Let B be a nonempty compact subset of X which is ϕ-positively invariant, i.e., $\phi(t,B) \subset B$ for all $t \in \mathbb{T}_0^+$. Then there exists a nonempty compact subset B_∞ of B which is ϕ-invariant, i.e., $\varphi(t,B_\infty) = B_\infty$ for all $t \in \mathbb{T}_0^+$.*

PROOF. Since B is compact and $\phi(t,\cdot)$ is continuous, the set $\phi(t,B)$ is compact for each $t \in \mathbb{T}_0^+$. Moreover, by the semi-group property

$$\phi(s+t,B) = \phi(s,\phi(t,B)) \subset \phi(s,B) \subset B \quad \text{for all } s,t \in \mathbb{T}_0^+ ,$$

so the family $(\phi(t,B))_{t\in\mathbb{T}_0^+}$ of nonempty compact subsets is nested. Hence, the set B_∞, defined by

$$B_\infty = \bigcap_{t\geq 0} \phi(t,B) ,$$

is a nonempty compact subset of B. Moreover, B_∞ is ϕ-invariant by the same argument to that in the proof of Theorem 1.23. In fact, the construction of the set B_∞ here is the same as that used there for the local attractor, since

$$\phi(s,B) = \overline{\bigcup_{t\geq s} \phi(t,B)}$$

by positive invariance. □

The negative invariant case is more complicated, so the discrete and continuous time cases will be considered separately.

LEMMA 1.29. *Let X be a metric space and $\phi : \mathbb{N}_0 \times X \to X$ be a discrete time semi-dynamical system which is induced by a continuous map $f : X \to X$ with $f(x) = \phi(1,x)$, and consider a nonempty compact set $B \subset X$ which is ϕ-negatively invariant, i.e., $B \subset f(B)$. Then there exists a nonempty compact subset B_∞ of B which is f-invariant, i.e., $f(B_\infty) = B_\infty$.*

PROOF. Define $B_0 = B$, and let B_{-1} be the maximal subset of B_0 such that $B_0 = f(B_{-1})$. Since B_0 is compact and f continuous, the set B_{-1} is compact. To see this, consider a sequence $b_n \in B_{-1}$ and define $a_n = f(b_n)$. Since B_0 is compact and both $a_n \in B_0$ and $b_n \in B_{-1} \subset B_0$, there are convergent subsequences $a_{n'} \to \bar{a} \in B_0$ and $b_{n'} \to \bar{b} \in B_0$ as $n' \to \infty$. Then by the continuity of f, $a_{n'} = f(b_{n'}) \to f(\bar{b})$ as $n' \to \infty$, so $f(\bar{b}) = \bar{a}$, that is $\bar{b} \in B_{-1}$, which means that B_{-1} is compact.
Repeating this procedure gives a nested sequence of nonempty compact subsets $B_{-n-1} \subset B_{-n} = f(B_{-n-1})$ for $n \in \mathbb{N}_0$. Hence, the set defined by

$$B_\infty = \bigcap_{n\geq 0} B_{-n}$$

is a nonempty compact subset of B. Moreover, B_∞ is f-invariant by a similar argument to that in the proof of Theorem 1.23. □

In the continuous time case, one has to ensure that constructed subsets remain in the original set B.

THEOREM 1.30. *Let X be a metric space and B be a nonempty compact subset of X which is negatively invariant for a continuous time semi-dynamical system $\phi : \mathbb{R}_0^+ \times X \to X$, i.e., $B \subset \phi(t, B)$ for all $t \in \mathbb{R}_0^+$. Then there exists a nonempty compact subset B_∞ of B which is ϕ-invariant.*

PROOF. The result of Lemma 1.29 will be applied to the discrete time system formed by the time-1 mapping $\phi(1, \cdot) : X \to X$. This yields a nonempty compact subset B_∞^1 of B which is the maximal $\phi(1, \cdot)$-invariant subset of B, i.e., $\phi\big(1, B_\infty^1\big) = B_\infty^1$. The problem is that $\phi\big(t, B_\infty^1\big)$ may not be a subset of B for all $t \in (0, 1)$. Therefore, the procedure will be repeated for the discrete time system formed by the time-$\frac{1}{2}$ mapping $\phi\big(\frac{1}{2}, \cdot\big) : X \to X$ in order to obtain a nonempty compact subset B_∞^2 of B which is the maximal $\phi\big(\frac{1}{2}, \cdot\big)$-invariant subset of B, i.e., with $\phi\big(\frac{1}{2}, B_\infty^2\big) = B_\infty^2$. By this and the semi-group property,

$$B_\infty^2 = \phi\big(\tfrac{1}{2}, B_\infty^2\big) = \phi\big(\tfrac{1}{2}, \phi\big(\tfrac{1}{2}, B_\infty^2\big)\big) = \phi\big(1, B_\infty^2\big)$$

so B_∞^2 is also a $\phi(1, \cdot)$-invariant subset of B. However, B_∞^1 is the maximal compact $\phi(1, \cdot)$-invariant subset of B, which means that $B_\infty^2 \subset B_\infty^1$.
Repeat this procedure with the discrete time system formed by the time-2^{-n} mapping $\phi(2^{-n}, \cdot) : X \to X$ to obtain a nonempty compact subset B_∞^n of B which is the maximal $\phi(2^{-n}, \cdot)$-invariant subset of B, and is thus also $\phi(2^{-n+1}, \cdot)$-invariant. Hence $B_\infty^n \subset B_\infty^{n-1}$ for $n \in \mathbb{N}$. This is a nested family of nonempty compact subsets, so the set defined by

$$B_\infty = \bigcap_{n \geq 1} B_\infty^n$$

is a nonempty compact subset of B. Moreover, by Lemma 1.29, the set B_∞ is $\phi(2^{-n}, \cdot)$-invariant for each $n \in \mathbb{N}_0$, i.e., $\phi(2^{-n}, B_\infty) = B_\infty$. From this and the semi-group property, it follows that $\phi(j2^{-n}, B_\infty) = B_\infty$ for all $n \in \mathbb{N}$ and $j \in \{0, \ldots, 2^n\}$, i.e., for all dyadic numbers in $[0, 1]$. By continuity,

$$h\big(\phi(t, B_\infty), \phi(\tau, B_\infty)\big) \to 0 \quad \text{as } \tau \to t$$

for dyadic $\tau \in [0, 1]$ with $\tau \to t \in [0, 1]$, where t is arbitrary. Here h is the Hausdorff metric. Finally, since $h\big(\phi(\tau, B_\infty), B_\infty\big) = 0$ by invariance, one obtains

$$h\big(\phi(t, B_\infty), B_\infty\big) \leq h\big(\phi(t, B_\infty), \phi(\tau, B_\infty)\big) + h\big(\phi(\tau, B_\infty), B_\infty\big),$$

which gives $\phi(t, B_\infty) = B_\infty$ for all $t \in [0, 1]$, and hence, this also holds for all $t \in \mathbb{R}_0^+$, since $h\big(\phi(\tau, B_\infty), B_\infty\big) = 0$ for all dyadic $\tau \in [0, 1]$.
This construction is then repeated on other time intervals $[n, n+1]$ with $n \in \mathbb{N}$. □

3. Global asymptotic behavior

3.1. Attractors. In contrast to local attractors, global attractors attract not only trajectories in a neighborhood but in the entire state space. The formal definition is given as follows.

DEFINITION 1.31 (Global attractor). A nonempty compact subset A of X is called a *global attractor* of a semi-dynamical system ϕ on X if it is ϕ-invariant and attracts bounded sets, i.e.,

$$\lim_{t \to \infty} \operatorname{dist}(\phi(t, D), A) = 0 \quad \text{for any bounded subset } D \subset X. \tag{1.7}$$

Usually, the qualifier *global* will be omitted. If an attractor exists, then it is unique.

THEOREM 1.32 (Uniqueness of global attractor). *There exists at most one attractor for a semi-dynamical system.*

PROOF. Let A_1 and A_2 be attractors of the semi-dynamical system ϕ. Since A_1 and A_2 are bounded, the relation (1.7) implies that both

$$\lim_{t\to\infty} \operatorname{dist}(\phi(t, A_2), A_1) = 0 \quad \text{and} \quad \lim_{t\to\infty} \operatorname{dist}(\phi(t, A_1), A_2) = 0\,.$$

In view of the invariance of A_1 and A_2 these reduce to

$$\operatorname{dist}(A_2, A_1) = 0 \quad \text{and} \quad \operatorname{dist}(A_1, A_2) = 0\,.$$

Hence, $h(A_1, A_2) = 0$, which implies $A_1 = A_2$, because A_1 and A_2 are compact. □

By the invariance of A, this attraction can be written equivalently as the convergence

$$\lim_{t\to\infty} \operatorname{dist}(\phi(t, D), \phi(t, A)) = 0\,,$$

which will often be called *forward convergence* to distinguish it from other kinds of convergence to be introduced later.

EXAMPLE 1.33. The continuous time dynamical system on $\mathbb{R}$ generated by the autonomous scalar ordinary differential equation

$$\dot{x} = x(1 - x^2)$$

has the attractor $A = [-1, 1]$. The dynamics within the attractor is quite simple, since all solutions are monotone increasing or decreasing.

EXAMPLE 1.34. The discrete time semi-dynamical system generated by the autonomous scalar difference equation (1.3) with mapping

$$f(x) := \max\{0, 4x(1 - x)\}$$

has the attractor $A = [0, 1]$. Here $f(x_0) \in A$ for all $x_0 \in \mathbb{R}$, so the dynamics are very simple outside of the attractor, but is chaotic within it.

As in the local case, the existence of global attractors follows from that of suitable absorbing sets.

DEFINITION 1.35 (Absorbing set). A nonempty compact subset B of X is called an *absorbing set* of a semi-dynamical system ϕ on X if for every bounded subset D of X, there exists a $T = T(D) \in \mathbb{T}^+$ such that $\phi(t, D) \subset B$ for all $t \geq T$.

Absorbing sets are often called *attracting sets* when they are also *positively invariant*, i.e., with $\phi(t, B) \subseteq B$ for all $t \in \mathbb{T}_0^+$. Attractors differ from attracting sets in that they consist entirely of limit points of the system and are thus strictly invariant in the sense of (1.4).

THEOREM 1.36 (Existence of global attractors). *Suppose that a semi-dynamical system ϕ on X has an absorbing set B. Then ϕ has a unique attractor $A \subset B$ given by*

$$A = \bigcap_{s\geq 0} \overline{\bigcup_{t\geq s} \phi(t, B)}\,,$$

or simply by $A = \bigcap_{t\geq 0} \phi(t, B)$ when B is positively invariant.

PROOF. See the proof of Theorem 1.23. □

REMARK 1.37. For an infinite-dimensional state space X, it is often easier to find a closed and bounded rather than compact set which is absorbing. To obtain the existence of an attractor in this set, one then needs to assume that the mappings $\phi(t, \cdot)$ of the semi-dynamical system are eventually compact, i.e., they map bounded subsets B into relatively compact subsets $\phi(t, B)$ for all $t \geq T$ with some $T = T(B) \geq 0$, or asymptotically compact (see Chapter 12).

In Example 1.33, the dynamics is reversible *inside* the attractor $[-1, 1]$, i.e., ϕ restricted to the space $[-1, 1]$ with $\phi(-t, x_0)$ defined by $\phi(-t, x_0) := y$, where $y = y(t, x_0)$ is the unique point in $[-1, 1]$ with $\phi(t, y) := x_0$, is a two-sided dynamical system rather than just a semi-dynamical system. In general, a semi-dynamical system can be extended backwards in time within a compact invariant set only as a set-valued dynamical system (see Chapter 9).

THEOREM 1.38 (Asymptotic stability of attractors). *A global attractor is* uniformly asymptotically stable *in the sense of Lyapunov, i.e., it is both Lyapunov stable and* uniformly Lyapunov attractive, *i.e., for any $\varepsilon > 0$ and every bounded subset D of X, there exists a $T > 0$ such that*

$$\operatorname{dist}(\phi(t, D), A) < \varepsilon \quad \textit{for all } t \geq T .$$

The asymptotic stability of attractors and that of attracting sets in general can be characterized by Lyapunov functions. These can be used to establish the existence of an absorbing set and hence that of a nearby maximal attractor in a nearby perturbed system such as in parameterized family of semi-dynamical systems, see Chapter 11.

3.2. Attractor-repeller pairs. If the phase space X is compact, the notion of a global attractor becomes trivial, because the attractor is then given by the phase space itself. However, the following theorem implies that in this case, a local attractor implies the existence of a local repeller and vice versa.

THEOREM 1.39 (Existence of attractor-repeller pairs). *Let $\phi : \mathbb{T} \times X \to X$ be a dynamical system on a compact metric space X, and let A be a local attractor. Then the set*

$$A^* := \{x \in X : \omega(x) \cap A = \emptyset\}$$

is a local repeller, and (A, A^*) is called an attractor-repeller pair. *In addition, if R is a local repeller of ϕ, then the set*

$$R^* := \{x \in X : \alpha(x) \cap R = \emptyset\}$$

is a local attractor, leading to the attractor-repeller pair (R^*, R).

PROOF. Only the first assertion will be proved, since the second one follows analogously by time-reversal. Since A is a local attractor, there exists an $\eta > 0$ such that

$$\lim_{t\to\infty} \operatorname{dist}\big(\phi(t, B_\eta(A)), A\big) = 0 .$$

This means that there exists a $T > 0$ such that $\operatorname{dist}\big(\phi(t, B_\eta(A)), A\big) < \frac{\eta}{2}$ for all $t \geq T$. This implies that $\phi(-t, X \setminus B_{\eta/2}(A)) \subset X \setminus B_\eta(A)$ for all $t \geq T$. Hence,

$B := X \setminus B_\eta(A)$ is locally repelling, and this means by Theorem 1.24 that the set $R = \alpha(B)$ is a local repeller. It remains to show that $A^* = R$. Firstly, choose some $x \notin R = \alpha(B)$. Then there exists a $\tau \leq 0$ such that $x \notin \overline{\bigcup_{t \leq \tau} \phi(t, B)}$, in particular, $x \notin \phi(\tau, B)$. This implies $\phi(-\tau, x) \in B_\eta(A)$, and hence, $\omega(\phi(-\tau, x)) = \omega(x) \in A$, which means that $x \notin A^*$. On the other hand, let $x \in R$. Since R is invariant (as an α-limit set) and $R \cap A = \emptyset$ (note that $R = \alpha(B) \subset B$), one obtains that $\omega(x) \cap A = \emptyset$, and thus, $x \in A^*$. □

This theorem can be applied in the noncompact case when a global attractor exists by restricting the system to the attractor, which yields a dynamical system on a compact space. This helps to understand the dynamics within an attractor.

EXERCISE 1.40. Show that the local repeller A^* of Theorem 1.39 is the maximal repeller outside of the local attractor A in the following sense: Every local repeller $R \subsetneq A^*$ fulfills $R \cap A \neq \emptyset$.

The next theorem provides a fundamental statement about the dynamics starting outside of such an attractor-repeller pair.

THEOREM 1.41 (Dynamics of attractor-repeller pairs). *Let $\phi : \mathbb{T} \times X \to X$ be a dynamical system on a compact metric space X, and let (A, A^*) be an attractor-repeller pair. Then for any $x \in X$ with $x \notin A \cup A^*$,*

$$\omega(x) \subset A \quad \textit{and} \quad \alpha(x) \subset A^* .$$

PROOF. Choose an $\eta > 0$ such that

$$\lim_{t \to \infty} \operatorname{dist}\big(\phi(t, B_\eta(A)), A\big) = 0 .$$

Since $x \notin A^*$, one has $\omega(x) \cap A \neq \emptyset$, and thus, there exists a $\tau > 0$ such that $\phi(\tau, x) \in B_\eta(A)$. This implies $\lim_{t \to \infty} \operatorname{dist}\big(\phi(t, x), A\big) = 0$, and hence, $\omega(x) \subset A$. The relation concerning the α-limit set can be shown analogously. □

EXAMPLE 1.42. Consider the autonomous linear system

$$\dot{x} = Bx$$

with a matrix $B \in \mathbb{R}^{d \times d}$. Assume that the spectrum of B, given by

$$\sigma(B) := \{\lambda \in \mathbb{C} : \lambda \text{ is an eigenvalue of } B\},$$

has a splitting $\sigma(B) = \sigma_- \cup \sigma_+$, where $\Re\lambda_- < \Re\lambda_+$ for all $\lambda_- \in \sigma_-$ and $\lambda_+ \in \sigma_+$. This means that there exists a *pseudo-stable* linear subspace, given by $\mathcal{W}^- := \bigoplus_{\lambda \in \sigma_-} \ker(B - \lambda \mathbb{1})^d$, and a *pseudo-unstable* linear subspace $\mathcal{W}^+ := \bigoplus_{\lambda \in \sigma_+} \ker(B - \lambda \mathbb{1})^d$. It is easy to see that the projection of the flow $e^{Bt}\xi$ defines a flow $\mathbb{P}e^{Bt}\xi$ in the projective space $\mathbb{P}^{d-1}$ (see Section 4 of Chapter 5 for the nonautonomous case). It is, however, technically involved to show that the projections of the linear spaces form an attractor-repeller pair $(\mathbb{P}\mathcal{W}^+, \mathbb{P}\mathcal{W}^-)$ in $\mathbb{P}^{d-1}$.

3.3. Morse decompositions. The notion of an attractor-repeller pair allows a dynamical description of the system for all points which are outside the local attractor and repeller. With the concept of a Morse decomposition, a more detailed analysis of the asymptotic behavior is possible.

DEFINITION 1.43 (Morse decomposition). Let $\phi : \mathbb{T} \times X \to X$ be a dynamical system on a compact metric space X. A set $\{M^1, \dots, M^n\}$ containing finitely many subsets of X is called a *Morse decomposition* of ϕ if there exists a finite sequence of local attractors $\emptyset = A^0 \subsetneq A^1 \subsetneq \dots \subsetneq A^n = X$ such that

$$M^i = A^i \cap A^{i-1,*} \quad \text{for all } i \in \{1, \dots, n\}.$$

The sets M^i, $i \in \{1, \dots, n\}$, are called *Morse sets*.

There always exists a Morse decomposition, because the sets $A^0 := \emptyset$ and $A^1 := X$ are local attractors. The corresponding Morse decomposition is then given by the single set $M^1 = X$.

The following theorem shows that Morse decompositions are indeed crucial for the global asymptotic behavior.

THEOREM 1.44 (Dynamics of Morse decompositions). *Let $\phi : \mathbb{T} \times X \to X$ be a dynamical system on a compact metric space X and $\{M^1, \dots, M^n\}$ be a Morse decomposition. Then the following two statements hold.*

(i) *The Morse sets M^i, $i \in \{1, \dots, n\}$, are pairwise disjoint.*
(ii) *For all $x \in X \setminus (M^1 \cup \dots \cup M^n)$, there exist $i \le j$ with*

$$\omega(x) \subset M^i \quad \text{and} \quad \alpha(x) \subset M^{j+1}.$$

PROOF. (i) One has

$$M^i \cap M^j = A^i \cap A^{i-1,*} \cap A^j \cap A^{j-1,*} = A^i \cap A^{j-1,*} \subset A^{j-1} \cap A^{j-1,*} = \emptyset$$

for $1 \le i < j \le n$.
(ii) Let $x \in X \setminus (M^1 \cup \dots \cup M^n)$. Because $A^n = X$ and $A^{0,*} = X$, there exists a minimal $i > 0$ with $\omega(x) \subset A^i$ and a maximal $j < n$ with $\alpha(x) \subset A^{j,*}$. Since i was chosen to be minimal, the relation $\omega(x) \not\subset A^{i-1}$ holds, and thus, Theorem 1.41 implies that $x \in A^{i-1,*}$. Because $A^{i-1,*}$ is invariant, one has $\omega(x) \subset A^{i-1,*}$. Analogously, the maximality of j implies that $x \in A^{j+1}$ and $\alpha(x) \subset A^{j+1}$. In addition, $i \le j$ holds, since the assumption $i > j$ leads to $x \in A^{i-1,*} \cap A^{j+1} \subset A^{i-1,*} \cap A^i = M^i$, which is a contradiction. Finally, both $\omega(x) \subset A^{i-1,*} \cap A^i = M^i$ and $\alpha(x) \subset A^{j,*} \cap A^{j+1} = M^{j+1}$ hold. □

The fact that each point of the phase space converges both in forwards and backwards time to a Morse set is not only a necessary but also a sufficient condition.

THEOREM 1.45 (Characterizations of Morse decompositions). *Let $\phi : \mathbb{T} \times X \to X$ be a dynamical system on a compact metric space X, and consider a set $M := \{M^1, \dots, M^n\}$ consisting of nonempty, compact, invariant and pairwise disjoint subsets of X. Then the following statements are equivalent.*

(i) *M is a Morse decomposition.*

(ii) *For all $x \in X$, there exist $i \leq j$ with*

$$\omega(x) \subset M^i \quad \textit{and} \quad \alpha(x) \subset M^j .$$

EXERCISE 1.46. Prove Theorem 1.45.

EXAMPLE 1.47. Consider the autonomous differential equation

$$\dot{x} = x(1-x)(2-x)$$

on the compact interval $[0,2]$, and let $\phi : \mathbb{R} \times [0,2] \to [0,2]$ be the corresponding dynamical system. First choose the attractor sequence: $A^0 = \emptyset \subsetneq A^1 = [0,1] \subsetneq A^2 = [0,2]$. This implies the sequence of local repellers $A^{0,*} = [0,2] \supsetneq A^{1,*} = \{2\} \supsetneq A^{2,*} = \emptyset$, and the Morse sets of the corresponding Morse decomposition are given by $M^1 = A^1 \cap A^{0,*} = [0,1]$ and $M^2 = A^2 \cap A^{1,*} = \{2\}$. Due to Theorem 1.44, the dynamical behavior of the system outside the Morse sets is known, i.e., in the interval $(1,2)$. This is improved by considering a longer sequence of local attractors $A^0 = \emptyset \subsetneq A^1 = \{1\} \subsetneq A^2 = [0,1] \subsetneq A^3 = [0,2]$. This implies the sequence of local repeller $A^{0,*} = [0,2] \supsetneq A^{1,*} = \{0,2\} \supsetneq A^{2,*} = \{2\} \supsetneq A^{3,*} = \emptyset$. The corresponding Morse sets are then given by the singletons containing the equilibria of the system: $M^1 = \{1\}$, $M^2 = \{0\}$, $M^3 = \{2\}$.

The second Morse decomposition of the previous example provides a finer description of the dynamics.

DEFINITION 1.48 (Fineness of Morse decompositions). A Morse decomposition $\{M^1, \dots, M^n\}$ is said to be *finer* than the Morse decomposition $\{\tilde{M}^1, \dots, \tilde{M}^m\}$ if, for all $j \in \{1, \dots, m\}$, there exists an $i \in \{1, \dots, n\}$ such that $M^i \subset \tilde{M}^j$.

The Morse decomposition of the following example is a finest Morse decomposition.

EXAMPLE 1.49. Consider again the autonomous linear system

$$\dot{x} = Bx$$

with a matrix $B \in \mathbb{R}^{d \times d}$ which was already discussed in Example 1.42. Repeated application of the example leads to a Morse decomposition in the projective space $\{\mathbb{P}\mathcal{W}_1, \dots, \mathbb{P}\mathcal{W}_n\}$, where $\sigma_1 > \sigma_2 > \dots > \sigma_n$ are the ordered elements of $\Re\sigma(B) = \{\Re\lambda : \lambda \text{ is an eigenvalue of } B\}$ and $\mathcal{W}_i := \bigoplus_{\lambda \in \sigma(B), \Re\lambda = \sigma_i} \ker(B - \lambda\mathbb{1})^d$ for $i \in \{1, \dots, n\}$.

In general, however, a finest Morse decomposition need not exist.

EXERCISE 1.50. Construct an example for which no finest Morse decomposition exists.

4. Dependence on parameters

It is analyzed in this section how (global) attractors depend on parameters.

4.1. Upper semi-continuity of attractors. Consider a family of semi-dynamical systems $\phi_\nu : \mathbb{T}_0^+ \times X \to X$ which depend on a parameter ν and each with a global attractor A^ν. For example, suppose that these systems are generated by the scalar autonomous differential equation

$$\dot{x} = -x\nu + x^3,$$

with a scalar parameter ν, which undergoes a supercritical bifurcation as ν increases through $\nu = 0$. In particular, the global attractor $A^\nu = \{0\}$ for $\nu \leq 0$ and $A^\nu = [-\sqrt{\nu}, \sqrt{\nu}]$ for $\nu > 0$. The set-valued mapping $\nu \mapsto A^\nu$ is continuous with respect to the Hausdorff metric, i.e.,

$$h\big(A^\nu, A^{\nu_0}\big) \to 0 \quad \text{as } \nu \to \nu_0$$

for all $\nu_0 \in \mathbb{R}$.

However, this result is not typical. A classical result in dynamical systems theory says that, under appropriate natural assumptions, the attractors A^ν depend *upper semi-continuously on the parameter* ν in the sense that

$$\lim_{\nu \to \nu_0} \operatorname{dist}\big(A^\nu, A^{\nu_0}\big) = 0\,.$$

A proof of this is given in the more general context of nonautonomous attractors in Chapter 3, see Theorem 3.34.

In general, this cannot be strengthened to continuous dependence, i.e., with the Hausdorff semi-distance dist on X replaced by the Hausdorff distance h on X. As a counterexample, consider the scalar autonomous differential equation

$$\dot{x} = -x\big(x^4 - 2x^2 + 1 - \nu\big)\,, \tag{1.8}$$

with a scalar parameter ν, which undergoes a subcritical bifurcation as ν increases. There are three parameter regimes for equilibrium solutions $\bar{x}_\nu$:

(i) $\bar{x}_\nu = 0$ for $\nu < 0$,
(ii) $\bar{x}_\nu = 0, \pm\sqrt{1+\sqrt{\nu}}, \pm\sqrt{1-\sqrt{\nu}}$ for $0 \leq \nu < 1$, and
(iii) $\bar{x}_\nu = 0, \pm\sqrt{1+\sqrt{\nu}}$ for $\nu \geq 1$.

The zero solution here loses linear stability at $\nu = 1$ in a subcritical bifurcation to the nonlocal solutions $\pm\sqrt{1+\sqrt{\nu}}$. Note, however, that these equilibria, as well as $\pm\sqrt{1-\sqrt{\nu}}$, first appear at $\nu = 0$. The equilibria $\pm\sqrt{1+\sqrt{\nu}}$ are asymptotically stable for $\nu > 0$, whereas the equilibria $\pm\sqrt{1-\sqrt{\nu}}$ are unstable in their existence interval $0 \leq \nu < 1$.

The attractor here is $A^\nu = \{0\}$ for $\nu < 0$ and

$$A^\nu = \left[-\sqrt{1+\sqrt{\nu}}, \sqrt{1+\sqrt{\nu}}\right] \quad \text{for all } \nu \geq \nu_0 = 0\,.$$

In particular, the set-valued mapping $\nu \mapsto A^\nu$ is not continuous at $\nu_0 = 0$, but only upper semi-continuous; it is, however, continuous at all other ν_0.

An inspection of the vector fields shows that the attractors are clearly not uniformly attracting for different parameter values in a neighborhood of $\nu_0 = 0$.

4.2. Equi-attraction and continuity of attractors. There are results in the literature which ensure that attractors depend continuously on a parameter rather than just upper semi-continuously. These usually assume something about the dynamics within the attractors, e.g., they are Morse–Smale attractors consisting of hyperbolic steady state solutions and their connecting heteroclinic trajectories. This information is, however, often unknown since the attractors themselves may not be exactly known.

The concept of equi-attraction provides an alternative test; it depends on the behavior of the systems outside of the attractors and is often provided by the vector field functions of the parameterized differential equations which generate the systems. The prefix *equi* is used to denote *uniformity* with respect to a family of systems. Specifically, the attractors A^ν are said to be *equi-attracting* if for every $\varepsilon > 0$ and every bounded subset B of X, there exists a $T = T(\varepsilon, B) \in \mathbb{R}^+$, independent of ν, such that

$$\operatorname{dist}\left(\phi_\nu(t, B), A^\nu\right) < \varepsilon \quad \text{for all } t \geq T\,.$$

To ensure that systems for nearby parameter values actually have attractors, it has to be assumed that the family of semi-dynamical systems is equi-dissipative and eventually equi-compact (the latter follows trivially from equi-dissipativity in finite-dimensional state spaces, and is only required in the infinite-dimensional case for this reason).

DEFINITION 1.51. Let (Λ, ρ) be a compact metric space. A family of semi-dynamical systems $(\phi_\nu)_{\nu\in\Lambda}$ on X is said to be

(i) *equi-dissipative* on X if there exists a bounded subset U of X so that for any bounded subset $B \subset X$, there exists a $T = T(B) > 0$, independent of $\nu \in \Lambda$, such that

$$\phi_\nu(t, B) \subset U \quad \text{for all } t \geq T\,,$$

(ii) *eventually equi-compact* if for any bounded subset B of X, there exists a $T = T(B) > 0$, independent of $\nu \in \Lambda$, such that $\bigcup_{\nu\in\Lambda} \phi_\nu(t, B)$ is relatively compact for any $t \geq T$.

The following theorem was first proved in LI & KLOEDEN [**67**]. The proof given here is adapted from KLOEDEN & PISKAREV [**137**].

THEOREM 1.52. *Let (Λ, ρ) be a compact metric space and suppose that a family of semi-dynamical systems $(\phi_\nu)_{\nu\in\Lambda}$ on X is equi-dissipative and eventually equi-compact and that A^ν is the attractor of ϕ_ν for each $\nu \in \Lambda$. In addition, suppose that*

(i) *for any $t \in \mathbb{T}_0^+$ fixed, $\phi_\nu(t, x)$ is jointly continuous in (x, ν) on $X \times \Lambda$.*
(ii) *$\phi_\nu(t, x)$ is equi-continuous in ν for (t, x) in any bounded subset of $\mathbb{T}_0^+ \times X$.*

Then $(A^\nu)_{\nu\in\Lambda}$ is equi-attracting if and only if the set-valued mapping $\nu \mapsto A^\nu$ is continuous with respect to Hausdorff distance.

PROOF. The dissipativity and compactness assumptions ensure the existence of the global attractors and, moreover, that the attractors A^ν all belong to the absorbing set U, so are uniformly bounded.

Each of the other assertions will be proved by contradiction.
Firstly, consider the upper semi-continuous convergence of the attractors. Assume that this does not hold, i.e., there is an $\varepsilon_0 > 0$ and a sequence $\nu_n \to \nu_0$ as $n \to \infty$ such that

$$\text{dist}\,(A^{\nu_n}, A^{\nu_0}) \geq \varepsilon_0 > 0 \quad \text{for all } n \in \mathbb{N}, \tag{1.9}$$

and let $x^*_{\nu_n} \in A^{\nu_n}$ (which exist due to the compactness of A^{ν_n}) be such that

$$\text{dist}\left(x^*_{\nu_n}, A^{\nu_0}\right) = \text{dist}\,(A^{\nu_n}, A^{\nu_0}) \quad \text{for all } n \in \mathbb{N}.$$

Since $\phi_{\nu_n}(t, A^{\nu_n}) = A^{\nu_n}$ for any $t > 0$, for each $T > 0$ there is a $x_{\nu_n} \in A^{\nu_n}$ such that $\phi_{\nu_n}(T, x_{\nu_n}) = x^*_{\nu_n}$. The points x_{ν_n} belong to the bounded set U by the eventual equi-dissipativity assumption and the set $\{x_{\nu_n}\}$ is pre-compact by the eventual equi-compactness assumption, so there is a subsequence $x_{\nu_{n'}} \to x_T \in \bar{U}$ as $n' \to \infty$. Now $\phi_{\nu_{n'}}(T, x_{\nu_{n'}}) \to \phi_{\nu_0}(T, x_T)$ as $n' \to \infty$ by the joint continuity assumption (i). Moreover, $\phi_{\nu_0}(T, x_T)$ belongs to the closed and bounded set $\bar{U}$ since the $\phi_{\nu_{n'}}(T, x_{\nu_{n'}})$ belong to the bounded absorbing set U. Hence, for any $\varepsilon > 0$, there is a $T(\varepsilon, \bar{U}) > 0$ with $\text{dist}\left(\phi_{\nu_0}(T, x_T), A^{\nu_0}\right) \leq \varepsilon$ for all $T \geq T(\varepsilon, \bar{U})$. Thus, the point $\phi_{\nu_0}(T, x_T)$ can be chosen arbitrarily close to A^{ν_0} by taking T large enough, so from

$$\text{dist}\left(x^*_{\nu_{n'}}, A^{\nu_0}\right) \leq d(\phi_{\nu_{n'}}(T, x_{\nu_{n'}}), \phi_{\nu_0}(T, x_T)) + \text{dist}\left(\phi_{\nu_0}(T, x_T), A^{\nu_0}\right),$$

it follows that $\text{dist}(x^*_{\nu_{n'}}, A^{\nu_0}) \to 0$ as $\nu_{n'} \to \nu_0$ in Λ, which contradicts (1.9). Hence, A^ν converges upper semi-continuously to A^{ν_0}.
Suppose now that the family of attractors $(A^\nu)_{\nu \in \Lambda}$ is equi-attracting, but the A^ν do not converge lower semi-continuously to A^{ν_0}. Then there exist sequences $\nu_n \to \nu_0$ as $n \to \infty$ and $x_n \in A^{\nu_0}$ such that

$$\text{dist}(x_n, A^{\nu_n}) \geq \varepsilon_0 > 0 \quad \text{for all } n \in \mathbb{N}. \tag{1.10}$$

Since A^{ν_0} is compact, there is a convergent subsequence $x_{n'} \to x^* \in A^{\nu_0}$ as $n' \to \infty$. Therefore,

$$\text{dist}(x_{n'}, A^{\nu_{n'}}) \leq \text{dist}(x^*, A^{\nu_{n'}}) + \frac{\varepsilon}{3} \quad \text{for all } n' \geq N(\varepsilon).$$

By the $\phi_{\nu_0}(T, \cdot)$-invariance of A^{ν_0}, there is an element $x_T \in A^{\nu_0}$ such that $\phi_{\nu_0}(T, x_T) = x^*$. Consider a sequence $y_n \in X$ with $y_n \to x_T$. Then $\phi_{\nu_n}(T, y_n) \to \phi_{\nu_0}(T, x_T)$ as $n \to \infty$ by the joint continuity assumption (i). Since A^{ν_0} is bounded, the sequence $\{y_n\}$ is also contained in a bounded ball B, i.e. $y_n \in B$ for each n. By the equi-attraction, there is a $T = T(B, \varepsilon/3)$ such that $\text{dist}(\phi_\nu(T, B), A^\nu) \leq \varepsilon/3$ holds for all $\nu \in \Lambda$. Finally, with $T = T(B, \varepsilon/3)$, it follows that

$$\begin{aligned}
\text{dist}(x_{n'}, A^{\nu_{n'}}) &\leq \text{dist}(x^*, A^{\nu_{n'}}) + \frac{\varepsilon}{3} \\
&\leq d(\phi_{\nu_0}(T, x_T), \phi_{\nu_{n'}}(T, x_{n'})) + \text{dist}(\phi_{\nu_{n'}}(T, x_{n'}), A^{\nu_{n'}}) + \frac{\varepsilon}{3} \\
&\leq \varepsilon
\end{aligned}$$

for n' large enough (depending on ε). Taking $\varepsilon < \varepsilon_0$ gives a contradiction to (1.10). Thus, the lower semi-continuous convergence of the attractors must hold. Since upper semi-continuous convergence always holds, the continuous convergence follows.
Finally, suppose that the A^ν converge continuously to A^{ν_0}, i.e., $h(A^\nu, A^{\nu_0}) \to 0$ as

$\nu \to \nu_0$, but the attractors are not equi-attracting. In particular, there is an $\varepsilon_0 > 0$, a sequence $\nu_n \to \nu_0$ as $n \to \infty$ and a bounded set B in X such that

$$\operatorname{dist}(\phi_{\nu_n}(T,B), A^{\nu_n}) \geq \varepsilon_0 > 0 \quad \text{for all } T > 0 \text{ and } n \in \mathbb{N}. \tag{1.11}$$

Then for all $T > 0$, there is a sequence $u_{\nu_n} \in B$ such that $\operatorname{dist}\left(\phi_{\nu_n}(T, u_{\nu_n}), A^{\nu_n}\right) \geq \varepsilon_0 > 0$ uniformly in $n \in \mathbb{N}$. By the eventual equi-compactness assumption, $\{\phi_{\nu_n}(T_0, u_{\nu_n})\}$ is pre-compact for some $T_0 > 0$, so there is a convergent subsequence $\phi_{\nu_{n'}}(T_0, u_{\nu_{n'}}) \to u^*$ as $n' \to \infty$. Hence, by the joint continuity assumption (i),

$$\phi_{\nu_{n'}}(T, u_{\nu_{n'}}) = \phi_{\nu_{n'}}(T - T_0, \phi_{\nu_{n'}}(T_0, u_{\nu_{n'}})) \to \phi_{\nu_0}(T - T_0, u^*) \quad \text{as } n' \to \infty.$$

Now $\operatorname{dist}\left(\phi_{\nu_0}(T - T_0, u^*), A^{\nu_0}\right)$ can be made arbitrarily small as $T - T_0$ becomes arbitrarily large. Hence,

$$\begin{aligned}\operatorname{dist}(\phi_{\nu_{n'}}(T, u_{\nu_{n'}}), A^{\nu_{n'}}) \leq\ & d(\phi_{\nu_{n'}}(T, u_{\nu_{n'}}), \phi_{\nu_0}(T - T_0, u^*)) \\ & + \operatorname{dist}(\phi_{\nu_0}(T - T_0, u^*), A^{\nu_0}) + \operatorname{dist}(A^{\nu_0}, A^{\nu_{n'}})\end{aligned}$$

can be made smaller ε_0 for n' and T large enough. However, this contradicts (1.11), so the family of attractors $(A^\nu)_{\nu \in \Lambda}$ is equi-attracting. □

Endnotes. There are many excellent textbooks and monographs on autonomous dynamical systems and their applications. A few, each with a different perspective, are Bhatia & Szegö [**23**], Brin & Stuck [**26**], Guckenheimer & Holmes [**86**], Hirsch, Smale & Devaney [**95**], Katok & Hasselblatt [**105**], Robinson [**200**], Sibirsky [**223**], Stuart & Humphries [**226**] and Wiggins [**235**]. See also Babin & Vishik [**16**], Hale [**90**], Ladyzhenskaya [**151**], Robinson [**201**], Sell & You [**220**], Temam [**230**] and Vishik [**233**] for infinite-dimensional systems. Bhatia & Szegö [**23**], Sibirsky [**223**] and Stuart & Humphries [**226**] are also useful sources on Lyapunov functions. Exercise 1.14 is taken from Bhatia & Szegö [**23**], see also Stuart & Humphries [**226**]. For attractor-repeller pairs and Morse decompositions, see Conley [**56**] and Selgrade [**215**] as well as Rasmussen [**194**]. The results on entire solutions and invariant sets are based on Kloeden & Marín-Rubio [**132**], while those on equi-attraction and continuous convergence of attractors with respect to a parameter in Subsection 4.2 are taken from Li Desheng & Kloeden [**67**] and Kloeden & Piskarev [**137**].

CHAPTER 2

Nonautonomous dynamical systems

The formulation of an autonomous dynamical system as a group or semi-group of mappings depends on the fact that such systems depend only on the elapsed time $t - t_0$ since starting and not directly on the current time t or starting time t_0 themselves. For a nonautonomous system both the current time t and starting time t_0 are important rather than just their difference.

The most natural generalization of a semi-group formalism to nonautonomous dynamical systems is the *two-parameter semi-group* or *process* formalism of a nonautonomous dynamical system, where both t and t_0 are the parameters. The process formulation will be treated in the first section of this chapter.

An alternative method includes an autonomous dynamical system as a driving mechanism which is responsible for, e.g., the temporal change of the vector field of a nonautonomous dynamical system. This leads to the *skew product flow* formalism of a nonautonomous dynamical system, which is discussed in the second part of this chapter.

1. Processes formulation

Consider an initial value problem for a nonautonomous ordinary differential equation in $\mathbb{R}^d$,

$$\dot{x} = f(t, x)\,, \qquad x(t_0) = x_0\,. \tag{2.1}$$

In contrast to the autonomous case, the solutions may now depend separately on the actual time t and the starting time t_0 rather than only on the elapsed time $t - t_0$ since starting. For example, the scalar initial value problem

$$\dot{x} = -2tx\,, \qquad x(t_0) = x_0\,,$$

has the explicit solution

$$x(t) = x(t, t_0, x_0) = x_0\, e^{-(t^2 - t_0^2)} \quad \text{for all } t, t_0, x_0 \in \mathbb{R}\,,$$

and $t^2 - t_0^2 = (t - t_0)^2 + 2(t - t_0)t_0$ cannot be expressed in terms of $t - t_0$ alone. Assuming global existence and uniqueness of solutions in forward time, the solutions form a continuous mapping $(t, t_0, x_0) \mapsto x(t, t_0, x_0) \in \mathbb{R}^d$ for $t \geq t_0$ with $t, t_0 \in \mathbb{R}$ and $x_0 \in \mathbb{R}^d$ fulfilling the initial value and evolution properties

(i) $x(t_0, t_0, x_0) = x_0$ for all $t_0 \in \mathbb{R}$ and $x_0 \in \mathbb{R}^d$,
(ii) $x(t_2, t_0, x_0) = x\big(t_2, t_1, x(t_1, t_0, x_0)\big)$ for all $t_0 \leq t_1 \leq t_2$ and $x_0 \in \mathbb{R}^d$.

The evolution property (ii) is a consequence of the causality principle that the solutions are determined uniquely by their initial values (for the given differential equation).

1.1. Definition. Solution mappings of nonautonomous ordinary differential equations are one of the main motivations for the process formulation of an abstract nonautonomous dynamical system on a metric state space (X, d) and time set $\mathbb{T}$, where $\mathbb{T} = \mathbb{R}$ for a continuous time process and $\mathbb{T} = \mathbb{Z}$ for a discrete time process.

The following definition originates from DAFERMOS [**62**] and HALE [**90**].

DEFINITION 2.1 (Process formulation). A *process* is a continuous mapping $(t, t_0, x_0) \mapsto \phi(t, t_0, x_0) \in X$ for $t, t_0 \in \mathbb{T}$ and $x_0 \in X$ with $t \geq t_0$, which satisfies the initial value and evolution properties

(i) $\phi(t_0, t_0, x_0) = x_0$ for all $t_0 \in \mathbb{T}$ and $x_0 \in X$,
(ii) $\phi(t_2, t_0, x_0) = \phi\big(t_2, t_1, \phi(t_1, t_0, x_0)\big)$ for all $t_0 \leq t_1 \leq t_2$ and $x_0 \in X$.

A process is often called a *two-parameter semi-group* on X in contrast with the *one-parameter semi-group* of an autonomous semi-dynamical system since it depends on both the initial time t_0 and the actual time t rather than just the elapsed time $t - t_0$.

1.2. Examples. The solution $x(t, t_0, x_0)$ of the nonautonomous differential equation (2.1) defines a continuous time process under the assumption of global existence and uniqueness of solutions. Indeed, this was the motivating example behind the definition of a process. Similarly, a nonautonomous difference equation generates a discrete time process.

EXAMPLE 2.2 (Nonautonomous difference equations as processes). Let $f_n : \mathbb{R}^d \to \mathbb{R}^d$, $n \in \mathbb{Z}$, be continuous mappings. Then the nonautonomous difference equation

$$x_{n+1} = f_n(x_n) \tag{2.2}$$

generates a discrete time process ϕ which is defined for all $x_0 \in \mathbb{R}^d$ and $n, n_0 \in \mathbb{Z}$ with $n > n_0$ by

$$\phi(n_0, n_0, x_0) := x_0\,, \qquad \phi(n, n_0, x_0) := f_{n-1} \circ \cdots \circ f_{n_0}(x_0)\,.$$

In particular, note that ϕ is continuous, since the variation in the non-discrete variable $x_0 \mapsto \phi(n, n_0, x_0)$ is continuous as composition of finitely many continuous mappings.

Not all examples of processes involve either differential or difference equations.

EXAMPLE 2.3 (Nonhomogeneous Markov chains as processes). Consider a nonhomogeneous Markov chain on a finite state space $\{1, \ldots, N\}$ with $d \times d$ probability transition matrices

$$P(t_0, t) = \big(p_{i,j}(t_0, t)\big)_{i,j=1,\ldots,d} \quad \text{for all } t_0, t \in \mathbb{T} \text{ with } t \geq t_0\,.$$

Such transition matrices satisfy $P(t_0, t_0) = \mathbb{1}$, the $d \times d$ identity matrix, for all $t_0 \in \mathbb{T}$. They also satisfy the so-called *Chapman–Kolmogorov property*

$$P(t_0, s)P(s, t) = P(t_0, t) \quad \text{for all } t_0 \leq s \leq t\,.$$

Let Σ_d denote the subset of $\mathbb{R}^d$ consisting of the N-dimensional probability row vectors, i.e., $p = (p_1, \dots, p_d) \in \Sigma_d$ satisfies $\sum_{i=1}^d p_i = 1$ with $0 \le p_i \le 1$ for $i = 1, \dots, d$. If the states of the Markov chain at time t_0 satisfy the probability vector $p(t_0) \in \Sigma_d$, then they are distributed according to a probability vector $p(t) = p(t_0)P(t_0, t)$ at time $t \ge t_0$.
Thus, the mapping ϕ defined by $\phi(t, t_0, p_0) := p_0 P(t_0, t)$ is a process on the state space Σ_d, which is in fact linear in the initial state component p_0 and thus continuous in this variable. Continuity in the time variables is trivial in the discrete time case and requires the additional assumption of continuity of the transition matrices in both of their variables in the continuous time case. The two-parameter semi-group property follows from the Chapman–Kolmogorov property.

1.3. Perturbed motions. In contrast to autonomous dynamical systems, every solution $\bar{x}(t) := \phi(t, \bar{t}_0, \bar{x}_0)$ of a process ϕ on a Banach space $(X, \|\cdot\|)$ can be transformed to a constant solution of a related process $\bar{\phi}$, which is defined by

$$\bar{\phi}(t, t_0, x_0) := \phi(t, t_0, x_0) - \bar{x}(t) .$$

The mapping $\bar{\phi}$ is also called the *process of perturbed motion*. In particular, such a description yields a notational advantage in the study of local properties of one fixed solution of a process. If the process ϕ is generated by a nonautonomous differential equation

$$\dot{x} = f(t, x) ,$$

then $\bar{\phi}$ is generated by the differential equation

$$\dot{x} = \tilde{f}(t, x) := f\big(t, x + \bar{x}(t)\big) - f\big(t, \bar{x}(t)\big) ,$$

which is called *differential equation of perturbed motion*. Analogously, let $(\bar{x}_n)_{n\in\mathbb{Z}}$ be a solution of the nonautonomous difference equation

$$x_{n+1} = f_n(x_n)$$

with continuous mappings $f_n : \mathbb{R}^d \to \mathbb{R}^d$, $n \in \mathbb{Z}$. Then the corresponding *difference equation of perturbed motion* is given by

$$x_{n+1} = \tilde{f}_n(x_n) := f_n(x_n + \bar{x}_n) - \bar{x}_{n+1} .$$

1.4. An interesting property of processes. A process can be reformulated as an autonomous semi-dynamical system, which has some interesting implications.

The extended phase space will be denoted by $\mathfrak{X} := \mathbb{T} \times X$, and define a mapping $\pi : \mathbb{T}_0^+ \times \mathfrak{X} \to \mathfrak{X}$ by

$$\pi(t, (t_0, x_0)) := \big(t + t_0, \phi(t + t_0, t_0, x_0)\big) \quad \text{for all } (t, (t_0, x_0)) \in \mathbb{T}_0^+ \times \mathfrak{X} .$$

Note that the variable t in $\pi(t, (t_0, x_0))$ is the time which has elapsed since starting at time t_0, while the actual time is $t + t_0$.

THEOREM 2.4. *π is an autonomous semi-dynamical system on $\mathfrak{X}$.*

PROOF. It is obvious that π is continuous in its variables and satisfies the initial condition

$$\pi(0, (t_0, x_0)) = \big(t_0, \phi(t_0, t_0, x_0)\big) = (t_0, x_0) .$$

It also satisfies the (one-parameter) semi-group property

$$\pi(s + t, (t_0, x_0)) = \pi\big(s, \pi(t, (t_0, x_0))\big) \quad \text{for all } s, t \in \mathbb{T}_0^+ ,$$

since, by the evolution property (ii) of the process,

$$\begin{aligned}\pi(s+t,(t_0,x_0)) &= \big(s+t+t_0,\phi(s+t+t_0,t_0,x_0)\big)\\ &= \big(s+t+t_0,\phi\big(s+t+t_0,t+t_0,\phi(t+t_0,t_0,x_0)\big)\big)\\ &= \pi\big(s,(t+t_0,\phi(t+t_0,t_0,x_0)\big)\\ &= \pi\big(s,\pi(t,(t_0,x_0))\big)\,.\end{aligned}$$

This finishes the proof of this theorem. □

The autonomous semi-dynamical system π on the extended state space $\mathfrak{X}$ generated by a process ϕ on the state space X has some unusual properties. In particular, π has no nonempty ω-limit sets and, indeed, no compact subset of $\mathfrak{X}$ is π-invariant. This is a direct consequence of the fact that time is a component of the state.

This has significant implications and means that many concepts for autonomous systems need to be modified appropriately to be of any use in the nonautonomous context. For example, note that a π-invariant subset $\mathcal{A}$ of $\mathfrak{X}$ has the form $\mathcal{A} = \bigcup_{t_0\in\mathbb{T}}(t_0, A_{t_0})$, where A_{t_0} is a nonempty subset of X for each $t_0 \in \mathbb{T}$. Then the invariance property $\pi(t,\mathcal{A}) = \mathfrak{X}$ for $t \in \mathbb{T}_0^+$ is equivalent to

$$\phi\big(t+t_0,t_0,A_{t_0}\big) = A_{t+t_0} \quad \text{for all } t\in\mathbb{T}_0^+ \text{ and } t_0\in\mathbb{T}\,.$$

This will be used in Chapter 3 to motivate the definition of ϕ-invariant sets for a process ϕ.

2. Skew product flow formulation

To motivate the concept of a skew product flow, first a triangular system of ordinary differential equations is considered in which the uncoupled component can be considered as the driving force in the equation for the other component.

Consider an autonomous system of ordinary differential equations of the form

$$\dot p = f(p)\,, \qquad \dot x = g(p,x)\,, \tag{2.3}$$

where $p \in \mathbb{R}^n$ and $x \in \mathbb{R}^m$, i.e., with a triangular structure.

Assuming global existence and uniqueness of solutions forwards in time, the system of differential equations (2.3) generates an autonomous semi-dynamical system π on $\mathbb{R}^{n+m}$ which will be written in component form as

$$\pi(t,p_0,x_0) = \big(p(t,p_0),x(t,p_0,x_0)\big)\,,$$

with initial condition $\pi(0,p_0,x_0) = (p_0,x_0)$.

There are two important points to observe here. Firstly, the p-component of the system is an independent autonomous system in its own right, i.e., its solution mapping $p = p(t,p_0)$ generates an autonomous semi-dynamical system on $\mathbb{R}^n$ and amongst other properties satisfies the semi-group property

$$p(s+t,p_0) = p(s,p(t,p_0)) \quad \text{for all } s,t\geq 0\,. \tag{2.4}$$

Secondly, the semi-group property for π on $\mathbb{R}^{n+m}$, i.e.,

$$\pi(s+t,p_0,x_0) = \pi(s,\pi(t,p_0,x_0))\,,$$

can be expanded out componentwise as

$$\begin{aligned}\pi(s+t,p_0,x_0) &= \left(p(s+t,p_0),x(s+t,p_0,x_0)\right)\\ &= \left(p(s,p(t,p_0)),x(s+t,p_0,x_0)\right),\end{aligned}$$

using (2.4), and

$$\pi(s,\pi(t,p_0,x_0)) = \left(p(s,p(t,p_0)),x(s,p(t,p_0),x(t,p_0,x_0))\right).$$

These are identical for all $s,t \geq 0$ and all $(p_0,x_0) \in \mathbb{R}^{n+m}$. Equating for the second components gives

$$x(s+t,p_0,x_0) = x\big(s,p(t,p_0),x(t,p_0,x_0)\big) \quad \text{for all } s,t \geq 0,$$

which is a generalization of the semi-group property and known as the *cocycle property*.

Given a solution $p = p(t,p_0)$ of the p-component of the triangular system (2.3), the x-component becomes a nonautonomous ordinary differential equation in the x variable on $\mathbb{R}^m$ of the form

$$\dot{x} = g(p(t,p_0),x), \quad \text{where } t \geq 0 \text{ and } x \in \mathbb{R}^n. \tag{2.5}$$

The function $p = p(t,p_0)$ can be considered as "driving" the nonautonomous system here, i.e., as being responsible for the changes in the vector field with the passage of time.

The solution $x(t) = x(t,p_0,x_0)$ with initial value $x(0) = x_0$ (which also depends on the choice of p_0 as a parameter through the driving solution $p(t,p_0)$) then satisfies the following.

(i) *Initial condition.* $x(0,p_0,x_0) = x_0$.
(ii) *Cocycle property.* $x(s+t,p_0,x_0) = x\big(s,p(t,p_0),x(t,p_0,x_0)\big)$.
(iii) *Continuity condition.* $(t,p_0,x_0) \to x(t,p_0,x_0)$ is continuous.

The mapping $x : \mathbb{R}_0^+ \times \mathbb{R}^n \times \mathbb{R}^m \to \mathbb{R}^m$ is called a *cocycle mapping.* It describes the evolution of the solution of the nonautonomous differential equation (2.5) with respect to the driving system. Note that the variable t here is the time since starting at the state x_0 with the driving system at state p_0.

As mentioned above, the product system π on $\mathbb{R}^n \times \mathbb{R}^m$ is an autonomous semi-dynamical system and is known as a *skew product flow* due to the asymmetrical roles of the two component systems. This motivates an alternative definition of a nonautonomous dynamical system, called the *skew product flow formalism*, where, for various reasons, the driving system p is usually taken to be a reversible dynamical system, i.e., forming a group rather than a semi-group. This will happen for example, if the driving differential equation is restricted to a compact invariant subset of $\mathbb{R}^n$. Driving systems which are only semi-groups or semi-dynamical systems will be considered in Chapter 10.

REMARK 2.5. Any nonautonomous differential equation $\dot{x} = f(t,x)$ can be written as the triangular autonomous system

$$\frac{\mathrm{d}}{\mathrm{d}t}\begin{pmatrix} t \\ x \end{pmatrix} = \begin{pmatrix} 1 \\ f(t,x) \end{pmatrix},$$

with an increase of the dimension of the state space by one. Note, however, that this system has no equilibrium points and bounded solutions, and thus, the theory of autonomous dynamical systems is of no use here.

The process formulation of a nonautonomous dynamical system defined by the solution mapping of a nonautonomous differential equation is quite intuitive. In contrast, the skew product flow formulation is more abstract, but it contains more information about how the system evolves in time, especially when the driving system is on a compact space P.

2.1. Definition. Let (X, d_X) and (P, d_P) be metric spaces. A *nonautonomous dynamical system* (θ, φ) is defined in terms of a cocycle mapping φ on a state space X which is driven by an autonomous dynamical system θ acting on a base or parameter space P and the time set $\mathbb{T} = \mathbb{R}$ or $\mathbb{Z}$.

Specifically, the *dynamical system* θ on P is a group of homeomorphisms $(\theta_t)_{t\in\mathbb{T}}$ under composition on P with the properties that

(i) $\theta_0(p) = p$ for all $p \in P$,
(ii) $\theta_{s+t} = \theta_s(\theta_t(p))$ for all $s,\, t \in \mathbb{T}$,
(iii) the mapping $(t, p) \mapsto \theta_t(p)$ is continuous,

and the *cocycle mapping* $\varphi : \mathbb{T}_0^+ \times P \times X \to X$ satisfies

(i) $\varphi(0, p, x) = x$ for all $(p, x) \in P \times X$,
(ii) $\varphi(t + s, p, x) = \varphi\big(t, \theta_s(p), \varphi(s, p, x)\big)$ for all $s, t \in \mathbb{T}_0^+$, $(p, x) \in P \times X$,
(iii) the mapping $(t, p, x) \mapsto \varphi(t, p, x)$ is continuous.

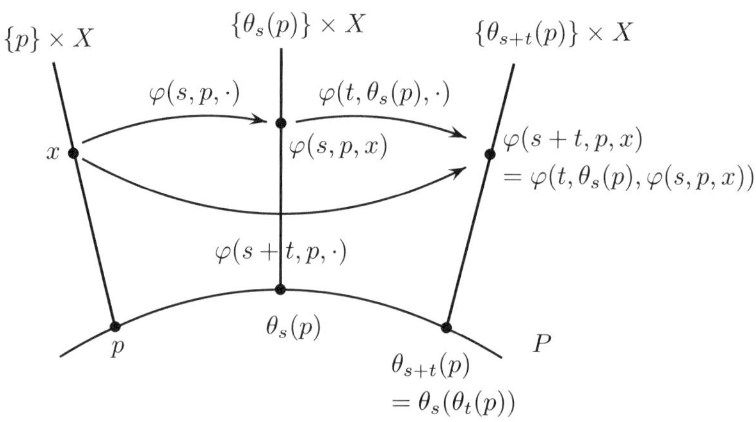

FIGURE 2.1. The cocycle property.

The mapping $\pi : \mathbb{T}_0^+ \times P \times X \to P \times X$ defined by

$$\pi(t, (p, x)) := \big(\theta_t(p), \varphi(t, p, x)\big) \tag{2.6}$$

forms an autonomous semi-dynamical system on $\mathfrak{X} = P \times X$.

DEFINITION 2.6 (Skew product formulation). The autonomous semi-dynamical system π on $\mathfrak{X} = P \times X$ defined by (2.6) is called the *skew product flow* associated with the nonautonomous dynamical system (θ, φ).

EXERCISE 2.7. Show that the mapping π defined by (2.6) defines a continuous time autonomous semi-dynamical system on $\mathfrak{X}$.

2.2. Examples. Nonautonomous difference equations and differential equations provide a rich source of examples for skew product flows.

EXAMPLE 2.8. The solution mapping $x(t) = x(t, t_0, x_0)$ of a nonautonomous differential equation (2.1) with initial value $x(t_0, t_0, x_0) = x_0$ at time t_0 defines a process. Theorem 2.4 shows that such a process can be reformulated as a skew product flow with the cocycle mapping φ on the state space $X = \mathbb{R}^d$ defined by $\varphi(t, t_0, x_0) := x(t_0 + t, t_0, x_0)$ and the driving system θ on the (noncompact) base space $P = \mathbb{R}$ defined by the shift operators $\theta_t(t_0) := t - t_0$. The disadvantages of this representation were discussed above.

The advantages of the skew product flow formulation reveals itself, for instance, when the generating nonautonomous differential equation is periodic or almost periodic in time, because the base space is then compact.

EXAMPLE 2.9. The skew product formulation of a nonautonomous differential equation (2.1) is based on the fact that whenever $x(t)$ is a solution of the differential equation, then the time-shifted solution $x_\tau(t) := x(\tau+t)$ (for some fixed τ) satisfies the nonautonomous differential equation

$$\dot{x}_\tau(t) = f_\tau(t, x_\tau(t)) := f(\tau + t, x(\tau + t)) .$$

Consider the set of functions $\{f_\tau(\cdot,\cdot) := f(\tau + \cdot, \cdot) : \tau \in \mathbb{R}\}$. Its closure $\mathcal{F}$ in an appropriate topology is called the *hull* of the vector field given by the nonautonomous differential equation (2.1). See SELL **[218]** for examples and typical topologies. For example, $\mathcal{F}$ is a compact metric space for periodic or almost periodic differential equations (see Exercise 2.12 below).
Introduce a group of shift operators $\theta_\tau : \mathcal{F} \mapsto \mathcal{F}$ by $\theta_\tau(f) := f_\tau$ for each $\tau \in \mathbb{R}$, define $\mathfrak{X} = \mathcal{F} \times \mathbb{R}^d$ and write $\varphi(t, f, x_0)$ for the solution of (2.1) with initial value x_0 at initial time $t_0 = 0$. Finally, define $\pi : \mathbb{R}^+ \times \mathfrak{X} \to \mathfrak{X}$ by $\pi(t, x_0, f) := \big(\theta_t(f), \varphi(t, f, x_0)\big)$.
Then, $\pi = (\theta, \varphi)$ is a continuous-time skew product flow on the state space $\mathfrak{X}$. To see this, observe that the second component of the semi-group identity $\pi(t + s, f, x_0) = \pi(t, \pi(s, f, x_0))$ expands out as the cocycle property

$$\varphi(t + s, f, x_0) = \varphi\big(t, \theta_s(f), \varphi(s, f, x_0)\big) .$$

Nonautonomous difference equations (2.2) generate discrete time skew product flows, the simplest coming from discrete time processes via Theorem 2.4 and have $\mathbb{Z}$ as their base space. When more is known about how the different mappings f_n vary with $n \in \mathbb{Z}$, it is often possible to have a compact base space.

EXAMPLE 2.10. Suppose that the mappings f_n in the nonautonomous difference equation (2.2) are chosen in some way from a finite number of continuous mappings $R_i : \mathbb{R}^d \to \mathbb{R}^d$ for $i \in \{1, \dots, r\}$. Then the difference equation has the form

$$x_{n+1} = R_{i_n}(x_n) ,$$

where the $i_n \in \{1, \ldots, r\}$ for all $n \in \mathbb{N}$. It generates a discrete time skew product flow over the parameter set $P = \{1, \ldots, r\}^{\mathbb{Z}}$ of bi-infinite sequences $p = (i_n)_{n\in\mathbb{Z}}$ in $\{1, \ldots, r\}$ with respect to the group of left shift operators $(\theta_m)_{m\in\mathbb{N}}$, where $\theta_m\big((i_n)_{n\in\mathbb{Z}}\big) = (i_{n+m})_{n\in\mathbb{Z}}$. The cocycle mapping $\varphi(n, \cdot, \cdot)$ is defined by

$$\varphi(0, p, x) := x \quad \text{and} \quad \varphi(n, p, x) := \big(R_{i_{n-1}} \circ \cdots \circ R_{i_0}\big)(x)$$

for all $n \in \mathbb{N}$, $x \in \mathbb{R}^d$ and $p = (i_n)_{n\in\mathbb{N}} \in P$. The parameter space $P = \{1, \ldots, r\}^{\mathbb{Z}}$ here is a compact metric space with the metric

$$d(p, p') = \sum_{n=-\infty}^{\infty} (r+1)^{-|n|} \left|i_n - i'_n\right|,$$

and the mappings $p \mapsto \theta_n(p)$ and $(p, x) \mapsto \varphi(n, p, x)$ are continuous for each $n \in \mathbb{N}$. To see this, note that $d(p, p') \leq \delta < 1$ requires $i_j = i_{j'}$ for $j = 0, \pm 1, \ldots, \pm N(\delta)$. Then take δ small enough corresponding to a given $\varepsilon > 0$ and fixed n.

More generally, the difference equation may involve a parameter $q \in Q$, which varies from iterate to iterate, by choice or randomly,

$$x_{n+1} = f(x_n, q_n)\,. \tag{2.7}$$

EXAMPLE 2.11. Consider the parametrically dependent difference equation (2.7) with the continuous mapping $f : \mathbb{R} \times [-1, 1] \to \mathbb{R}$, given by

$$f(x, q) = f_q(x) := \nu x + q\,,$$

where $\nu \in [0, 1)$ and $q \in [-1, 1]$. Let $P = [-1, 1]^{\mathbb{Z}}$ be the space of bi-infinite sequences $p = (q_n)_{n\in\mathbb{Z}}$ taking values in $[-1, 1]$, which is a compact metric space with the metric

$$d(p, p') = \sum_{n=-\infty}^{\infty} 2^{-|n|}\left|q_n - q'_n\right|,$$

and let $(\theta_n)_{n\in\mathbb{Z}}$ be the group of the left shift operators on this sequence space (cf. Example 2.10). Finally, define the mappings $\varphi(n, \cdot, \cdot)$ by

$$\varphi(0, p, x_0) := x \quad \text{and} \quad \varphi(n, p, x) := \big(f_{q_{n-1}} \circ \cdots \circ f_{q_0}\big)(x)$$

for all $n \in \mathbb{N}$, $x \in \mathbb{R}$ and $p = (q_n)_{n\in\mathbb{N}} \in P$. Specifically,

$$\varphi(n, p, x) = \nu^n x + \sum_{j=0}^{n-1} \nu^{n-1-j} q_j$$

for all $n \in \mathbb{N}$.

The mappings $p \mapsto \theta_n(p)$ and $(p, x) \mapsto \varphi(n, p, x)$ are obviously continuous here for each $n \in \mathbb{N}$. Thus, (θ, φ) is a discrete time skew product flow on $\mathbb{R}$ with the compact base space P.

Skew product flows need not be generated by either differential equations or difference equations. The reader is invited to invent an example.

EXERCISE 2.12. Show using the Theorem of Arzelà–Ascoli that the hull of the cosine function $\cos t$ is the compact subset $\big\{\cos(\tau + \cdot) : \tau \in [0, 2\pi]\big\}$ of the Banach space $C(\mathbb{R}, \mathbb{R})$ of all uniformly continuous functions $f : \mathbb{R} \to \mathbb{R}$, which is equipped with the supremum norm $\|f\|_\infty = \sup_{t\in\mathbb{R}} |f(t)|$.

3. Entire solutions and invariant sets

The definition of an entire solution of a nonautonomous dynamical system is an obvious generalization of the autonomous case.

DEFINITION 2.13 (Entire solution of a process). An *entire solution* of a process ϕ on a metric space (X, d) with time set $\mathbb{T}$ is a mapping $\xi : \mathbb{T} \to X$ such that

$$\xi(t) = \phi(t, \tau, \xi(\tau)) \quad \text{for all } t, \tau \in \mathbb{T} \text{ with } t \geq \tau\,. \tag{2.8}$$

The discussion following Theorem 2.4, in which a process ϕ on X was formulated as a skew product flow on the extended state space $\mathbb{R} \times X$, suggests that it is more appropriate to consider the invariance of a family of time-dependent subsets rather than of a single set. This motivates the following definition.

DEFINITION 2.14 (Invariant families for processes). Let ϕ be a process on a metric space (X, d). A family $\mathcal{A} = (A_t)_{t\in\mathbb{T}}$ of nonempty subsets of X is said to be *invariant* with respect to ϕ, or *ϕ-invariant*, if

$$\phi\,(t, t_0, A_{t_0}) = A_t \quad \text{for all } t \geq t_0\,.$$

A simple example of a ϕ-invariant family $\mathcal{A} = (A_t)_{t\in\mathbb{T}}$ is given by an entire solution of ϕ, i.e., having the singleton subsets $A_t = \{\xi(t)\}$ for each $t \in \mathbb{T}$. In fact, ϕ-invariant families consist of entire solutions.

LEMMA 2.15. *Let $\mathcal{A} = (A_t)_{t\in\mathbb{T}}$ be a nonempty ϕ-invariant family of subsets of X of a process ϕ. Then for any $t_0 \in \mathbb{T}$ and $a_0 \in A_{t_0}$, there exists an entire solution ξ through a_0 which is contained in $\mathcal{A}$, i.e., with $\xi(t_0) = a_0$ and $\xi(t) \in A_t$ for all $t \in \mathbb{T}$.*

PROOF. This is easy to see in the forward direction by defining $\xi(t) := \phi(t, t_0, a_0)$ for $t \geq t_0$, so in particular $\xi(t_0) = a_0$. For negative times, note that, since $\phi(t_0, t_0 - 1, A_{t_0-1}) = A_{t_0}$, there is a point $a_{-1} \in A_{t_0-1}$ such that $\phi(t_0, t_0 - 1, a_{-1}) = a_0$ and define $\xi(t) := \phi(t, t_0 - 1, a_{-1})$ for $t \in [t_0 - 1, t_0] \cap \mathbb{T}$, which is in $\mathcal{A}$ by the invariance of $\mathcal{A}$. Then repeat this construction on each subinterval $[t_{t_0} - n - 1, t_0 - n]$ for $n \in \mathbb{N}$ by defining $\xi(t) := \phi(t, t_0 - n - 1, a_{-n+1})$ for $t \in [t_0 - n - 1, t_0 - n]$, where $a_{-n-1} \in A_{t_0-n-1}$ is chosen so that $\phi(t_0 - n, t_0 - n - 1, a_{-n-1}) = a_{-n}$. It follows from the semi-group property that ξ satisfies (2.8) and from the ϕ-invariance of $\mathcal{A}$ that $\xi(t) \in A_t$ for all $t \in \mathbb{T}$. □

When the subsets in a ϕ-invariant family are compact, it follows from the continuity of a continuous time process that the set-valued mapping $t \mapsto A_t$ is continuous in $t \in \mathbb{R}$ with respect to the Hausdorff metric h_X, since

$$h_X(A_t, A_{t_0}) = h_X\big(\phi(t, t_0, A_{t_0}), \phi(t_0, t_0, A_{t_0})\big) \to 0 \quad \text{as } t \to t_0$$

by the continuity of the process ϕ in its first variable. The proof requires the result of the following exercise (see also ROXIN [**205**]).

EXERCISE 2.16. Let $f : \mathbb{R} \times X \to X$ be continuous, and define

$$F(t) := f(t, A) := \bigcup_{a\in A} \{f(t, a)\} \quad \text{for all } t \in \mathbb{R}\,,$$

where A is a compact subset of a metric space (X, d). Show that the sets $F(t)$ are compact subsets of X and that the setvalued mapping $t \mapsto F(t)$ is continuous with respect to the Hausdorff distance h.

Similar definitions hold for positive and negative invariant families of sets.

DEFINITION 2.17 (Positive and negative invariance). Let ϕ be a process on a metric space (X, d). A family $\mathcal{A} = (A_t)_{t\in\mathbb{T}}$ of nonempty subsets of X is said to be *positive invariant* with respect to ϕ, or *ϕ-positive invariant*, if

$$\phi\,(t, t_0, A_{t_0}) \subset A_t \quad \text{for all } t \geq t_0\,,$$

and *negative invariant* with respect to ϕ, or *ϕ-negative invariant*, if this holds with "$\supset$" instead of "$\subset$".

The corresponding definitions for skew product flows are stated here for completeness and later use.

DEFINITION 2.18 (Entire solution of a skew product flow). An *entire solution* of a skew product flow (θ, φ) on a metric phase space (X, d) and a base set P with time set $\mathbb{T}$ is a mapping $\xi : P \to X$ such that

$$\xi(\theta_t(p)) = \varphi(t - s, \theta_s(p), \xi(\theta_s(p))) \quad \text{for all } p \in P \text{ and } s, t \in \mathbb{T} \text{ with } s \leq t\,.$$

DEFINITION 2.19 (Invariant families for skew product flows). Let (θ, φ) be a skew product flow on a metric phase space (X, d) and a base set P. A family $\mathcal{A} = (A_p)_{p\in P}$ of nonempty sets of X is said to be *invariant* with respect to (θ, φ), or *φ-invariant*, if

$$\varphi(t, p, A_p) = A_{\theta_t(p)} \quad \text{for all } t \geq 0 \text{ and } p \in P\,.$$

For positive and negative invariant families, replace "$=$" here by "$\subset$" or "$\supset$", respectively.

The compact set-valued mapping $t \mapsto A_{\theta_t(p)}$ induced by a φ-invariant family $(A_p)_{p\in P}$ of compact subsets is continuous in $t \in \mathbb{R}$ with respect to the Hausdorff metric for each fixed $p \in P$.

3.1. Invariant subfamily of positive invariant family of sets. Analogously to the autonomous case, there is an invariant family of sets of a positively invariant family of compact sets of nonautonomous dynamical systems, in the sense that its component sets are subsets of the component sets of the positive invariant family. It then follows from Lemma 2.15 that the positive invariant family contains entire solutions.

This will be proved here for processes, which is notationally simpler than for skew product flows.

LEMMA 2.20. *Let $\mathcal{A} = (A_t)_{t\in\mathbb{T}}$ be a family of nonempty compact subsets of X which is positively invariant for the process ϕ, i.e., $\phi(t, t_0, A_{t_0}) \subset A_t$ for all $t_0 \in \mathbb{T}$ and $t \geq t_0$. Then there exists a family of nonempty compact subsets $\mathcal{A}^\infty = (A_t^\infty)_{t\in\mathbb{T}}$ of $\mathcal{A}$, which is ϕ-invariant, i.e., $\phi(t, t_0, A_{t_0}^\infty) = A_t^\infty$ for all $t \geq t_0$.*

PROOF. Since A_{t_0} is compact and the process ϕ is continuous, the set $\phi(t, t_0, A_{t_0})$ is compact for all $t \geq t_0$. Moreover, by the two-parameter semi-group property

$$\phi(t, s_0, A_{s_0}) = \phi\big(t, t_0, \phi(t_0, s_0, A_{s_0})\big) \subset \phi(t, t_0, A_{t_0}) \subset A_t$$

for all $s_0 \leq t_0 \leq t$, so for fixed $t \in \mathbb{T}$, the sets $\phi(t, t_0, A_{t_0})$ for $t_0 \leq t$ are a nested family of nonempty compact subsets of A. Hence, the set defined by

$$A_t^\infty = \bigcap_{t_0 \leq t} \phi(t, t_0, A_{t_0})$$

is a nonempty compact subset of A_t for each $t \in \mathbb{T}$. It remains to prove that $\mathcal{A}^\infty = (A_t^\infty)_{t\in\mathbb{T}}$ is ϕ-invariant, i.e., $\phi(t, t_0, A_{t_0}^\infty) = A_t^\infty$.
($\subset$) Let $\bar{a} \in A_{t_0}^\infty$. Then $\bar{a} \in \phi(t_0, s_0, A_{s_0})$ for all $s_0 \leq t_0$. Hence,

$$\phi(t, t_0, \bar{a}) \in \phi(t, t_0, \phi(t_0, s_0, A_{s_0})) = \phi(t, s_0, A_{s_0})$$

for any $t \geq t_0$ and $s_0 \leq t_0$, and

$$\phi(t, t_0, A_{t_0}) = \phi(t, s_0, \phi(s_0, t_0, A_{t_0})) \subset \phi(t, s_0, A_{s_0})$$

for any $t_0 \leq s_0 \leq t$, so

$$\phi(t, t_0, \bar{a}) \in \bigcap_{s_0 \leq t_0} \phi(t, s_0, A_{s_0}) = \bigcap_{s_0 \leq t} \phi(t, s_0, A_{s_0}) = A_t^\infty .$$

It follows that $\phi(t, t_0, A_{t_0}^\infty) \subset A_t^\infty$.
($\supset$) Let $\bar{a} \in A_t^\infty$. Then $\bar{a} \in \phi(t, s_n, A_{s_n}) = \phi(t, t_0, \phi(t_0, s_n, A_{s_n}))$ for all $s_n \leq t_0 \leq t$. Hence, there exist $b_n \in \phi(t_0, s_n, A_{s_n}) \subset A_{t_0}$ for all $n \in \mathbb{N}$ such that $\phi(t, t_0, b_n) = \bar{a}$. Now $b_n \in A_{t_0}$ for all $n \in \mathbb{N}$, and A_{t_0} is compact, so there exists a convergent subsequence $b_{n_j} \to \bar{b}$ in A_{t_0}. Moreover, one can choose the s_n so that $s_n \to -\infty$. In fact, $\bar{b} \in A_{t_0}^\infty$, since

$$\operatorname{dist}\left(\bar{b}, A_{t_0}^\infty\right) \leq \operatorname{dist}\left(\bar{b}, b_{n_j}\right) + \operatorname{dist}\left(\phi\left(t_0, s_{n_j}, A_{s_{n_j}}\right), A_{t_0}^\infty\right) \to 0 \tag{2.9}$$

as $j \to \infty$. Finally, by continuity, $\bar{a} = \phi(t, t_0, b_{n_j}) \to \phi(t, t_0, \bar{b})$, so $\bar{a} = \phi(t, t_0, \bar{b})$, which means that $A_t^\infty \subset \phi(t, t_0, A_{t_0}^\infty)$.
The convergence of the second term in (2.9) to zero follows from Lemma 1.27. □

3.2. Invariant subfamily of negative invariant family of sets. A negative invariant family of sets also contains an invariant subfamily, but the proof is somewhat more complicated than in the positive invariant case just considered. It will be given first for discrete time processes.

LEMMA 2.21. *Let $\mathcal{A} = (A_n)_{n\in\mathbb{Z}}$ be a family of nonempty compact subsets of X which is ϕ-negatively invariant for a discrete time process ϕ, i.e., $A_n \subset \phi(n, n_0, A_{n_0})$ for all $n \geq n_0$. Then there exists a family of nonempty compact subsets $\mathcal{A}^\infty = (A_n^\infty)_{n\in\mathbb{Z}}$ of $\mathcal{A}$, which is ϕ-invariant, i.e., $\phi(n, n_0, A_{n_0}^\infty) = A_n^\infty$ for all $n \geq n_0$.*

PROOF. Define $B_{n,0} := A_n$ for all $n \in \mathbb{Z}$, and for a fixed $n \in \mathbb{Z}$, define $B_{n,-1}$ to be the maximal subset of $B_{n-1,0}$ such that

$$B_{n,0} = \phi\big(n, n-1, B_{n,-1}\big) .$$

To see that the set $B_{n,-1}$ is compact, consider a sequence $(b_k)_{k\in\mathbb{N}}$ in $B_{n,-1}$ and define $a_k = \phi(n, n-1, b_k)$ for all $k \in \mathbb{N}$. Since $B_{n,0}$ and $B_{n-1,0}$ are compact and both $a_k \in B_{n,0}$ and $b_k \in B_{n,-1} \subset B_{n-1,0}$, there are convergent subsequences

$a_{k_j} \to \bar{a} \in B_{n,0}$ and $b_{k_j} \to \bar{b} \in B_{n-1,0}$. Then by the continuity of the mapping $\phi(n, n-1, \cdot)$, one has $a_{k_j} = \phi(n, n-1, b_{k_j}) \to \phi(n, n-1, \bar{b})$, so $\phi(n, n-1, \bar{b}) = \bar{a}$. This means that $\bar{b} \in B_{n,-1}$, which implies compactness of $B_{n,-1}$. Repeating this procedure gives a sequence of nonempty compact subsets $B_{n,-j}$ for $j \in \mathbb{N}$ such that $B_{n,-j} = \phi\big(n-j, n-j-1, B_{n,-j-1}\big)$. This means that

$$B_{n,0} = \phi\big(n, n-j, B_{n,-j}\big) \quad \text{for all } j \in \mathbb{N}_0 .$$

It is proved now that $B_{n+k+1,-k-1} \subset B_{n+k,-k}$ for each $k \in \mathbb{N}$, which yields a nested family of nonempty compact subsets of A_n. To see this, consider the case $k = 1$ and recall that $\phi(n+1, n, B_{n+1,-1}) = A_{n+1}$. Using the two-parameter semi-group property, this implies that

$$\begin{aligned} A_{n+2} = \phi(n+2, n, B_{n+2,-2}) &\subset \phi(n+2, n+1, A_{n+1}) \\ &= \phi\big(n+2, n+1, \phi(n+1, n, B_{n+1,-1})\big) \\ &= \phi(n+2, n, B_{n+1,-1}) . \end{aligned}$$

The set A_n^∞, defined by

$$A_n^\infty = \bigcap_{k \in \mathbb{N}} B_{n+k,-k} \quad \text{for all } n \in \mathbb{N}$$

is a nonempty compact subset of A_n. It remains to prove that the family of nonempty compact subsets $\mathcal{A}^\infty = (A_n^\infty)_{n \in \mathbb{Z}}$ is ϕ-invariant, i.e., $\phi\big(n, n_0, A_{n_0}^\infty\big) = A_n^\infty$ for all $n \geq n_0$.

($\subset$) Let $\bar{a} \in A_{n_0}^\infty$. Then $\bar{a} \in B_{n_0+k,-k}$ and $\phi(n, n_0, \bar{a}) \in \phi\big(n, n_0, B_{n_0+k,-k}\big)$ for all $k \in \mathbb{N}$. Moreover, for $k \geq n - n_0$ and $\ell = k - n + n_0 \geq 0$, one has

$$B_{n_0+k,-k} = B_{n_0+(\ell+n-n_0),-(\ell+n-n_0)} = B_{n+\ell,-\ell-n+n_0} .$$

However,

$$\phi\big(n, n_0, B_{n+\ell,-\ell-n+n_0}\big) = B_{n+\ell,-\ell}$$

by construction, which means that

$$\phi(n, n_0, \bar{a}) \in \phi\big(n, n_0, B_{n_0+k,-k}\big) = \phi\big(n, n_0, B_{n+\ell,-\ell-n+n_0}\big) = B_{n+\ell,-\ell} .$$

Hence,

$$\phi(n, n_0, \bar{a}) \in \bigcap_{\ell \in \mathbb{N}} B_{n+\ell,-\ell} = A_n^\infty ,$$

which implies that $\phi\left(n, n_0, A_{n_0}^\infty\right) \subset A_n^\infty$.

($\supset$) Let $\bar{a} \in A_n^\infty$. Then one obtains $\bar{a} \in B_{n+\ell,-\ell}$ for all $\ell \in \mathbb{N}$. However,

$$\begin{aligned} B_{n+\ell,-\ell} &= B_{n_0+(\ell+n-n_0),-\ell} \\ &= B_{n_0+k,-(k-n+n_0)} \quad \text{for all } k = \ell + n - n_0 \geq n - n_0 . \end{aligned}$$

Moreover,

$$\phi\big(n, n_0, B_{n_0+k,-k}\big) = B_{n_0+k,-(k-n+n_0)} ,$$

so $\bar{a} \in \phi\big(n, n_0, B_{n_0+k,-k}\big)$. Hence, there exist $b_k \in B_{n_0+k,-k} \subset A_{n_0}$ such that $\phi(n, n_0, b_k) = \bar{a}$. Note that $b_k \in A_{n_0}$ for all $k \in \mathbb{N}$, which is a compact set, so there exists a convergent subsequence $b_{k_j} \to \bar{b}$ in A_{n_0}. In fact, $\bar{b} \in A_{n_0}^\infty$, since

$$\operatorname{dist}\left(\bar{b}, A_{n_0}^\infty\right) \leq \operatorname{dist}\left(\bar{b}, b_{k_j}\right) + \operatorname{dist}\left(B_{n_0+k_j,-k_j}, A_{n_0}^\infty\right) \to 0 \quad \text{as } j \to \infty .$$

It follows by continuity that $\bar{a} = \phi(n, n_0, b_{k_j}) \to \phi(n, n_0, \bar{b})$ as $j \to \infty$, so $\bar{a} = \phi(n, n_0, \bar{b})$, which finally means that $A_n^\infty \subset \phi(n, n_0, A_{n_0}^\infty)$. □

THEOREM 2.22. *Let $\mathcal{A} = (A_t)_{t\in\mathbb{R}}$ be a family of nonempty compact subsets of X which is ϕ-negatively invariant for a continuous time process ϕ, i.e., $A_t \subset \phi(t, t_0, A_{t_0})$ for all $t \geq t_0$. Then there exists a family of nonempty compact subsets $\mathcal{A}^\infty = (A_t^\infty)_{t\in\mathbb{R}}$ of $\mathcal{A}$, which is ϕ-invariant, i.e., $\phi(t, t_0, A_{t_0}^\infty) = A_t^\infty$ for all $t \geq t_0$.*

PROOF. To simplify the notation, the proof will be given only for the special case that $A_t \equiv A$ for all $t \in \mathbb{R}$. First consider the process restricted the dyadic numbers in $\mathbb{R}$. Let $\mathbb{T}_0 = \mathbb{Z}$ and $\mathbb{D}_n = \{j2^{-n} : j = 0, 1, \ldots, 2^n\}$, and define

$$\mathbb{T}_n := \mathbb{Z} + \mathbb{D}_n = \{k + t_j^n : k \in \mathbb{Z} \text{ and } t_j^n \in \mathbb{D}_n\} \quad \text{for all } n \in \mathbb{N}$$

and apply the result of Lemma 2.21 to the discrete time system formed by the restriction $\phi|_{\mathbb{T}_0}$ of the mapping ϕ to the time set $\mathbb{T}_0$. This gives a family $\mathcal{A}^0 = (A_t^0)_{t\in\mathbb{T}_0}$ of nonempty compact subsets of A which is the maximal $\phi|_{\mathbb{T}_0}$-invariant family of subsets of A, i.e., with $\phi(n+1, n, A_n^0) = A_{n+1}^0$ for all $n \in \mathbb{Z}$.
A difficulty is that the $\phi(n + t, n, A_n^0)$ may not be a subset of A for all $t \in (0,1)$. Therefore, the procedure will be repeated for the discrete time system formed by the restriction $\phi|_{\mathbb{T}_1}$ of the mapping ϕ to the time set $\mathbb{T}_1$, and one obtains a family $\mathcal{A}^1 = (A_t^1)_{t\in\mathbb{T}_1}$ of nonempty compact subsets of A which is the maximal $\phi|_{\mathbb{T}_1}$-invariant family of subsets of A, i.e., with

$$\phi\left(t_{j+1}^1, t_j^1, A_{t_j^1}^1\right) = A_{t_{j+1}^1}^1$$

for every $t_j^1, t_{j+1}^1 \in \mathbb{T}_1$ with $t_{j+1}^1 - t_j^1 = \frac{1}{2}$. By this and the semi-group property, one has

$$\begin{aligned}
A_{m+1}^1 &= \phi(m+1, m+\tfrac{1}{2}, A_{m+1/2}^1) \\
&= \phi(m+1, m+\tfrac{1}{2}, \phi(m+\tfrac{1}{2}, m, A_m^1)) \\
&= \phi(m+1, m, A_m^1) \quad \text{for all } m \in \mathbb{Z},
\end{aligned}$$

so $\mathcal{A}^1$ is also a $\phi|_{\mathbb{T}_0}$-invariant family of compact subsets of A. But since $\mathcal{A}^0$ is the maximal $\phi|_{\mathbb{T}_0}$-invariant family of compact subset of A, one has $A_t^1 \subset A_t^0$ for all $t \in \mathbb{T}_0$.
Now repeat this procedure with the discrete time system formed by the restriction $\phi|_{\mathbb{T}_n}$ of the mapping ϕ to the time set $\mathbb{T}_n$ and obtain a family $\mathcal{A}^n = (A_t^n)_{t\in\mathbb{T}_n}$ of nonempty compact subsets of A, which is the maximal $\phi|_{\mathbb{T}_n}$-invariant family of subsets of A. Note that this family is also $\phi|_{\mathbb{T}_{n-1}}$-invariant. Hence, $A_t^n \subset A_t^{n-1}$ for all $t \in \mathbb{T}_{n-1} \cap \mathbb{T}_n = \mathbb{T}_{n-1}$ and $n \in \mathbb{N}$.
Thus for each $t_\ell \in \mathbb{T}_\ell$ for an arbitrary $\ell \in \mathbb{N}$, the subsets $A_{t_\ell}^n$ for $n \geq \ell$ are nonempty, compact and nested. Hence, the set defined by

$$A_{t_\ell}^\infty = \bigcap_{n\geq\ell} A_{t_\ell}^n$$

is a nonempty compact subset of A. In this way, one obtains a family $\mathcal{A}^\infty = (A_{t_d}^\infty)_{t_d\in\bigcup_{\ell\geq 0}\mathbb{T}_\ell}$ of nonempty compact subsets of A.
Moreover, by Lemma 2.21, the family $\mathcal{A}^\infty$ is $\phi|_{\mathbb{T}_n}$-invariant for each $n \in \mathbb{N}$, i.e.,

$$\phi\left(t_{j+1}^n, t_j^n, A_{t_j^n}^\infty\right) = A_{t_{j+1}^n}^\infty$$

for every $t_j^n, t_{j+1}^n \in \mathbb{T}_n$ with $t_{j+1}^n - t_j^n = 2^{-n}$. From this and the semi-group property, it follows that $\phi(t_1, t_0, A_{t_0}^\infty) = A_{t_1}^\infty$ for all dyadic numbers $t_0 \leq t_1$ in $\mathbb{R}$. Finally, for non-dyadic t, one defines A_t^∞ by

$$A_t^\infty = \phi\big(t, t_0, A_{t_0}^\infty\big)\,,$$

where $t_0 < t$ is an arbitrary dyadic number (note that this definition is independent of the choice of t_0 by the semi-group property for the dyadic numbers). It follows that A_t^∞ is a nonempty compact subset of A. By continuity of ϕ in its first variable,

$$h\big(A_t^\infty, A_{t_n}^\infty\big) \le h\big(\phi(t, t_0, A_{t_0}^\infty), \phi(t_n, t_0, A_{t_0}^\infty)\big) \to 0 \quad \text{as } t_n \to t$$

for dyadic $t_n \ge t_0$ with $t_n < t$.
Finally, extend that definition of $\mathcal{A}^\infty$ by $\mathcal{A}^\infty = (A_t^\infty)_{t\in\mathbb{R}}$. It remains to show that $A_t^\infty = \phi(t, s, A_s^\infty)$ for all $s < t$ in $\mathbb{R}$. From above, the only remaining case to show is for s non-dyadic. The desired result follows from the definition of A_s^∞ and the semi-group property, i.e.,

$$\phi(t, s, A_s^\infty) = \phi\big(t, s, \phi\big(s, t_0, A_{t_0}^\infty\big)\big) = \phi\big(t, t_0, A_{t_0}^\infty\big) = A_t^\infty\,,$$

where $t_0 < s$ is dyadic but otherwise arbitrary. Thus, $\mathcal{A}^\infty$ is ϕ-invariant, which finishes the proof of this theorem. □

Endnotes. The concept of a process is due to Dafermos [**62**] and Hale [**90**]. Skew product flows originated in ergodic theory and were extensively studied in connection with ordinary differential equations by Sell [**217, 218**]. See also Wenxian Shen & Yingfei Yi [**221**] as well as the monographs Carvalho, Langa & Robinson [**35**], Cheban [**38**], Chepyzhov & Vishik [**43**], Fink [**77**] and Kato, Martynyuk & Shestakov [**104**]. Section 1.3 on entire solutions and invariant sets is based on Kloeden & Marín-Rubio [**132**], and see Kloeden & Rodrigues [**139**] for the dynamics of a class of differential equations which are more general than almost periodic. The figures in this chapter were made by van Geene [**231**].

CHAPTER 3

Attractors

Simple generalizations of concepts for autonomous dynamical systems to nonautonomous dynamical systems are not always adequate or appropriate. For instance, it was seen in the last chapter that for nonautonomous dynamical systems, it is often too restrictive to consider the invariance of just a single set and that instead a family of subsets is more appropriate. A similar situation also applies to attractors, which are the most important examples of invariant sets.

Attractors of autonomous dynamical systems are given by ω-limit sets, which are invariant sets. Since the solution of a process φ depends both on initial time t_0 and initial value x_0, ω-limit sets for a process will also depend on both of these two parameters, specifically

$$\omega(t_0, x_0) = \left\{ x \in X : \lim_{n\to\infty} \varphi(t_n, t_0, x_0) = x \text{ for some sequence } t_n \to \infty \right\}.$$

As in the autonomous case, one can show that $\omega(t_0, x_0)$ is a nonempty compact set when, for example, the forward trajectory $\bigcup_{t\geq t_0} \{\varphi(t, t_0, x_0)\}$ is precompact. However, unlike its autonomous counterpart, a nonautonomous ω-limit set $\omega(t_0, x_0)$ may not be invariant for the process.

As an example, consider the nonautonomous scalar differential equation

$$\dot{x} = -x + e^{-t},$$

which can be solved with the variation of constants formula to give the explicit solution

$$x(t, t_0, x_0) = e^{-(t-t_0)}x_0 + (t - t_0)\, e^{-t}.$$

This implies that

$$\lim_{t\to\infty} x(t, t_0, x_0) = 0 \quad \text{for all } (t_0, x_0) \in \mathbb{R} \times \mathbb{R},$$

so the nonautonomous ω-limit set is given by

$$\omega(t_0, x_0) = \{0\} \quad \text{for all } (t_0, x_0) \in \mathbb{R} \times \mathbb{R}.$$

However, $x(t, t_0, 0) = (t - t_0)\, e^{-t} \neq 0$ for all $t > t_0$, i.e., the ω-limit set here is not invariant in the sense of autonomous systems.

Nonautonomous sets. Let ϕ be a process on a metric space (X, d), and consider a family $\tilde{\mathcal{M}} = (M_t)_{t\in\mathbb{T}}$ of subsets of X. For a more compact and elegant formulation, such families $\tilde{\mathcal{M}}$ will henceforth be viewed equivalently as subsets of the extended phase space $\mathbb{T} \times X$, and the translation is as follows. The family $\tilde{\mathcal{M}}$ induces a subset $\mathcal{M} \subset \mathbb{T} \times X$, defined by

$$\mathcal{M} := \{(t, x) : x \in M_t\}$$

and, conversely, a subset $\mathcal{M}$ of the extended phase space $\mathbb{T} \times X$ leads to a family $\tilde{\mathcal{M}} = (M_t)_{t\in\mathbb{T}}$ of subsets of X with

$$M_t := \{x \in X : (t,x) \in \mathcal{M}\} \quad \text{for all } t \in \mathbb{T}.$$

The advantage of the new formulation is that $\mathcal{M}$ is a set, which makes the direct use of set-valued operations possible, and the notation becomes easier to read. Such sets $\mathcal{M}$ are called *nonautonomous sets* in the following. It will become clear soon that all interesting objects in the nonautonomous context are nonautonomous sets, i.e., subsets of the *extended* phase space, whereas the main interest focusses on subsets of the phase space in an autonomous setting. The precise definition of a nonautonomous set is given as follows.

DEFINITION 3.1 (Nonautonomous set of a process). Let ϕ be a process on a metric space (X,d). A subset $\mathcal{M}$ of the extended phase space $\mathbb{T} \times X$ is called a *nonautonomous set*, and for each $t \in \mathbb{T}$, the set

$$M_t := \{x \in X : (t,x) \in \mathcal{M}\}$$

is called the *t-fiber of* $\mathcal{M}$. A nonautonomous set $\mathcal{M}$ is said to be *invariant* if $\phi(t,t_0,M_{t_0}) = M_t$ for all $t \geq t_0$. In general, $\mathcal{M}$ is said to have a topological property (such as compactness or closedness) if each fiber of $\mathcal{M}$ has this property.

The notion of a nonautonomous set will also be used in the setting of skew product flows.

DEFINITION 3.2 (Nonautonomous set of a skew product flow). Let (θ,φ) be a skew product flow on a base set P and a metric phase space (X,d). A subset $\mathcal{M}$ of the extended phase space $P \times X$ is called a *nonautonomous set*, and for each $p \in P$, the set

$$M_p := \{x \in X : (p,x) \in \mathcal{M}\}$$

is called *p-fiber of* $\mathcal{M}$. A nonautonomous set $\mathcal{M}$ is said to be *invariant* if $\varphi(t,p,M_p) = M_{\theta_t(p)}$ for all $t \geq 0$ and $p \in P$. In general, $\mathcal{M}$ is said to have a topological property (such as compactness or closedness) if each fiber of $\mathcal{M}$ has this property.

1. Attractors of processes

There are basically two ways to define attraction of a compact and invariant nonautonomous set $\mathcal{A}$ for a process ϕ on a metric space (X,d) with time set $\mathbb{T}$. The first, and perhaps more obvious, corresponds to the attraction in Lyapunov asymptotic stability, is called *forward attraction* and involves a moving target, while the latter, called *pullback attraction*, involves a fixed target set with progressively earlier starting time. In general, these two types of attraction are independent concepts, while for the autonomous case, they are equivalent.

DEFINITION 3.3 (Nonautonomous attractivity). Let ϕ be a process. A nonempty, compact and invariant nonautonomous set $\mathcal{A}$ is said to be

(i) *forward attracting* if

$$\lim_{t\to\infty} \operatorname{dist}\big(\phi(t,t_0,x_0), A_t\big) = 0 \quad \text{for all } x_0 \in X \text{ and } t_0 \in \mathbb{T},$$

(ii) and *pullback attracting* if

$$\lim_{t_0 \to -\infty} \operatorname{dist}\big(\phi(t, t_0, x_0), A_t\big) = 0 \quad \text{for all } x_0 \in X \text{ and } t \in \mathbb{T}.$$

Moreover, if the forward attraction in (i) is uniform with respect to $t_0 \in \mathbb{T}$, or equivalently, if the pullback attraction in (ii) is uniform with respect to $t \in \mathbb{T}$, then $\mathcal{A}$ is called *uniformly attracting.*

Figure 3.1 and Figure 3.2 illustrate forward and pullback attraction, respectively, of a nonautonomous set with singleton sets as fibers $A_t = \{\rho(t)\}$, i.e., an entire solution of the process.

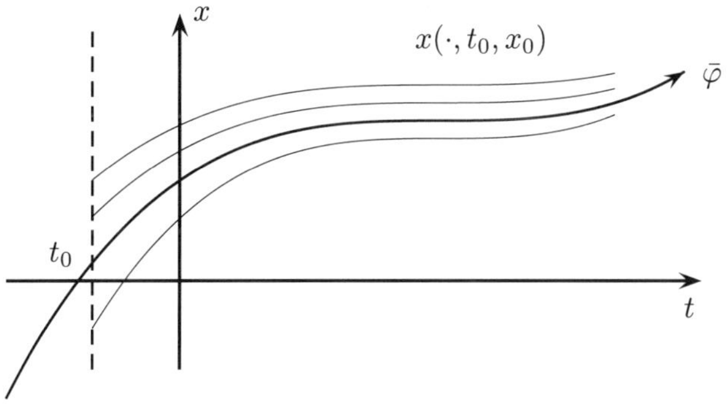

FIGURE 3.1. Forward attraction.

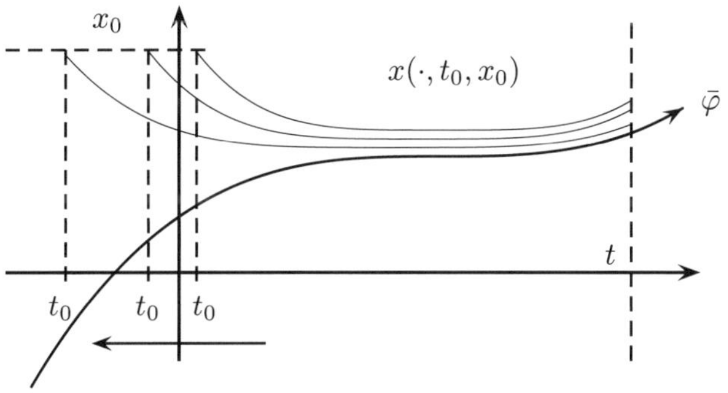

FIGURE 3.2. Pullback attraction.

In an autonomous system, the solutions depend only on the elapsed time $t - t_0$. Moreover, the limit relation $t - t_0 \to \infty$ either holds when $t \to \infty$ with t_0 fixed or as $t_0 \to -\infty$ with t fixed, so pullback and forward convergence are equivalent for an autonomous system.

Two types of nonautonomous attractors for processes are possible, depending which of the above types of attraction is used. It is required that the component subsets of such attractors are compact and that they attract bounded subsets D of initial values in X (rather than just individual points), in the sense that

$$\operatorname{dist}\big(\phi(t, t_0, D), A_t\big) \to 0 \begin{cases} \text{as } t \to \infty \text{ with } t_0 \text{ fixed (forward case),} \\ \text{as } t_0 \to -\infty \text{ with } t \text{ fixed (pullback case).} \end{cases}$$

Compare this with Definition 1.31 for the definition of an autonomous global attractor.

DEFINITION 3.4 (Nonautonomous attractors). Let ϕ be a process. A nonempty and invariant nonautonomous set $\mathcal{A}$ is called

(i) a *forward attractor* if it forward attracts bounded subsets of X,
(ii) a *pullback attractor* if it pullback attracts bounded subsets of X, and
(iii) a *uniform attractor* if it uniformly attracts bounded subsets of X.

Forward and pullback attractors will be discussed in more detail and generality in the context of skew product flows in the following sections. In general, they are independent concepts and one can exist without the other.

EXAMPLE 3.5. The nonautonomous set $\mathbb{R}\times\{0\}$, i.e., the trivial solution, is a forward attractor but not a pullback attractor of the system

$$\dot{x} = -2tx \tag{3.1}$$

with the general solution $x(t, t_0, x_0) = x_0 e^{-(t^2-t_0^2)}$, and a pullback attractor but not a forward attractor of the system

$$\dot{x} = 2tx \tag{3.2}$$

with the general solution $x(t, t_0, x_0) = x_0 e^{t^2-t_0^2}$.

This example demonstrates that forward attraction can be seen as an attraction concept for the *future* of the system, since the coefficient $-2t$ of (3.1) is negative for $t > 0$. On the other hand, pullback attraction means basically attraction for the *past* of the system, see the negativity of $2t$ for $t < 0$ in (3.2). The concept of a uniform attractor, however, is concerned with attractivity for the *entire time.*

EXAMPLE 3.6. Consider the nonautonomous scalar ordinary differential equation

$$\dot{x} = -x + 2\sin t\,. \tag{3.3}$$

If $x_1(t)$ and $x_2(t)$ are any two solutions, then their difference $z(t) = x_1(t) - x_2(t)$ satisfies the homogeneous linear differential equation

$$\dot{z} = -z$$

with the explicit solution $z(t) = z(t_0)e^{-(t-t_0)}$, so

$$\big|x_1(t) - x_2(t)\big| = \big|x_1(t_0) - x_2(t_0)\big|e^{-(t-t_0)} \to 0 \quad \text{as } t \to \infty\,,$$

from which it follows that all solutions converge to each other in time. What do they converge to?

The explicit solution of the nonautonomous differential equation (3.3) with initial value $x(t_0) = x_0$ is

$$\begin{aligned} x(t, t_0, x_0) &= x_0 e^{-(t-t_0)} + 2e^{-t}\int_{t_0}^{t} e^s \sin s\,\mathrm{d}s \\ &= \big(x_0 - (\sin t_0 - \cos t_0)\big)\, e^{-(t-t_0)} + (\sin t - \cos t)\,, \end{aligned}$$

from which it is clear that the forward limit $\lim_{t\to\infty} x(t, t_0, x_0)$ does not exist. On the other hand, the pullback limit does exist for all t and x_0, i.e.,

$$\lim_{t_0\to-\infty} x(t, t_0, x_0) = \sin t - \cos t =: \rho(t)\,,$$

and is independent of x_0, i.e.,

$$\lim_{t_0 \to -\infty} \left| x(t, t_0, x_0) - \rho(t) \right| = 0 .$$

Hence, the nonautonomous set $\mathcal{A}$ having the singleton fibers $A_t := \{\rho(t)\}$ is pullback attracting for the solution process.

Moreover, it is easily shown that $\rho(t)$ is a solution of the nonautonomous differential equation (3.3) and since all solutions converge to each other forward in time, the forward convergence

$$\lim_{t \to \infty} \left| x(t, t_0, x_0) - \rho(t) \right| = 0$$

also holds.

EXERCISE 3.7. The nonautonomous set $\mathcal{A}$ in Example 3.6 is both a pullback and forward attractor of the nonautonomous differential equation (3.3). Find other forward attractors of (3.3) which are not pullback attractors.

This exercise demonstrates that forward attractors can be nonunique, and this is quite typical for forward attractors. For pullback attractors, however, the following uniqueness result can be proved.

PROPOSITION 3.8 (Uniqueness of pullback attractors). *Suppose that a process ϕ has two pullback attractors $\mathcal{A}$ and $\bar{\mathcal{A}}$ such that both $\bigcup_{t \le 0} A_t$ and $\bigcup_{t \le 0} \bar{A}_t$ are bounded. Then $\mathcal{A} = \bar{\mathcal{A}}$.*

PROOF. The boundedness of $\bigcup_{t \le 0} A_t$ implies for all $t \in \mathbb{T}$ that

$$\begin{aligned} \operatorname{dist}\left(A_t, \bar{A}_t\right) &= \lim_{t_0 \to -\infty} \operatorname{dist}\left(\phi(t, t_0, A_{t_0}), \bar{A}_t\right) \\ &\le \lim_{t_0 \to -\infty} \operatorname{dist}\left(\phi(t, t_0, \textstyle\bigcup_{\tau \le 0} A_\tau), \bar{A}_t\right) = 0 . \end{aligned}$$

Analogously, one shows that $\operatorname{dist}\left(\bar{A}_t, A_t\right) = 0$, which finishes the proof, since both the sets A_t and $\bar{A}_t$ are compact. □

EXERCISE 3.9. Demonstrate by using a concrete example that an analogous statement of Proposition 3.8 for forward attractors is not possible.

EXERCISE 3.10. An invariant nonautonomous set $\mathcal{A}$ (such as a pullback attractor) consists of entire solutions, see Lemma 2.15. Give an example of a process ϕ which has entire solutions that are not contained in the pullback attractor.

2. Attractors of skew product flows

Let (θ, φ) be a skew product flow on a base space P and a state space X with time set $\mathbb{T} = \mathbb{R}$ or $\mathbb{Z}$, where (P, d_P) and (X, d_X) are metric spaces. Then $\pi = (\theta, \varphi)$ is an autonomous semi-dynamical system on the extended state space $\mathfrak{X} := P \times X$. The definition of a global attractor for an autonomous semi-dynamical system π was given in Chapter 1. Specifically, a nonempty compact subset $\mathcal{A}$ of $\mathfrak{X}$ which is π-invariant, i.e., which satisfies $\pi(t, \mathcal{A}) = \mathcal{A}$ for all $t \in \mathbb{T}_0^+$ is called a *global attractor* of π if

$$\lim_{t \to \infty} \operatorname{dist}_{\mathfrak{X}}\left(\pi(t, D), \mathcal{A}\right) = 0$$

for every nonempty bounded subset D of $\mathcal{A}$. Suppose that P is compact. Then the global attractor $\mathcal{A}$ of π has the form

$$\mathcal{A} = \bigcup_{p \in P} \{(p, x) : x \in A_p\},$$

where A_p is a nonempty compact subset of X for each $p \in P$, and the π-invariance property $\pi(t, \mathcal{A}) = \mathcal{A}$ for $t \in \mathbb{T}_0^+$ is equivalent to the φ-invariance property $\varphi(t, p, A_p) = A_{\theta_t(p)}$ for all $t \in \mathbb{T}_0^+$ and $p \in P$.

A global attractor of the autonomous system π is a possible candidate for an attractor of the nonautonomous dynamical system described by the skew product flow. A disadvantage of this definition is that the extended state space $\mathcal{X}$ includes the base space P as a component, which often does not have the same physical significance as the state space X.

Other types of attractors consisting of a family of nonempty compact subsets of the state space X have also been proposed for a skew product flow (θ, φ). These are analogues of the forward and pullback attractors of a process.

DEFINITION 3.11 (Pullback and forward attractor for skew product flows). Let (θ, φ) be a skew product flow. A nonempty, compact and invariant nonautonomous set $\mathcal{A}$ is called a *pullback attractor* of (θ, φ) if the *pullback convergence*

$$\lim_{t \to \infty} \operatorname{dist}_X(\varphi(t, \theta_{-t}(p), D), A_p) = 0$$

holds for every nonempty bounded subset D of X and $p \in P$, and is called a *forward attractor* if the *forward convergence*

$$\lim_{t \to \infty} \operatorname{dist}_X\left(\varphi(t, p, D), A_{\theta_t(p)}\right) = 0$$

holds for every nonempty bounded subset D of X and $p \in P$.

As for processes, the concepts of forward and pullback attractors for skew products are generally independent of each other, and one can exist without the other existing. If the above limit is replaced by $\lim_{t\to\infty} \sup_{p \in P} \operatorname{dist}_X(\cdot, \cdot)$, then the attractors are called *uniform pullback* and *uniform forward* attractors, respectively. If either of the limits is uniform in this sense, then so is the other and the attractor is both a uniform pullback and a uniform forward attractor, which will be called simply a *uniform attractor.*

The relationship between these different kinds of nonautonomous attractors will be discussed in some detail in Section 4 of this chapter.

EXERCISE 3.12. Formulate and prove the corresponding statement about entire solutions and pullback attractors in Exercise 3.10 for skew product flows.

EXAMPLE 3.13. Reconsider the nonautonomous scalar ordinary differential equation (3.3), now writing $p(t)$ instead of $\sin t$,

$$\dot{x} = -x + 2p(t),$$

with the initial condition $x(0) = x_0$. In the spirit of skew product flows as introduced in Section 2 of Chapter 2, the general solution of this equation depends on both p and x_0. The initial value problem has thus the explicit solution

$x(t) = x(t, p, x_0)$ given by

$$x(t) = x_0 e^{-t} + 2e^{-t} \int_0^t e^s p(s)\, \mathrm{d}s\,.$$

Introduce shift operators on the space $P = \{p(t + \cdot) : 0 \le t \le 2\pi\}$ defined by $\theta_t(p(\cdot)) = p(t + \cdot)$ and consider the solution corresponding to the driving term $\theta_{-\tau}(p(\cdot)) = p(-\tau + \cdot)$ at time τ, i.e.,

$$\begin{aligned} x\big(\tau, \theta_{-\tau}(p(\cdot)), x_0\big) &= x_0 e^{-\tau} + 2e^{-\tau} \int_0^\tau e^s \theta_{-\tau}(p(s))\, \mathrm{d}s \\ &= x_0 e^{-\tau} + 2e^{-\tau} \int_0^\tau e^s p(s-\tau)\, \mathrm{d}s \\ &= x_0 e^{-\tau} + 2 \int_0^\tau e^{s-\tau} p(s-\tau)\, \mathrm{d}s \\ &= x_0 e^{-\tau} + 2 \int_{-\tau}^0 e^t p(t)\, \mathrm{d}t\,, \end{aligned}$$

where the substitution $t := s - \tau$ has been used. The pullback limit as $\tau \to \infty$ gives

$$\lim_{\tau \to \infty} x\big(\tau, \theta_{-\tau}(p(\cdot)), x_0\big) = \alpha(p(\cdot)) := 2 \int_{-\infty}^0 e^t p(t)\, \mathrm{d}t$$

for x_0 in an arbitrary bounded subset D consists of singleton fibers $A_p = \{\alpha(p(\cdot))\}$, $p \in P$, and corresponds to the entire solution

$$\rho(t) := \sin t - \cos t$$

in the process version of this differential equation in Example 3.6, i.e., $\rho(t) = \alpha\big(\theta_t(p(\cdot))\big)$ for all $t \in \mathbb{R}$.

The pullback attraction in this example is uniform, and the pullback attractor is also a forward attractor, and hence a uniform attractor. Moreover, the autonomous semi-dynamical system $\pi = (\theta, \varphi)$ on the extended state space $P \times \mathbb{R}$ has a global attractor given by

$$\mathcal{A} = \bigcup_{p(\cdot) \in P} \{(p(\cdot), \alpha(p(\cdot)))\}\,.$$

REMARK 3.14. The analysis of this system as a skew product flow is somewhat more complicated and less transparent than its analysis as a process in Example 3.6. This is typical and is why the process formulation will often be used in subsequent examples (whenever it is possible).

EXAMPLE 3.15. The difference equation in Example 2.11 generates a discrete time skew product flow with cocycle mapping

$$\varphi(n, p, x) = \nu^n x + \sum_{j=0}^{n-1} \nu^{n-1-j} q_j \quad \text{for all } n \in \mathbb{N}\,, \tag{3.4}$$

on the state space $X = \mathbb{R}$. The base space $P = [-1, 1]^{\mathbb{Z}}$ is the space of bi-infinite sequences $p = (q_n)_{n \in \mathbb{N}}$ taking values in $[-1, 1]$ and θ is the left shift operator on this sequence space.

Replacing p by $\theta_{-n}(p)$ in (3.4) implies

$$\varphi\big(n, \theta_{-n}(p), x\big) = \nu^n x + \sum_{j=0}^{n-1} \nu^{n-1-j} q_{-n+j} \quad \text{for all } n \in \mathbb{N},$$

which can be reindexed as

$$\varphi\big(n, \theta_{-n}(p), x\big) = \nu^n x + \sum_{k=-n}^{-1} \nu^{-k-1} q_k \quad \text{for all } n \in \mathbb{N}.$$

Taking pullback convergence gives

$$\lim_{n\to\infty} \varphi\big(n, \theta_{-n}(p), x\big) = \alpha(p) := \sum_{k=-\infty}^{-1} \nu^{-k-1} q_k .$$

The pullback attractor $\mathcal{A}$ thus consists of singleton fibers $A_p = \{\alpha(p)\}$ for $p = (q_n)_{n\in\mathbb{N}} \in P$. Since the pullback convergence here is in fact uniform in $p \in P$, the pullback attractor is also a uniform forward attractor, and hence a uniform attractor. Moreover, the corresponding subset $\mathbb{A}$ of $\mathfrak{X}$ formed from $\mathcal{A}$ is also the global attractor of the autonomous semi-dynamical system $\pi = (\theta, \varphi)$ on the extended state space $P \times \mathbb{R}$.

3. Existence of pullback attractors

There are generalizations of Theorem 1.23 on the existence of an attractor for autonomous systems in Chapter 1 to pullback attractors for processes and skew product flows. These are also based on the supposed existence of an absorbing set, which is now absorbing in the pullback sense.

DEFINITION 3.16 (Pullback absorbing set for processes). Let ϕ be a process on a metric space (X, d). A nonempty compact subset B of X is called *pullback absorbing* if for each $t \in \mathbb{T}$ and every bounded subset D of X, there exists a $T = T(t, D) > 0$ such that

$$\phi(t, t_0, D) \subset B \quad \text{for all } t_0 \in \mathbb{T} \text{ with } t_0 \leq t - T .$$

DEFINITION 3.17 (Pullback absorbing set for skew product flows). Let (θ, ϕ) be a skew product flow on a metric space (X, d). A nonempty compact subset B of X is called *pullback absorbing* if for each $p \in P$ and every bounded subset D of X, there exists a $T = T(p, D) > 0$ such that

$$\varphi\big(t, \theta_{-t}(p), D\big) \subset B \quad \text{for all } t \geq T .$$

The existence theorems will be presented here under basic but restricted assumptions, which will then be relaxed and generalized.

3.1. Existence of pullback attractors for processes.

The following theorem is a simple generalization of Theorem 1.23 for attractors of autonomous semi-dynamical systems given in Chapter 1.

THEOREM 3.18 (Existence of pullback attractors for processes). *Let ϕ be a process on a complete metric space X with a compact pullback absorbing set B such that*

$$\phi(t, t_0, B) \subset B \quad \textit{for all } t \geq t_0 .$$

Then there exists a pullback attractor $\mathcal{A}$ with fibers in B uniquely determined by

$$A_t = \bigcap_{\tau \geq 0} \overline{\bigcup_{t_0 \leq -\tau} \phi(t, t_0, B)} \quad \text{for all } t \in \mathbb{T}. \tag{3.5}$$

A proof will not be given here since it is identical to that of Lemma 2.20, in which an invariant family of subsets contained in a positively invariant family was constructed, once the dynamics has entered the positively invariant pullback absorbing set. The proof is also similar to the proof of the corresponding theorem below, Theorem 3.20, for skew product flows.

The formula (3.5) is a kind of a nonautonomous ω-limit set of the set B. As seen in the introduction of this chapter, a naive definition of a nonautonomous ω-limit set leads to a set which is not positively invariant. However, the pullback construction used in (3.5) gives an invariant set and can be regarded as a proper version of a nonautonomous ω-limit set.

EXAMPLE 3.19. Consider a nonautonomous dynamical system in $\mathbb{R}^d$ given by

$$\dot{x} = f(t, x), \tag{3.6}$$

where f is continuously differentiable and satisfies the uniform dissipative condition

$$\langle x, f(t,x) \rangle \leq K - L\|x\|^2 \quad \text{for all } x \in \mathbb{R}^d \text{ and } t \in \mathbb{R} \tag{3.7}$$

with positive constants K and L. These assumptions ensure that the differential equation (3.6) generates a process.

Moreover, any solution $x(t)$ of (3.6) satisfies

$$\begin{aligned} \frac{\mathrm{d}}{\mathrm{d}t}\|x(t)\|^2 &= 2\langle x(t), \dot{x}(t) \rangle \\ &= 2\langle x(t), f(t, x(t)) \rangle \leq 2K - 2L\|x(t)\|^2, \end{aligned}$$

from which, on integrating, it follows that

$$\|x(t)\|^2 \leq \|x(t_0)\|^2 e^{-2L(t-t_0)} + \frac{K}{L}\left(1 - e^{-2L(t-t_0)}\right).$$

Suppose that for a bounded subset D of $\mathbb{R}^d$ with $\|D\| := \sup_{d \in D}\|d\| > 1$, one has $x(t_0) \in D$, and define

$$T := t_0 + \frac{1}{2L}\ln\left(L\|D\|^2\right).$$

Then

$$\|x(t)\|^2 \leq \frac{1}{L} + \frac{K}{L} = \frac{K+1}{L}$$

for $x(t_0) \in D$ and $t_0 \leq t - T$. Thus, the closed ball

$$\overline{B_{\sqrt{(K+1)/L}}(0)} = \left\{x \in \mathbb{R}^d : \|x\|^2 \leq (K+1)/L\right\}$$

is pullback absorbing and positively invariant.

From Theorem 3.18, it follows that the process generated by the differential equation (3.6) has a pullback attractor in $\mathbb{R}^d$ with components subsets $A_t \subset B$.

3.2. Existence of pullback attractors for skew product flows. The counterpart of Theorem 3.18 for skew product flows is the first part of the following theorem. The second part provides some information about a form of forwards convergence of the cocycle mapping, which is different from that in the definition of a forward attractor.

THEOREM 3.20 (Existence of pullback attractors). *Let (θ, φ) be a skew product flow on a complete metric space X with a compact pullback absorbing set B such that*

$$\varphi(t,p,B) \subset B \quad \text{for all } t \geq 0 \text{ and } p \in P. \tag{3.8}$$

Then there exists a unique pullback attractor $\mathcal{A}$ with fibers in B uniquely determined by

$$A_p = \bigcap_{\tau \geq 0} \overline{\bigcup_{t \geq \tau} \varphi(t, \theta_{-t}(p), B)} \quad \text{for all } p \in P. \tag{3.9}$$

If, in addition, (P, d_P) is a compact metric space, then

$$\lim_{t\to\infty} \sup_{p\in P} \operatorname{dist}\big(\varphi(t,p,D), A(P)\big) = 0 \tag{3.10}$$

for any bounded subset D of X, where $A(P) := \overline{\bigcup_{p\in P} A_p} \subset B$.

PROOF. The proof generalizes the proof of Theorem 1.23 for autonomous semi-dynamical systems. It will be divided into two parts, where in the first part, the existence of a pullback attractor is proved, and in the second part, the assertion concerning the compact base set P is treated.

Part 1. Let B be a pullback absorbing set satisfying (3.8), and let A_p be defined as in (3.9) for this absorbing set B.

(i) Firstly, it will be shown for any $p \in P$ that

$$\lim_{t\to\infty} \operatorname{dist}\big(\varphi(t, \theta_{-t}(p), B), A_p\big) = 0. \tag{3.11}$$

Assume to the contrary that there exist sequences $t_j \to \infty$ and $x_j \in \varphi\big(t_j, \theta_{-t_j}(p), B\big) \subset B$ such that $\operatorname{dist}(x_j, A_p) > \varepsilon$ for all $j \in \mathbb{N}$. The set $\{x_j : j \in \mathbb{N}\} \subset B$ is relatively compact, so there is a point $x_0 \in B$ and an index subsequence $j' \to \infty$ such that $x_{j'} \to x_0$. Now $x_{j'} \in \bigcup_{t\geq\tau} \varphi\big(t, \theta_{-t}(p), B\big)$ for all $\tau \geq 0$ with $t_{j'} \geq \tau$, which implies that

$$x_0 \in \overline{\bigcup_{t\geq\tau} \varphi\big(t, \theta_{-t}(p), B\big)} \quad \text{for all } \tau \geq 0.$$

Hence, $x_0 \in A_p$, and this contradiction proves the original assertion (3.11).

(ii) By (3.11), for every $\varepsilon > 0$ and $p \in P$, there exists a $T = T(\varepsilon, p) \geq 0$ such that

$$\operatorname{dist}\big(\varphi(T, \theta_{-T}(p), B), A_p\big) < \varepsilon.$$

Let D be a bounded subset of X. The fact that B is an absorbing set implies that $\varphi\big(t, \theta_{-t-T}(p), D\big) \subset B$ for all sufficiently large t. Hence, by the cocycle property, one has

$$\begin{aligned}\varphi\big(t+T, \theta_{-t-T}(p), D\big) &= \varphi\big(T, \theta_{-T}(p), \varphi(t, \theta_{-t-T}(p), D)\big)\\ &\subset \varphi\big(T, \theta_{-T}(p), B\big).\end{aligned}$$

(iii) The φ-invariance of the nonautonomous set $\mathcal{A}$ will now be shown. By (3.8), the set $F_\tau(p) := \bigcup_{s\geq\tau} \varphi\big(s, \theta_{-s}(p), B\big)$ is contained in B for every $\tau \geq 0$, and by definition, $A_{\theta_{-t}(p)} = \bigcap_{\tau\geq 0} \overline{F_\tau\left(\theta_{-t}(p)\right)}$. Firstly, it will be shown that

$$\varphi\Big(t, \theta_{-t}(p), \Big(\bigcap\nolimits_{\tau\geq 0} \overline{F_\tau(\theta_{-t}(p))}\Big)\Big) = \bigcap\nolimits_{\tau\geq 0} \varphi\big(t, \theta_{-t}(p), \overline{F_\tau(\theta_{-t}(p))}\big)\,, \tag{3.12}$$

and one sees directly that "$\subset$" holds. To prove "$\supset$", let x be contained in the set on the right side. Then for any $\tau \geq 0$, there exists an $x^\tau \in \overline{F_\tau(\theta_{-t}(p))} \subset B$ such that $x = \varphi\big(t, \theta_{-t}(p), x^\tau\big)$. Since the family $\overline{F_\tau(\theta_{-t}(p))}$ is monotonically decreasing with increasing τ, the set $\{x^\tau : \tau \geq 0\}$ has a limit point $\hat{x} \in \bigcap_{\tau\geq 0} \overline{F_\tau(\theta_{-t}(p))}$. By the continuity of $\varphi\big(t, \theta_{-t}(p), \cdot\big)$, it follows that $x = \varphi\big(t, \theta_{-t}(p), \hat{x}\big)$, and thus, $x \in \varphi\big(t, \theta_{-t}(p), \bigcap_{\tau\geq 0} \overline{F_\tau(\theta_{-t}(p))}\big) = \varphi\big(t, \theta_{-t}(p), A_{\theta_{-t}(p)}\big)$.
Hence, equation (3.12), the compactness of $\overline{F_\tau(\theta_{-t}(p))}$ and the continuity of $\varphi\big(t, \theta_{-t}(p), \cdot\big)$ imply that

$$\begin{aligned}
\varphi\big(t, \theta_{-t}(p), A_{\theta_{-t}(p)}\big) &= \bigcap_{\tau\geq 0} \varphi\big(t, \theta_{-t}(p), \overline{F_\tau(\theta_{-t}(p))}\big) \\
&\supset \bigcap_{\tau\geq 0} \overline{\varphi\big(t, \theta_{-t}(p), F_\tau(\theta_{-t}(p))\big)} \\
&= \bigcap_{\tau\geq 0} \overline{\bigcup_{s\geq\tau} \varphi\big(t, \theta_{-t}(p), \varphi(s, \theta_{-t-s}(p), B)\big)} \\
&= \bigcap_{\tau\geq 0} \overline{\bigcup_{s\geq\tau} \varphi\big(t+s, \theta_{-t-s}(p), B\big)} \\
&= \bigcap_{\tau\geq t} \overline{\bigcup_{s\geq\tau} \varphi\big(s, \theta_{-s}(p), B\big)} \supset A_p\,,
\end{aligned}$$

which means that

$$A_p \subset \varphi\left(t, \theta_{-t}(p), A_{\theta_{-t}(p)}\right) \quad \text{for all } t \geq 0 \text{ and } p \in P\,. \tag{3.13}$$

Replacing p by $\theta_{-\tau}(p)$ in (3.13) and using the cocycle property gives

$$\begin{aligned}
\varphi\big(\tau, \theta_{-\tau}(p), A_{\theta_{-\tau}(p)}\big) &\subset \varphi\big(\tau, \theta_{-\tau}(p), \varphi\big(t, \theta_{-\tau-t}(p), A_{\theta_{-\tau-t}(p)}\big)\big) \\
&= \varphi\big(t, \theta_{-t}(p), \varphi\big(\tau, \theta_{-\tau-t}(p), A_{\theta_{-\tau-t}(p)}\big)\big) \\
&\subset \varphi\big(t, \theta_{-t}(p), \varphi\big(\tau, \theta_{-\tau-t}(p), B\big)\big) \\
&\subset \varphi\big(t, \theta_{-t}(p), B\big) \subset U_\varepsilon(A_p)
\end{aligned}$$

for all ε-neighborhoods $U_\varepsilon(A_p)$ of A_p, $\varepsilon > 0$, provided that $t = t(\varepsilon)$ is sufficiently large. Hence,

$$\varphi\big(\tau, \theta_{-\tau}(p), A_{\theta_{-\tau}(p)}\big) \subset A_p \quad \text{for all } \tau \geq 0 \text{ and } p \in P\,.$$

With τ replaced by t, this yields with (3.13) the φ-invariance of the family $(A_p)_{p\in P}$.
(iv) It remains to observe that the sets A_p, $p \in P$, are uniformly bounded, because they are subsets of a common compact set B for all $p \in P$. They thus form a pullback attractor, the uniqueness of which follows by Proposition 3.8.
Part 2. Suppose now that the metric space (P, d_P) is compact, and assume to the contrary that the convergence (3.10) does not hold. Then there exist an $\varepsilon > 0$ and sequences $t_n \to \infty$, $\hat{p}_n \in P$ and $x_n \in B$ such that

$$\operatorname{dist}\big(\varphi(t_n, \hat{p}_n, x_n), A(P)\big) > \varepsilon\,. \tag{3.14}$$

Set $p_n = \theta_{t_n}(\hat{p}_n)$. By the compactness of P, there exists a convergent subsequence $p_{n'} \to p_0 \in P$. Because of the pullback attraction, there exists a $\tau > 0$ such that

$$\operatorname{dist}\left(\varphi(\tau, \theta_{-\tau}(p_0), B), A_{p_0}\right) < \frac{\varepsilon}{2}.$$

The cocycle property gives

$$\varphi\left(t_n, \theta_{-t_n}(p_n), x_n\right) = \varphi\left(\tau, \theta_{-\tau}(p_n), \varphi\left(t_n - \tau, \theta_{-t_n}(p_n), x_n\right)\right)$$

for any $t_n > \tau$. Now, by the positive invariance of B, it follows that

$$\varphi\left(t_n - \tau, \theta_{-t_n}(p_n), x_n\right) \subset B,$$

and since B is compact, there is also a further index subsequence n'' of n' (depending on τ) such that

$$s_{n''} := \varphi\left(t_{n''} - \tau, \theta_{-t_{n''}}(p_{n''}), x_{n''}\right) \to s_0 \in B.$$

The continuity of the skew product flow implies

$$\left\|\varphi\left(\tau, \theta_{-\tau}(p_{n''}), s_{n''}\right) - \varphi\left(\tau, \theta_{-\tau}(p_0), s_0\right)\right\| < \frac{\varepsilon}{2} \quad \text{when } n'' > n(\varepsilon).$$

Therefore,

$$\begin{aligned}\varepsilon > \operatorname{dist}\left(\varphi\left(t_{n''}, \theta_{-t_{n''}}(p), x_{n''}\right), A_{p_0}\right) &= \operatorname{dist}\left(\varphi(t_{n''}, \hat{p}_{n''}, x_{n''}), A_{p_0}\right) \\ &\geq \operatorname{dist}\left(\varphi(t_{n''}, \hat{p}_{n''}, x_{n''}), A(P)\right),\end{aligned}$$

which contradicts (3.14), and thus, the asserted convergence (3.10) must be true. □

3.3. A continuous time example. Consider a nonautonomous dynamical system in $\mathbb{R}^d$ given by

$$\dot{x} = f(p, x) \tag{3.15}$$

with the driving system θ on a compact metric space (P, d_P). Suppose that f is regular enough to ensure that the differential equation (3.15) generates a skew product flow.

In addition, suppose that f satisfies the uniform dissipative condition

$$\left\langle x, f(p, x)\right\rangle \leq K - L\|x\|^2 \quad \text{for all } p \in P \text{ and } x \in \mathbb{R}^d \tag{3.16}$$

with positive constants K and L. Then, similarly to Example 3.19, a solution $x(t)$ satisfies the differential inequality

$$\frac{\mathrm{d}}{\mathrm{d}t}\|x(t)\|^2 \leq K - L\|x(t)\|^2,$$

which implies that the closed ball

$$B := B\left[0, \sqrt{(K+1)/L}\right] := \left\{x \in \mathbb{R}^d : \|x\|^2 \leq (K+1)/L\right\}$$

is pullback absorbing and positively invariant. From Theorem 3.20, it follows that the skew product flow has a pullback attractor in $\mathbb{R}^d$ with components subsets $A_p \subset B$, $p \in P$.

Suppose instead that the vector field f satisfies the uniform one-sided dissipative Lipschitz conditions

$$\left\langle x_1 - x_2, f(p, x_1) - f(p, x_2)\right\rangle \leq -L\|x_1 - x_2\|^2 \tag{3.17}$$

for all $p \in P$ and $x_1, x_2 \in \mathbb{R}^d$ with some constant $L > 0$. Then f satisfies the uniform dissipative condition (3.16) with constants

$$K' = \frac{2}{L} \sup_{p \in P} \|f(0)\| \quad \text{and} \quad L' = \frac{L}{2},$$

and the closed ball $B' := B_d\big(0, \sqrt{(K'+1)/L'}\big)$ is pullback absorbing and positively invariant. Thus, the skew product flow has a pullback attractor with component subsets A_p in this ball.

In fact, the fibers of the pullback attractor are singleton sets. The proof uses the fact that due to the uniform one-sided dissipative Lipschitz condition (3.17), the system satisfies

$$\|x_1(t) - x_2(t)\| \le e^{-Lt} \|x_{0,1} - x_{0,2}\| \tag{3.18}$$

for any pair of solutions with the same initial value $p \in P$ of the driving system. This follows from

$$\begin{aligned} \frac{\mathrm{d}}{\mathrm{d}t} \big\|x_1(t) - x_2(t)\big\|^2 &= \frac{\mathrm{d}}{\mathrm{d}t} \big\langle x_1(t) - x_2(t), x_1(t) - x_2(t) \big\rangle \\ &= 2\big\langle x_1(t) - x_2(t), \dot{x}_1(t) - \dot{x}_2(t) \big\rangle \\ &= 2\big\langle x_1(t) - x_2(t), f\big(\theta_t p, x_1(t)\big) - f\big(\theta_t p, x_2(t)\big) \big\rangle \\ &\le -2L \big\|x_1(t) - x_2(t)\big\|^2, \end{aligned}$$

which is integrated to give

$$\|x_1(t) - x_2(t)\|^2 \le e^{-2Lt} \|x_{0,1} - x_{0,2}\|^2.$$

Taking square roots yields the desired result.

THEOREM 3.21. *The pullback attractor $\mathcal{A}$ of the skew product flow (θ, φ) generated by the differential equation (3.15) consists of singleton fibers $A_p = \{a_p\}$ for each $p \in P$ when the vector field f satisfies the uniform one-sided dissipative Lipschitz condition (3.17). Moreover, $t \mapsto a_{\theta_t(p)}$, $t \in \mathbb{R}$, is an entire solution of (3.15) for each $p \in P$.*

PROOF. Since $A_p \subset B'$ for all $p \in P$, it follows that $\|A_p\| := \max_{a \in A_p} \|a\| \le R := (K'+1)/L'$ for each $p \in P$. Now consider a fixed $p \in P$, and suppose that there exists an $\varepsilon_0 > 0$ and points a_1, $a_2 \in A_p$ such that $\|a_1 - a_2\| = \varepsilon_0$. Moreover, choose $T > 0$ such that $2Re^{-LT} = \varepsilon_0$. The φ-invariance of the pullback attractor gives $\varphi\big(T, \theta_{-T}(p), A_{\theta_{-T}(p)}\big) = A_p$, which means that there exist $a_1', a_2' \in A_{\theta_{-T}p}$ such that

$$\varphi\big(T, \theta_{-T}(p), a_1'\big) = a_1 \quad \text{and} \quad \varphi\big(T, \theta_{-T}(p), a_2'\big) = a_2 \,.$$

Then, from the inequality in (3.18), it follows that

$$\begin{aligned} 0 < \varepsilon_0 = \|a_1 - a_2\| &= \big\|\varphi\big(T, \theta_{-T}(p), a_1'\big) - \varphi\big(T, \theta_{-T}(p), a_2'\big)\big\| \\ &\le e^{-LT} \|a_1' - a_2'\| \le Re^{-LT} = \frac{1}{2}\varepsilon_0 \,, \end{aligned}$$

which is not possible. Hence, $a_1 = a_2$.

Finally, from the φ-invariance of the pullback attractor, $\varphi(t, p, a_p) = a_{\theta_t(p)}$ for all $t \in \mathbb{R}$ and $p \in P$, so the singleton sets forming the pullback attractor define an entire solution of the system. It follows from inequality (3.18) that this entire solution also forward attracts all other solutions with the same initial value of the driving system, so the pullback attractor is also a forward attractor. (There will be more than one such entire solution when P is not a minimal subset for the autonomous dynamical system θ). □

The above theorem generalizes the following autonomous result from STUART & HUMPHRIES [**226**].

COROLLARY 3.22. *An autonomous differential equation with a vector field f which satisfies a one-sided dissipative Lipschitz condition such as* (3.17) *(i.e., without the p-variable) has a unique globally asymptotically stable equilibrium point.*

3.4. A discrete time example. Consider the parametrically dependent difference equation

$$x_{n+1} = f(x_n, q_n)\,.$$

with the continuous mapping $f : \mathbb{R} \times \left[\frac{1}{2}, 1\right] \to \mathbb{R}$ given by

$$f(x,q) := f_q(x) := \frac{|x| + q^2}{1+q}\,.$$

Let $P = \left[\frac{1}{2}, 1\right]^{\mathbb{Z}}$ be the space of bi-infinite sequences $p = (q_n)_{n\in\mathbb{Z}}$ taking values in $\left[\frac{1}{2}, 1\right]$, which is a compact metric space with the metric

$$d(p,p') = \sum_{n=-\infty}^{\infty} 2^{-|n|} \left|q_n - q_n'\right|,$$

and let $\{\theta_n : n \in \mathbb{Z}\}$ be the group generated by the left shift operator θ on this sequence space (analogously to Examples 2.10 and 2.11). Then the family of mappings $\varphi(n,\cdot,\cdot)$ defined by

$$\varphi(0,p,x) := \{x\} \quad \text{and} \quad \varphi(n,p,x) := f_{q_{n-1}} \circ \cdots \circ f_{q_0}(x)$$

for all $n \in \mathbb{N}$, $x \in \mathbb{R}$ and $p = (q_n)_{n\in\mathbb{Z}} \in P$ is a discrete time skew product flow on $\mathbb{R}$. Moreover, the mappings $p \mapsto \theta_n(p)$ and $(p,x) \mapsto \varphi(n,p,x)$ are continuous for each $n \in \mathbb{N}$.

Since

$$|f(x,q)| \le \frac{1}{1+q}|x| + \frac{q^2}{1+q} \le \frac{2}{3}|x| + \frac{2}{3}\,,$$

this discrete time skew product flow has an absorbing set $B = [-2, 2]$, which is positively invariant. Theorem 3.20 applies here which means that there exists a pullback attractor. Moreover, the sets A_p of the pullback attractor are singleton sets, since

$$\left|f(x,q) - f(y,q)\right| \le \frac{1}{1+q}|x-y| \le \frac{2}{3}|x-y|\,,$$

and it follows that solutions with the same p but different initial values converge to each other uniformly in the forward sense, so the pullback attractor is also a forward attractor. The sets $A_p = \{\alpha(p)\}$ are given by

$$\alpha(p) := \sum_{n=1}^{\infty} \frac{q_{-n}^2}{(1+q_{-1})(1+q_{-2})\cdot\ldots\cdot(1+q_{-n})}\,,$$

where $p = (q_n)_{n\in\mathbb{Z}} \in \left[\frac{1}{2}, 1\right]^{\mathbb{Z}}$.

3.5. Pullback attractors for absorbing families and attraction universes. To take into account nonuniformities that are ubiquitous in nonautonomous dynamical systems, greater generality can be attained in the definition of a pullback attractor by considering arbitrary nonautonomous sets $\mathcal{B}$ and $\mathcal{D}$ instead just a single compact absorbing set B and single attracted bounded set D. This allows local as well as global attraction to be handled at the same time. The skew product flows (θ, φ) in this subsection are on a metric state space (X, d_X) with a metric base space (P, d_P) and a time set $\mathbb{T}$.

DEFINITION 3.23 (Attraction universe). An *attraction universe* $\mathfrak{D}$ of a skew product flow (θ, φ) is a collection of bounded nonautonomous sets $\mathcal{D}$, which is closed in the sense that if $\emptyset \subsetneq \mathcal{D}' \subseteq \mathcal{D}$ for some $\mathcal{D}, \mathcal{D}' \in \mathfrak{D}$, then $\mathcal{D}' \in \mathfrak{D}$.

The definitions of pullback convergence and pullback attractor need to be extended accordingly.

DEFINITION 3.24 (Pullback attractor with respect to an attraction universe). Let (θ, φ) be a skew product flow on $P \times X$. A nonempty, compact and invariant nonautonomous set $\mathcal{A}$ is called *pullback attractor* with respect to an attraction universe $\mathfrak{D}$ if the *pullback convergence*

$$\lim_{t\to\infty} \operatorname{dist}\big(\varphi\big(t, \theta_{-t}(p), D_{\theta_{-t}(p)}\big), A_p\big) = 0$$

holds for all $p \in P$ and $\mathcal{D} \in \mathfrak{D}$.

EXERCISE 3.25. Show that a pullback attractor is unique within a given attraction universe $\mathfrak{D}$.

The pullback absorbing property now depends on the attraction universe $\mathcal{D}$ under consideration.

DEFINITION 3.26 (Pullback absorbing set with respect to an attraction universe). Let $\mathfrak{D}$ be an attraction universe of a skew product flow (θ, φ) on $P \times X$. A nonempty and compact nonautonomous set $\mathcal{B} \in \mathfrak{D}$ is called *pullback absorbing* with respect to $\mathfrak{D}$ if for each $\mathcal{D} \in \mathfrak{D}$ and $p \in P$, there exists a $T = T(p, D) > 0$ such that

$$\varphi\big(t, \theta_{-t}(p), D_{\theta_{-t}(p)}\big) \subset B_p \quad \text{for all } t \geq T\,.$$

Theorem 3.20 on the existence of a pullback attractor assuming that of a pullback absorbing set generalizes to attraction universes and pullback absorbing families.

THEOREM 3.27 (Existence of a pullback attractor with respect to an attraction universe). *Let (θ, φ) be a skew product flow on $P \times X$, and suppose that the compact nonautonomous set $\mathcal{B}$ is pullback absorbing with respect to an attraction universe $\mathfrak{D}$. Then (θ, φ) has a pullback attractor $\mathcal{A}$ with respect to $\mathfrak{D}$, where the fibers A_p are defined for each $p \in P$ by*

$$A_p = \bigcap_{s>0} \overline{\bigcup_{t>s} \varphi\big(t, \theta_{-t}(p), B_{\theta_{-t}(p)}\big)}\,. \tag{3.19}$$

The proof is a direct modification of that of Theorem 3.20.

REMARK 3.28. The assumption that the absorbing sets in Theorem 3.20 and Theorem 3.27 are compact is no restriction in a state space such as $\mathbb{R}^d$, which is locally compact, and thus, closed and bounded subsets are equivalently compact. This is not true for a general state space. In particular, for infinite-dimensional spaces, compact subsets are "thin" and it is much easier to determine an absorbing property for a closed and bounded subset, such as a unit ball, rather than a compact subset. Counterparts of Theorem 3.20 and Theorem 3.27 then hold, if the cocycle mapping is assumed to be compact, i.e., the mapping $\varphi(t,p,\cdot) : X \to X$ maps bounded subsets into precompact subsets for all $t > 0$ and $p \in P$, or more generally, asymptotically compact. These generalizations will be considered in Chapter 12 on infinite-dimensional dynamical systems, i.e., with an infinite-dimensional state space X.

EXERCISE 3.29. Formulate corresponding definitions of an attraction universe, a pullback absorbing family and a pullback attractor for a process.

4. Relationship between nonautonomous attractors

Simple examples show that a pullback attractor need not be a forward attractor and vice versa. However, Example 3.5 involves processes, and the associated skew product flows as defined in Theorem 2.4 of Chapter 2, when considered as autonomous semi-dynamical systems, do not have global attractors since the base space P of the driving system is the noncompact set $\mathbb{R}$. Much more can be said, however, about the relationships between the various kinds of nonautonomous attractors when the skew product flow has a compact base space P.

In this section, let (θ,φ) be a skew product flow on metric state space (X,d_X) with compact metric base space (P,d_P) and let the metric on the extended phase space $\mathcal{X} = P \times X$ be defined as

$$d_{\mathcal{X}}\big((p_1,x_1),(p_2,x_2)\big) = d_P(p_1,p_2) + d_X(x_1,x_2).$$

PROPOSITION 3.30. *Suppose that $\mathcal{A}$ is a uniform attractor (i.e., uniform in both the forward and pullback senses) of a skew product flow (θ,φ) and that $\bigcup_{p\in P} A_p$ is precompact in X. Then $\mathbb{A} := \bigcup_{p\in P}\{p\}\times A_p$ is the global attractor of the autonomous semi-dynamical system π associated with a skew product flow (θ,φ).*

PROOF. The π-invariance of $\mathbb{A}$ follows from the φ-invariance of $\mathcal{A}$, and the θ-invariance of P via

$$\begin{aligned}&\pi(t,\mathbb{A})\\ &= \bigcup_{p\in P}\{\theta_t(p)\}\times\varphi(t,p,A_p) = \bigcup_{p\in P}\{\theta_t(p)\}\times A_{\theta_t(p)} = \bigcup_{q\in P}\{q\}\times A_q = \mathbb{A}\,.\end{aligned}$$

Since $\mathcal{A}$ is also a pullback attractor and $\bigcup_{p\in P} A_p$ is precompact in X (and P is compact too) by Theorem 3.34, the set-valued mapping $p \mapsto A_p$ is upper semi-continuous, which means that $p \mapsto F(p) := \{p\}\times A_p$ is also upper semi-continuous. Hence, $F(P) = \mathbb{A}$ is a compact subset of $\mathcal{X}$, cf. Example 2.16. Moreover, the definition of the metric $d_{\mathcal{X}}$ on $\mathcal{X}$ implies that

$$\begin{aligned}
d_{\mathcal{X}}\big(\pi(t,(p,x)),\mathcal{A}\big) &= d_{\mathcal{X}}\big((\theta_t(p),\varphi(t,p,x)),\mathcal{A}\big)\\
&\le d_{\mathcal{X}}\big((\theta_t(p),\varphi(t,p,x)),\{\theta_t(p)\}\times A_{\theta_t(p)}\big)\\
&= d_P\big(\theta_t(p),\theta_t(p)\big)+\operatorname{dist}_X\left(\varphi(t,p,x),A_{\theta_t(p)}\right)\\
&= \operatorname{dist}_X\left(\varphi(t,p,x),A_{\theta_t(p)}\right),
\end{aligned}$$

where $\pi(t,(p,x)) = \big(\theta_t(p),\varphi(t,p,x)\big)$. The desired attraction to $\mathcal{A}$ with respect to π then follows from the forward attraction of $\mathcal{A}$ with respect to φ. □

Without uniform attraction as in Proposition 3.30 a pullback attractor need not give a global attractor, see Example 3.33 below, but the following result does hold.

PROPOSITION 3.31. *If $\mathcal{A}$ is a pullback attractor for a skew product flow (θ,φ) and $\bigcup_{p\in P} A_p$ is precompact in X, then $\mathcal{A} := \bigcup_{p\in P}\{p\}\times A_p$ is the maximal invariant compact set of the associated autonomous semi-dynamical system π.*

PROOF. The compactness and π-invariance of $\mathcal{A}$ are proved in the same way as in first part of the proof of Proposition 3.30. To prove that the compact invariant set $\mathcal{A}$ is maximal, let $\mathcal{C}$ be any other compact invariant set of the autonomous semi-dynamical system π associated with the skew product flow. Then $\mathcal{A}$ is a compact and invariant nonautonomous set, and by pullback attraction, one has

$$\begin{aligned}
\operatorname{dist}_X(C_p,A_p) &= \operatorname{dist}_X\left(\varphi\big(t,\theta_{-t}(p),C_{\theta_{-t}(p)}\big),A_p\right)\\
&\le \operatorname{dist}_X\left(\varphi\big(t,\theta_{-t}(p),K\big),A_p\right)\to 0 \quad \text{as } t\to\infty\,,
\end{aligned}$$

where $K := \overline{\bigcup_{p\in P} C_p}$ is compact. Hence, $C_p \subseteq A_p$ for every $p\in P$, i.e., $\mathcal{C}\subseteq\mathcal{A}$, which finally means that $\mathcal{A}$ is a maximal π-invariant set. □

The set $\mathcal{A}$ here need not be the global attractor of π. In the opposite direction, the global attractor of the associated autonomous semi-dynamical system always forms a pullback attractor of the skew product flow.

PROPOSITION 3.32. *If $\mathcal{A}$ is the global attractor of the associated autonomous semi-dynamical system π, then $\mathcal{A}$ is a pullback attractor for the skew product flow (θ,φ).*

PROOF. The set $K = \bigcup_{p\in P} A_p$ is compact by the compactness of $\mathcal{A}$. Moreover, $\mathcal{A}\subset P\times K$, which is a compact set. Now

$$\begin{aligned}
\operatorname{dist}_X\left(\varphi(t,p,x),K\right) &= \operatorname{dist}_P\left(\theta_t(p),P\right)+\operatorname{dist}_X\left(\varphi(t,p,x),K\right)\\
&= \operatorname{dist}_{\mathcal{X}}\left((\theta_t(p),\varphi(t,p,x)),P\times K\right)\\
&\le \operatorname{dist}_{\mathcal{X}}\left(\pi(t,(p,x)),P\times K\right)\\
&\le \operatorname{dist}_{\mathcal{X}}\left(\pi(t,P\times D),\mathcal{A}\right)\to 0 \quad \text{as } t\to\infty
\end{aligned}$$

for all $(p,x)\in P\times D$ and every arbitrary compact subset D of X, since $\mathcal{A}$ is the global attractor of π. Hence, replacing p by $\theta_{-t}(p)$ implies

$$\lim_{t\to\infty}\operatorname{dist}_X\left(\varphi(t,\theta_{-t}(p),D),K\right)=0\,.$$

Then the system is pullback asymptotic compact (see Definition 12.10) and by Theorem 12.14 in Chapter 12, this is a sufficient condition for the existence of a pullback attractor $\mathcal{A}'$ with $\bigcup_{p\in P} A'_p \subset K$. From Proposition 3.31, $\mathcal{A}'$ is the maximal π-invariant subset of $\mathcal{X}$, but so is the global attractor $\mathcal{A}$, which means that $\mathcal{A}' = \mathcal{A}$. Thus, $\mathcal{A}$ is a pullback attractor of the skew product flow (θ, φ). □

The following counterexample is taken from CHEBAN, KLOEDEN & SCHMALFUSS [**41**].

EXAMPLE 3.33. Let $f : \mathbb{R} \to \mathbb{R}$ be defined by

$$f(t) := -\left(\frac{1+t}{1+t^2}\right)^2 \quad \text{for all } t \in \mathbb{R}\,,$$

and let θ be the autonomous dynamical system on $P = H(f)$, the hull of f in $C(\mathbb{R}, \mathbb{R})$, formed by the shift operators $\theta_t f(\cdot) := f(t + \cdot)$ for $t \in \mathbb{R}$. Note that P is compact with respect to the supremum norm on $C(\mathbb{R}, \mathbb{R})$. Moreover,

$$P = H(f) = \bigcup_{h\in\mathbb{R}} \{f(\cdot + h)\} \cup \{0\}.$$

Finally, let E be the *evaluation functional* on $C(\mathbb{R}, \mathbb{R})$, i.e., $E(p) := p(0)$ for $p \in C(\mathbb{R}, \mathbb{R})$. From a straightforward calculation, it follows that the functional

$$\gamma(p) = -\int_0^\infty e^{-\tau} E(\theta_\tau(p))\, \mathrm{d}\tau = -\int_0^\infty e^{-\tau} p(\tau)\, \mathrm{d}\tau$$

is well-defined and continuous on P, and that the function

$$t \mapsto \gamma(\theta_t(p)) = -e^t \int_t^\infty e^{-\tau} p(\tau)\, \mathrm{d}\tau = \begin{cases} \dfrac{1}{1+(t+h)^2} & : \ p = \theta_h(f)\,, \\ 0 & : \ p = 0\,, \end{cases}$$

is the unique solution of the scalar ordinary differential equation

$$\dot{x} = x + E(\theta_t(p)) = x + p(t)\,,$$

which exists and is bounded for all $t \in \mathbb{R}$. Now consider the nonautonomous scalar ordinary differential equation

$$\dot{x} = g(\theta_t(p), x)\,, \tag{3.20}$$

where

$$g(p, x) := \begin{cases} -x - E(p)x^2 & : \ p \neq 0\,,\ 0 \le x\gamma(p) \le 1\,, \\ -\frac{1}{\gamma(p)}\left(1 + \frac{E(p)}{\gamma(p)}\right) & : \ p \neq 0\,,\ 1 < x\gamma(p)\,, \\ -x & : \ p = 0\,,\ 0 \le x\,. \end{cases}$$

It is easily shown that this differential equation has a unique solution defined on $\mathbb{R}$ passing through any point $x_0 \in X = \mathbb{R}^+$ at time $t = 0$. These solutions define a cocycle mapping

$$\varphi(t, p, x_0) = \begin{cases} \frac{x_0}{e^t(1 - x_0\gamma(p)) + x_0\gamma(\theta_t p)} & : \ p \neq 0\,,\ 0 \le x_0\gamma(p) \le 1\,, \\ x_0 + \frac{1}{\gamma(\theta_t(p))} - \frac{1}{\gamma(p)} & : \ p \neq 0\,,\ 1 < x_0\,\gamma(p)\,, \\ e^{-t}x_0 & : \ p = 0\,,\ 0 \le x_0\,. \end{cases} \tag{3.21}$$

From this construction, it follows that $\mathcal{A} := P \times \{0\}$ is the only compact invariant nonautonomous set. To see that $\mathcal{A}$ is a pullback attractor, observe that

$$\varphi(t, \theta_{-t}(p), x_0) = \begin{cases} \frac{x_0}{e^t(1-x_0\gamma(\theta_{-t}(p)))+x_0\gamma(p)} & : \; p \neq 0\,,\; 0 \leq x_0\gamma(\theta_{-t}(p)) \leq 1\,, \\ x_0 + \frac{1}{\gamma(p)} - \frac{1}{\gamma(\theta_{-t}(p))} & : \; p \neq 0\,,\; 1 < x_0\,\gamma(\theta_{-t}(p))\,, \\ e^{-t}x_0 & : \; p = 0\,,\; 0 \leq x_0\,. \end{cases}$$

In particular, note that $\gamma(\theta_t(p))^{-1}$ is a solution of the nonautonomous differential equation (3.20). Since $\gamma(\theta_{-t}(p))^{-1}$ tends to ∞ subexponentially fast for $t \to \infty$, it follows that

$$\varphi(t, \theta_{-t}(p), x_0) \leq \frac{1}{2} L\, e^{-\frac{1}{2}t}$$

for any $x_0 \in [0, L]$ for any $L \geq 0$ and $p \in P$, provided that t is taken sufficiently large. Consequently, $\mathcal{A}$ is a pullback attractor for the skew product flow (θ, φ). In view of (3.21), the stable set

$$W^s(\mathcal{A}) := \Big\{(p, x) \in \mathfrak{X} : \lim_{t\to\infty} \operatorname{dist}_{\mathfrak{X}} \big(\pi(t, (p, x)), \mathcal{A}\big) = 0\Big\}$$

of $\mathcal{A} = P \times \{0\}$, i.e., the set of all points in in $\mathfrak{X} = P \times X$ that are attracted to $\mathcal{A}$ by the associated semi-dynamical system π, is given by

$$W^s(\mathcal{A}) = \big\{(p, x) \in \mathfrak{X} : p \in P \text{ and } x \geq 0 \text{ with } x\gamma(p) < 1\big\} \subsetneq \mathfrak{X}\,.$$

In summary, the skew product flow (θ, φ) in the above example has a pullback attractor which is not a forward attractor and also not a global attractor of the associated autonomous semi-dynamical system. One can show, however, that it is a local attractor. The situation of local attractivity in the nonautonomous case is discussed in Section 8.

5. Upper semi-continuous dependence on parameters

Analogously to autonomous attractors, pullback attractors also depend, in general, upper semi-continuously on parameters. This also holds for the base space "parameter" $p \in P$. Throughout this subsection, it is assumed that the state and base spaces of the skew product flow are metric spaces (X, d_X) and (P, d_P).

THEOREM 3.34 (Upper semi-continuity of pullback attractors). *Let (θ, φ) be a skew product flow with a pullback attractor $\mathcal{A}$ such that $A(P) := \overline{\bigcup_{p\in P} A_p}$ is compact. Then the setvalued mapping $p \mapsto A_p$ is upper semi-continuous.*

PROOF. Suppose that this is not true. Then there exists an $\varepsilon_0 > 0$ and a sequence $p_n \to p_0$ in P such that $\operatorname{dist}_X\big(A_{p_n}, A_{p_0}\big) \geq 3\varepsilon_0$ for all $n \in \mathbb{N}$. Since the sets A_{p_n} are compact, there exists a sequence $a_n \in A_{p_n}$ such that

$$\operatorname{dist}_X\big(a_n, A_{p_0}\big) = \operatorname{dist}_X\big(A_{p_n}, A_{p_0}\big) \geq 3\varepsilon_0 \quad \text{for all } n \in \mathbb{N}\,. \tag{3.22}$$

By pullback attraction, $\operatorname{dist}_X\big(\varphi(\tau, \theta_{-\tau}(p_0), B), A_{p_0}\big) \leq \varepsilon_0$ for $\tau > 0$ large enough, and by the φ-invariance of the pullback attractor, there exist $b_n \in A_{\theta_{-\tau}(p_n)} \subset A(P)$, $n \in \mathbb{N}$ such that $\varphi(\tau, \theta_{-\tau}(p_n), b_n) = a_n$. Since $A(P)$ is compact, there is a convergent subsequence $b_{n'} \to \bar{b} \in A(P)$. Finally, by the continuity of $\theta_{-\tau}(\cdot)$ and of the cocycle mapping,

$$d_X\big(\varphi(\tau, \theta_{-\tau}(p_{n'}), b_{n'}), \varphi(\tau, \theta_{-\tau}(p_0), \bar{b})\big) \leq \varepsilon_0 \quad \text{for } n' \text{ large enough.}$$

Thus,

$$\begin{aligned}\operatorname{dist}_X\left(a_{n'}, A_{p_0}\right) &= \operatorname{dist}_X\left(\varphi(\tau, \theta_{-\tau}(p_{n'}), b_{n'}), A_{p_0}\right) \\ &\le d_X\left(\varphi(\tau, \theta_{-\tau}(p_{n'}), b_{n'}), \varphi(\tau, \theta_{-\tau}(p_0), \bar{b})\right) \\ &\qquad + \operatorname{dist}_X\left(\varphi(\tau, \theta_{-\tau}(p_0), \bar{b}), A_{p_0}\right) \le 2\varepsilon_0\,,\end{aligned}$$

which contradicts (3.22). □

The upper semi-continuous dependence of the component subsets A_p cannot, in general, be strengthened to continuous dependence as the following simple counterexample shows.

EXAMPLE 3.35. Consider the autonomous scalar differential equation

$$\dot{x} = -x\left(x^4 - 2x^2 + 1 - p\right), \quad \text{where } p \in P := [-1, 1]\,, \tag{3.23}$$

for which there are three parameter regimes for equilibrium solutions $\bar{x}_p$:

(i) $\bar{x}_p = 0$ for $p < 0$
(ii) $\bar{x}_p = 0,\ \pm\sqrt{1+\sqrt{\nu}},\ \pm\sqrt{1-\sqrt{p}}$ for $0 \le p < 1$, and
(iii) $\bar{x}_p = 0,\ \pm\sqrt{1+\sqrt{p}}$ for $p \ge 1$.

The zero solution here loses linear stability at $p = 1$ in a subcritical bifurcation to the nonlocal solutions $\pm\sqrt{1+\sqrt{p}}$. Note, however, that these equilibria, as well as $\pm\sqrt{1-\sqrt{p}}$, first appear at $p = 0$. The equilibria $\pm\sqrt{1+\sqrt{p}}$ are asymptotically stable for $p > 0$, whereas the equilibria $\pm\sqrt{1-\sqrt{p}}$ are unstable in their existence interval $0 \le p < 1$. The global (autonomous) attractors here are $A_p = \{0\}$ for $p < 0$ and

$$A_p = \left[-\sqrt{1+\sqrt{p}}, \sqrt{1+\sqrt{p}}\right] \quad \text{for } p \ge 0\,.$$

In particular, the set-valued mapping $p \mapsto A_p$ is not continuous at $p = 0$ (being only upper semi-continuous there), but is continuous elsewhere, for example, at $p = 1$.

Now consider (3.23) as a nonautonomous differential equation with a driving system θ on $P = [-1, 1]$ such that $\theta_t(p) \equiv p$ for all $t \in \mathbb{R}$ and all $p \in P$, i.e., the driving system just remains at its initial value. Then $\mathcal{A}$ is a pullback (and forward) attractor for the resulting nonautonomous dynamical system for which the set-valued mapping $p \mapsto A_p$ is upper semi-continuous but not continuous at $p = 0$.

A similar result holds when the cocycle mappings of a family of skew product flows and their pullback attractors depend on a parameter, and the proof of this result is similar.

THEOREM 3.36. *Let $(\theta, \varphi^\lambda)$ be a family of skew product flows on a common state space X and base space P for which the cocycle mappings depend on a parameter $\lambda \in \Lambda$, where (Λ, d_Λ) is a compact metric space and (X, d_X) and (P, d_P) are metric spaces. Suppose that each skew product flow has a pullback attractor $\mathcal{A}^\lambda$ and that there is a common compact subset B of X such that $A_p^\lambda \subset B$ for all $p \in P$ and $\lambda \in \Lambda$. If the mapping $(t, p, x, \lambda) \mapsto \varphi^\lambda(t, p, x)$ is continuous, then the set-valued mapping $\lambda \mapsto A_p^\lambda$ is upper semi-continuous for each $p \in P$.*

PROOF. Suppose that this is not true. Then there exists $p \in P$, $\varepsilon_0 > 0$ and a sequence $\lambda_n \to \lambda_0$ in P such that $\operatorname{dist}_X\left(A_p^{\lambda_n}, A_p^{\lambda_0}\right) \geq 3\varepsilon_0$ for all $n \in \mathbb{N}$. Since the sets $A_p^{\lambda_n}$ are compact, there exists a sequence $a_n \in A_p^{\lambda_n}$ such that

$$\operatorname{dist}_X\left(a_n, A_p^{\lambda_0}\right) = \operatorname{dist}_X\left(A_p^{\lambda_n}, A_p^{\lambda_0}\right) \geq 3\varepsilon_0 \quad \text{for all } n \in \mathbb{N}\,. \tag{3.24}$$

By pullback attraction, $\operatorname{dist}_X\left(\varphi(\tau, \theta_{-\tau}(p), B), A_p^{\lambda_0}\right) \leq \varepsilon_0$ for $\tau > 0$ large enough, and by the φ-invariance of the pullback attractor, there exists $b_n \in A^{\lambda_n}_{\theta_{-\tau}(p)} \subset B$ for $n \in \mathbb{N}$ such that $\varphi\big(\tau, \theta_{-\tau}(p), b_n\big) = a_n$. Since B is compact, there is a convergent subsequence $b_{n'} \to \bar{b} \in B$. In addition, by the continuity of the cocycle mapping in x and λ, one obtains

$$d_X\big(\varphi^{\lambda_n}\big(\tau, \theta_{-\tau}(p), b_{n'}\big), \varphi^{\lambda_0}\big(\tau, \theta_{-\tau}(p), \bar{b}\big)\big) \leq \varepsilon_0 \quad \text{for } n' \text{ large enough.}$$

Thus,

$$\begin{aligned}
\operatorname{dist}_X\left(a_{n'}, A_p^{\lambda_0}\right) &= \operatorname{dist}_X\left(\varphi^{\lambda_{n'}}\big(\tau, \theta_{-\tau}(p), b_{n'}\big), A_p^{\lambda_0}\right) \\
&\leq d_X\big(\varphi^{\lambda_{n'}}\big(\tau, \theta_{-\tau}(p), b_{n'}\big), \varphi^{\lambda_0}\big(\tau, \theta_{-\tau}(p), \bar{b}\big)\big) \\
&\quad + \operatorname{dist}_X\left(\varphi^{\lambda_0}\big(\tau, \theta_{-\tau}(p), \bar{b}\big), A_p^{\lambda_0}\right) \leq 2\varepsilon_0\,,
\end{aligned}$$

which contradicts (3.24). □

REMARK 3.37. An analogue of Theorem 1.52 on the equivalence of equi-attraction and continuity with respect to parameters also holds for pullback attractors, see LI & KLOEDEN [**68**].

6. Parametrically inflated pullback attractors

A pullback attractor $\mathcal{A}$ of a skew product flow (θ, φ) on $P \times X$ need not, in general, be forward attracting. However, by Theorem 3.20, one has under the assumption that (P, d_P) is a compact metric space that

$$\lim_{t\to\infty} \sup_{p\in P} \operatorname{dist}\big(\varphi(t, p, D), A(P)\big) = 0$$

for any bounded subset D of $\mathbb{R}^d$, where $A(P) := \overline{\bigcup_{p\in P} A_p}$. This is a form of forward attraction, but $A(P)$ is not φ-invariant and is thus not an attractor. The concept of a parametrically inflated pullback attractor, which was introduced in WANG, LI & KLOEDEN [**237**], provides a more precise description of this forward attraction.

DEFINITION 3.38 (Parametrically inflated pullback attractor). Let $\mathcal{A}$ be a pullback attractor of a skew product flow (θ, φ) on $P \times X$, and let $\varepsilon_0 > 0$. Then the nonautonomous set $\mathcal{A}(\varepsilon_0) = (A_p[\varepsilon_0])_{p\in P}$ with fibers defined by

$$A_p[\varepsilon_0] := \bigcup_{d_P(q,p)\leq\varepsilon_0} A_q$$

is called the *parametrically inflated pullback attractor*.

It is assumed throughout this section that (X, d_X) is a complete metric space and (P, d_P) is a compact metric space.

LEMMA 3.39. *The fiber sets $A_p[\varepsilon_0]$ of a parametrically inflated pullback attractor $\mathcal{A}(\varepsilon_0)$ are nonempty compact subsets of X.*

PROOF. The set $A_p[\varepsilon_0]$ is nonempty since it contains the nonempty set A_p. Compactness follows from the compactness of the subsets A_p, the upper semi-continuity of the mapping $p \mapsto A_p$ and, using Exercise 2.16, the fact that the closed ball $\overline{B_{\varepsilon_0}(p)}$ is compact in the compact space P. □

A parametrically inflated pullback attractor is not an attractor since it is not invariant but only φ-negative invariant, i.e., satisfies

$$A_{\theta_t(p)}[\varepsilon_0] \subset \varphi\big(t, p, A_p[\varepsilon_0]\big) \quad \text{for all } t \geq 0 \text{ and } p \in P. \tag{3.25}$$

EXERCISE 3.40. Prove (3.25).

However, in view of Theorem 3.41 below, a parametrically inflated pullback attractor is a forward attracting set. This was proved in [**237**] under weaker assumptions. It uses the fact that a skew product flow is an autonomous semi-dynamical system π on the product space $\mathcal{X} = P \times X$ defined by $\pi(t,(p,x)) = (\theta_t(p), \varphi(t,p,x))$ with respect to the metric $d_{\mathcal{X}}$ on $\mathcal{X}$ defined by

$$d_{\mathcal{X}}\big((p,x),(q,y)\big) = d_P(p,q) + d_X(x,y) \quad \text{for all } (p,x),(q,y) \in \mathcal{X}.$$

Clearly, $(\mathcal{X}, d_{\mathcal{X}})$ is complete.

THEOREM 3.41. *Assume that a skew product flow (θ,φ) with compact base space P has a compact positively invariant set $B \subset X$ which is* forward absorbing uniformly *in $p \in P$, i.e., for all bounded sets $D \subset X$, there exists a $T = T(D) > 0$ independently of $p \in P$ such that*

$$\varphi(t,p,D) \subset B \quad \textit{for all } t \geq T \textit{ and } p \in P, \tag{3.26}$$

and let $\mathcal{A}$ be the corresponding pullback attractor given by (3.19). *Then for any fixed $\varepsilon_0 > 0$, the parametrically inflated pullback attractor $\mathcal{A}[\varepsilon_0] = (A_p[\varepsilon_0])_{p\in P}$ forward attracts each bounded subset D of X uniformly in $p \in P$, i.e., for any $\gamma > 0$, there is a $\tau = \tau(D,\gamma) > 0$ independent of $p \in P$ such that*

$$\operatorname{dist}_X\big(\varphi(t,p,D), A_{\theta_t p}[\varepsilon_0]\big) < \gamma \quad \textit{for all } t \geq \tau \textit{ and } p \in P.$$

REMARK 3.42. Since $T = T(D)$ is independent of $p \in P$, (3.26) is equivalent to $\varphi\big(t, \theta_{-t}(p), D\big) \subset B$ for all $t \geq T$ and $p \in P$, which corresponds to pullback absorption.

PROOF. Note that Theorem 3.20 holds here, so the skew product flow (θ,φ) has a unique pullback attractor $\mathcal{A}$, and let $\varepsilon_0 > 0$. It needs to be shown that for any bounded set $D \subset X$ and $\gamma > 0$ (it can be assumed that $\gamma < \varepsilon_0$), there is a $\tau = \tau(D,\varepsilon) > 0$ which is independent of $p \in P$ such that

$$\operatorname{dist}_X\big(\varphi(t,p,D), A_{\theta_t p}[\varepsilon_0]\big) < \gamma \quad \text{for all } t \geq \tau \text{ and } p \in P.$$

It suffices to show this for the compact set B since any bounded set D is absorbed by B in finite time and remains there since B is positively invariant. Because of the uniformity in $p \in P$, the compact subset $P \times B$ is a positively invariant absorbing set for the autonomous semi-dynamical system π on the product space $P \times X$ associated with the skew product flow (θ,φ). Thus, π has a global attractor $\mathbb{A}$ in $P \times X$, which has the form

$$\mathbb{A} = \bigcup_{p\in P} \{p\} \times \tilde{A}_p,$$

where $\tilde{A}_p$ is a nonempty compact subset of B. Due to Proposition 3.31, the so-generated nonautonomous set $\tilde{A}$ is a pullback attractor for the skew product flow (θ, φ). Then $\tilde{A}_p = A_p$ for all $p \in P$, since the pullback attractor $\mathcal{A}$ of this skew product flow is unique. Thus,

$$\mathcal{A} = \bigcup_{p \in P} \{p\} \times A_p \,.$$

Since $\mathcal{A}$ attracts $P \times B$ under π, there exists a $T_1 = T_1(P \times B, \gamma) > 0$ such that

$$\mathrm{dist}_{\mathcal{X}}(\pi_t(p,x), \mathcal{A}) < \frac{\gamma}{2} \quad \text{for all } t \geq T_1 \text{ and } (p,x) \in P \times B\,,$$

i.e.,

$$\inf_{v \in \mathcal{A}} d_{\mathcal{X}}\big((\theta_t(p), \varphi(t,p,x)), v\big) < \frac{\gamma}{2} \quad \text{for all } t \geq T_1 \text{ and } (p,x) \in P \times B\,. \tag{3.27}$$

Replacing $\theta_t(p) = q$ in (3.27) yields

$$\inf_{v \in \mathcal{A}} d_{\mathcal{X}}\big((q, \varphi(t, \theta_{-t}(q), x), v\big) < \frac{\gamma}{2} \quad \text{for all } t \geq T_1 \text{ and } (p,x) \in P \times B\,.$$

For each q, partition $\mathcal{A}$ into two parts $\mathcal{A} = \mathcal{A}_q^1 \cup \mathcal{A}_q^2$, where

$$\mathcal{A}_q^1 = \bigcup_{\substack{p \in P,\\ d_P(p,q) \leq \varepsilon_0}} \{p\} \times A_p \quad \text{and} \quad \mathcal{A}_q^2 = \bigcup_{\substack{p \in P,\\ d_P(p,q) > \varepsilon_0}} \{p\} \times A_p\,.$$

Now suppose that $(q,x) \in P \times B$ and $t \geq T_1$. Then by the definition of $d_{\mathcal{X}}$, one has

$$d_{\mathcal{X}}\big((q, \varphi(t, \theta_{-t}(q), x)), v\big) \geq d_P(q,p) > \varepsilon_0 > \gamma \quad \text{for all } v = (p,y) \in \mathcal{A}_q^2\,,$$

so

$$\inf_{v \in \mathcal{A}_q^1} d_{\mathcal{X}}\big((q, \varphi(t, \theta_{-t}(q), x)), v\big) < \frac{\gamma}{2}\,.$$

Thus, there exists a point $v' = (p', y') \in \mathcal{A}_q^1$ such that

$$d_{\mathcal{X}}\big((q, \varphi(t, \theta_{-t}(q), x)), v'\big) \leq \frac{2}{3}\gamma\,.$$

Then $y' \in A_{p'} \subset A_q[\varepsilon_0]$, since $d_P(p', q) \leq \varepsilon_0$. From this and the definition of $d_{\mathcal{X}}$, it follows that

$$\begin{aligned} \mathrm{dist}_X\big(\varphi(t, \theta_{-t}(q), x), A_{p'}\big) &\leq d_X\big(\varphi(t, \theta_{-t}(q), x), y'\big) \\ &\leq d_{\mathcal{X}}\big((q, \varphi(t, \theta_{-t}(q), x)), (p', y')\big) < \frac{2}{3}\gamma\,, \end{aligned}$$

and hence, one has

$$\mathrm{dist}_X\big(\varphi(t, \theta_{-t}(q), x), A_q[\varepsilon_0]\big) \leq \mathrm{dist}_X\big(\varphi(t, \theta_{-t}(q), x), A_{p'}\big) < \frac{2}{3}\gamma\,.$$

Since $q \in P$, $x \in B$ and $t \geq T_1$ are otherwise arbitrary, this means that

$$\mathrm{dist}_X\big(\varphi(t, \theta_{-t}(q), B), A_q[\varepsilon_0]\big) < \frac{2}{3}\gamma < \gamma \quad \text{for all } t \geq T_1 \text{ and } q \in P\,.$$

Finally, writing $p = \theta_{-t}(q)$, this gives

$$\mathrm{dist}_X\big(\varphi(t, p, B), A_{\theta_t p}[\varepsilon_0]\big) < \gamma \quad \text{for all } t \geq T_1 \text{ and } p \in P\,,$$

since T_1 is independent of $p \in P$. $\square$

The following theorem shows the robust stability of pullback attraction of each fiber A_p of the pullback attractor $\mathcal{A}$ with respect to perturbations in p. It will be used in the next section.

THEOREM 3.43. *Suppose that a skew product flow (θ, φ) satisfies the assumptions of Theorem 3.41 and let $\mathcal{A}$ be its pullback attractor. Then for any $p_0 \in P$ and $\varepsilon > 0$, there exists $\delta = \delta(p_0, \varepsilon) > 0$ such that for any bounded set $D \subset X$, one can find a $T > 0$ such that*

$$\varphi(t, \theta_{-t}(p), D) \subset B_\varepsilon(A_{p_0}) \quad \textit{for all } t \geq T \textit{ and } p \in B_\delta(p_0)\,. \tag{3.28}$$

PROOF. Since the compact set B is forward absorbing uniformly in $p \in P$, it is also pullback absorbing, i.e., for any bounded subset D of X, there exists a $T_0 = T_0(D) > 0$ independent of $p \in P$ such that

$$\varphi(t, \theta_{-t}(p), D) \subset B \quad \text{for all } t \geq T_0 \text{ and } p \in P\,.$$

To prove the theorem, it thus suffices to check that (3.28) holds true for $D = B$. Firstly, one has

$$\varphi(t, \theta_{-t}(p), B) \subset B \quad \text{for all } t \geq 0 \text{ and } p \in P \tag{3.29}$$

by the positive invariance of B. Secondly, by the definition of pullback attraction, there is a time $T_1 \geq 0$, such that

$$\varphi(t, \theta_{-t}(p_0), B) \subset B_\varepsilon(A_{p_0}) \quad \text{for all } t \geq T_1\,.$$

Then, since $p \mapsto \varphi(T_1, p, x)$ is continuous uniformly in $x \in B$, given $\varepsilon > 0$, there exists a $\delta = \delta(p_0, \varepsilon) > 0$ such that

$$\varphi(T_1, \theta_{-T_1}(p), B) \subset B_\varepsilon(A_{p_0}) \quad \text{for all } \in B_\delta(p_0)\,. \tag{3.30}$$

Finally, each $t \geq T = T_1$ can be rewritten as $t = nT_1 + s$, where $s \in [0, T_1]$, so, by (3.29) and (3.30), it follows that

$$\begin{aligned}\varphi(t, \theta_{-t}(p), B) &= \varphi\big(T_1, \theta_{-T_1}(p), \varphi((n-1)T_1 + s, \theta_{-nT_1 - s}(p), B)\big) \\ &\subset \varphi\big(T_1, \theta_{-T_1}(p), B\big) \subset B_\varepsilon(A_{p_0})\end{aligned}$$

for all $p \in B_\delta(p_0)$. □

7. Pullback attractors with continuous fibers

In general, the fibers of a pullback attractor $\mathcal{A}$ of a skew product flow (θ, φ) are only upper semi-continuous in the parameter p. However, in special cases, they are also lower semi-continuous and hence continuous with respect to the Hausdorff metric. This happens, for example when the fibers are singleton sets.

As in the previous section, it is also assumed in this section that (X, d) is a complete metric space and (P, d_P) is a compact metric space.

THEOREM 3.44. *Suppose that a skew product flow (θ, φ) satisfies the assumptions of Theorem 3.41, and let $\mathcal{A}$ be its pullback attractor. In addition, suppose that the set-valued mapping $p \mapsto A_p$ is lower semi-continuous in p, i.e.,*

$$\lim_{p \to p_0} \operatorname{dist}_X\big(A_{p_0}, A_p\big) = 0 \quad \textit{for all } p_0 \in P\,.$$

Then $\mathcal{A}$ is a uniform attractor of (θ, φ).

PROOF. It suffices to show that the rate of pullback attraction of $\mathcal{A}$ is uniform with respect to $p \in P$, i.e., for any bounded set $D \subset X$ and $\varepsilon > 0$, there exists a time $T = T(D, \varepsilon) > 0$, which is independent of $p \in P$, such that

$$\operatorname{dist}_X \left(\varphi(t, \theta_{-t}(p), D), A_p\right) < \varepsilon \quad \text{for all } t \geq T \text{ and } p \in P.$$

Assume on the contrary that this is not true. Then there are sequences $t_n \in \mathbb{R}^+$, $x_n \in D$, and $p_n \in P$ with $t_n \to \infty$ as $n \to \infty$ such that

$$d\big(\varphi(t_n, \theta_{-t_n}(p_n), x_n), A_{p_n}\big) \geq \varepsilon .$$

Since P is compact, there is a subsequence of p_n which will also be labeled as p_n such that $p_n \to p_0$ as $n \to \infty$. By lower semi-continuity of A_p in p, one has

$$\operatorname{dist}_X \left(A_{p_0}, A_{p_n}\right) < \frac{\varepsilon}{2} \tag{3.31}$$

for n sufficiently large. On the other hand, by Theorem 3.43,

$$\operatorname{dist}_X \left(\varphi\big(t_n, \theta_{-t_n}(p_n), D\big), A_{p_0}\right) < \frac{\varepsilon}{2} \tag{3.32}$$

for n sufficiently large. Combining (3.31) and (3.32), it follows that

$$\operatorname{dist}_X \left(\varphi\big(t_n, \theta_{-t_n}(p_n), x_n\big), A_{p_n}\right) < \varepsilon$$

for n sufficiently large, which is a contradiction. This finishes the proof of this theorem. □

7.1. Periodic and almost periodic driving systems. When the driving system is periodic or almost periodic, the fibers of the pullback attractor are continuous in the parameter. This is easy to show in the periodic case.

LEMMA 3.45. *Suppose that the driving system θ of a skew product flow (θ, φ) is periodic with minimal period $T > 0$ and $P = \{\theta_t(\bar{p}) : 0 \leq t \leq T\}$ for some $\bar{p} = \theta_T(\bar{p}) \in P$. Then for any pullback attractor $\mathcal{A}$, the set-valued mapping $p \mapsto A_p$ is continuous in the Hausdorff metric.*

PROOF. Let $q_n \to \bar{q}$ in P. Then $q_n = \theta_{t_n}(\bar{p})$ and $\bar{q} = \theta_{\bar{t}}(\bar{p})$ for some $t_n, \bar{t} \in [0, T]$. Assume that $t_n - \bar{t} \in [0, T]$ (otherwise replace t_n by $t_n + T$) and that $t_n \to \bar{t}$ as $n \to \infty$. Then $q_n = \theta_{t_n}(\bar{p}) = \theta_{t_n - \bar{t}}(\theta_{\bar{t}}(\bar{p})) = \theta_{t_n - \bar{t}}(\bar{q})$ and

$$A_{q_n} = A_{\theta_{t_n - \bar{t}}(\bar{q})} = \varphi(t_n - \bar{t}, \bar{q}, A_{\bar{q}}) \to \varphi(0, \bar{q}, A_{\bar{q}}) = A_{\bar{q}} \quad \text{as } n \to \infty,$$

since the set-valued mapping $t \mapsto A_{\theta_t(p)} = \varphi(t, p, A_p)$ is continuous in the Hausdorff metric. □

The almost periodic case is similar, but the proofs are more complicated. See the Appendix for the definition and basic properties of almost periodic functions. Let M be a complete metric space with metric d_M and let $C_b(\mathbb{R}, M)$ be the space of uniformly continuous bounded functions $f : \mathbb{R} \to M$, which is complete under the metric $d_\infty(f, g) = \sup_{t \in \mathbb{R}} d_M(f(t), g(t))$. Recall that the hull of a function $f \in C_b(\mathbb{R}, M)$ is defined as the closure of the subset $\{\theta_t(\bar{p}) : t \in \mathbb{R}\}$, where θ_t is defined to be the left shift operator, $\theta_t(f(\cdot)) = f(\cdot + t)$. By classical results, the hull of an almost periodic function is a compact subset of $C_b(\mathbb{R}, M)$. Similar results hold for other function spaces and topologies, see SELL [**218**]. In addition, let $(\mathcal{K}, h_X)$ be the complete metric space of all nonempty compact subsets of X with the Hausdorff metric h_X.

THEOREM 3.46. *Let P be the hull of an almost periodic function $\bar{p} \in C_b(\mathbb{R}, M)$ in the space $(C_b(\mathbb{R}, M), d_\infty)$, and let $\mathcal{A}$ be the pullback attractor of a (θ, φ) on $P \times X$. If $p \mapsto A_p$ is continuous, then $t \mapsto A(t) := A_{\theta_t(\bar{p})}$ is almost periodic as a mapping in $C(\mathbb{R}, \mathcal{K})$.*

PROOF. Since $p \mapsto A_p$ is continuous and P is compact due to the assumed almost periodicity, the set-valued mapping $p \mapsto A_p$ is uniformly continuous in p. Therefore, for every $\varepsilon > 0$, there exists a $\delta(\varepsilon) > 0$ such that

$$h_X(A_p, A_q) < \varepsilon \quad \text{for all } p, q \in P \text{ with } d_P(p, q) < \delta .$$

Since $\bar{p}$ is almost periodic, there is a relatively dense set $E \subset \mathbb{R}$ such that

$$d_P\big(\theta_{t+\tau}(\bar{p}), \theta_t(\bar{p})\big) < \delta \quad \text{for all } \tau \in E \text{ and } t \in \mathbb{R} .$$

Then

$$h_X\big(A_{\theta_{t+\tau}(\bar{p})}, A_{\theta_t(\bar{p})}\big) < \varepsilon \quad \text{for all } t \in \mathbb{R} \text{ and } \tau \in E .$$

Hence, the set-valued mapping $t \mapsto A(t)$ is almost periodic. □

A simple and interesting case is when the A_p are singleton sets for each p. Then, by Theorem 3.41 and Theorem 3.46, $\mathcal{A}$ is both a uniform pullback attractor and a uniform forward attractor of (θ, φ), hence a uniform attractor. Moreover, $t \mapsto A_{\theta_t(p)}$ is almost periodic. Singleton fibers occur in the following case.

THEOREM 3.47. *Let P be the hull of an almost periodic function $\bar{p} \in C_b(\mathbb{R}; M)$. Suppose that the skew product flow (θ, φ) has a pullback attractor $\mathcal{A}$ and is uniformly asymptotically stable, i.e., for any bounded subset D of X and $\varepsilon > 0$, there exists a $T = T(D, \varepsilon) > 0$ such that*

$$d_X\big(\varphi(t, p, x), \varphi(t, p, y)\big) < \varepsilon \quad \textit{for all } t \geq T, p \in P \textit{ and } x, y \in D .$$

Then the following statements hold.

(i) *A_p is a singleton set for each $p \in P$,*
(ii) *$p \mapsto A_p$ is continuous,*
(iii) *the single-valued function $\gamma_p(t) := A_{\theta_t(p)}$ is almost periodic.*

PROOF. Only the first conclusion (i) needs to be verified. This follows easily from the uniform asymptotic stability of (θ, φ) and the φ-invariance of $\mathcal{A}$. □

8. Local attractors and repellers

The attractors discussed in this chapter so far are *global* in the sense that they attract all bounded subsets of the phase space. The aim of this section is to provide suitable concepts of *local* attractivity for both forward and pullback convergences, meaning that only sets within a certain neighborhood of the attractor need to be attracted.

DEFINITION 3.48 (Local attractivity). Let ϕ be a process on a metric space (X, d). A compact and invariant nonautonomous set $\mathcal{A}$ is called

(i) a *local forward attractor* if there exists an $\eta > 0$ such that

$$\lim_{t\to\infty} \operatorname{dist}\big(\phi(t, t_0, B_\eta(A_{t_0})), A_t\big) = 0 \quad \text{for all } t_0 \geq 0 ,$$

(ii) a *local pullback attractor* if there exists an $\eta > 0$ such that

$$\lim_{t_0 \to -\infty} \operatorname{dist}\big(\phi(t, t_0, B_\eta(A_{t_0})), A_t\big) = 0 \quad \text{for all } t \le 0\,,$$

(iii) a *local uniform attractor* if it is a local forward or local pullback attractor such that the attraction is uniform with respect to $t_0 \in \mathbb{T}$ or $t \in \mathbb{T}$, respectively, i.e., there exists an $\eta > 0$ such that

$$\lim_{t \to \infty} \sup_{t_0 \in \mathbb{T}} \operatorname{dist}\big(\phi(t_0 + t, t_0, B_\eta(A_{t_0})), A_{t_0+t}\big) = 0\,.$$

The supremum over all $\eta > 0$ for which the above relations hold is called the *forward (pullback, uniform, respectively) radius of attraction* of $\mathcal{A}$.

These definitions allow the empty set to be a local uniform attractor, and hence, also a local forward and pullback attractor. In addition, if the phase space X is compact, then the entire extended phase space $\mathcal{A} = \mathbb{T} \times X$ is also a local uniform attractor.

EXERCISE 3.49. Show that the concept of a local pullback attractor fits into Definition 3.24, i.e., a local pullback attractor is a pullback attractor with respect to an appropriate attraction universe.

EXAMPLE 3.50. Consider again the linear inhomogeneous differential equation (3.3) from the Examples 3.6 and 3.13, which is given by

$$\dot{x} = -x + 2\sin t\,.$$

It was shown that the entire solution $t \mapsto \rho(t) = \sin t - \cos t$ gives rise to both a pullback and forward attractor $\mathcal{A}$ with $A_t = \{\rho(t)\}$, which is also a uniform attractor. It follows from the exercise below that the set $\mathcal{A}$ is also a local uniform attractor, and hence, also a local pullback and forward attractor.

EXERCISE 3.51. Consider a process ϕ with a (global) pullback or uniform attractor $\mathcal{A}$, respectively, and assume that $\bigcup_{t\in\mathbb{R}} A_t$ is compact. Show that $\mathcal{A}$ is also a local pullback or uniform attractor, respectively. Formulate and prove a corresponding statement for local forward attractors under weaker assumptions.

In addition to attractivity, also corresponding concepts of repulsivity will be treated in the following, where repulsivity means attraction in backward time. This means that the notion of repulsivity requires that of invertibility for processes, which needs to be defined. Let (X, d) be a complete metric state space, and consider a time set $\mathbb{T} = \mathbb{R}$ or $\mathbb{T} = \mathbb{Z}$. A process $(t, t_0, x_0) \mapsto \phi(t, t_0, x_0)$ is said to be *invertible* if it is not only defined for all $t \ge t_0$, but also for $t < t_0$, so an invertible process satisfies both the initial value and evolution property

(i) $\phi(t_0, t_0, x_0) = x_0$ for all $t_0 \in \mathbb{T}$ and $x_0 \in X$,
(ii) $\phi(t_2, t_0, x_0) = \phi\big(t_2, t_1, \phi(t_1, t_0, x_0)\big)$ for all $t_0, t_1, t_2 \in \mathbb{T}$ and $x_0 \in X$.

Invertibility is satisfied if the process comes from an ordinary differential equation restricted to an invariant subset in contrast to the discrete case of difference equations.

DEFINITION 3.52 (Local repulsivity). Let ϕ be an invertible process on a metric space (X, d). A compact and invariant nonautonomous set $\mathcal{R}$ is called

(i) a *local forward repeller* if there exists an $\eta > 0$ such that

$$\lim_{t\to-\infty} \operatorname{dist}\big(\phi(t, t_0, B_\eta(R_{t_0})), R_t\big) = 0 \quad \text{for all } t_0 \le 0\,,$$

(ii) a *local pullback repeller* if there exists an $\eta > 0$ such that

$$\lim_{t_0\to\infty} \operatorname{dist}\big(\phi(t, t_0, B_\eta(R_{t_0})), R_t\big) = 0 \quad \text{for all } t \ge 0\,,$$

(iii) a *local uniform repeller* if it is a local forward or pullback repeller such that the repulsion is uniform with respect to $t_0 \in \mathbb{T}$ or $t \in \mathbb{T}$, respectively, i.e., there exists an $\eta > 0$ such that

$$\lim_{t\to\infty} \sup_{t_0\in\mathbb{T}} \operatorname{dist}\big(\phi(t_0 - t, t_0, B_\eta(R_{t_0})), R_{t_0-t}\big) = 0\,.$$

The supremum over all $\eta > 0$ for which the above relations hold is called the *forward (pullback, uniform, respectively) radius of repulsion* of $\mathcal{R}$.

It was already seen in Section 1 that forward attraction concerns the *future* of the system, pullback attraction the *past* and uniform attraction the *entire time*. The situation is different for the notions of repulsivity. It can be seen directly from the definitions that a local forward repeller is a repeller for the past, a local pullback repeller concerns the future, and uniform repulsivity is a concept for the entire time. In particular, this point of view will be important in the next chapter, where the interplay of attractor and repeller is discussed with respect to the different time domains.

The above notions of attractivity and repulsivity will be used in particular also for invariant nonautonomous sets with singleton fibers. These are given as graphs of solutions $t \mapsto x(t) := \phi(t, t_0, x_0)$ for fixed initial time t_0 and initial value x_0. A solution x is called

(i) *locally pullback attractive* if $\operatorname{graph} x$ is a local pullback attractor,
(ii) *locally forward attractive* if $\operatorname{graph} x$ is a local forward attractor,
(iii) *locally uniformly attractive* if $\operatorname{graph} x$ is a local uniform attractor,
(iv) *locally pullback repulsive* if $\operatorname{graph} x$ is a local pullback repeller,
(v) *locally forward repulsive* if $\operatorname{graph} x$ is a local forward repeller,
(vi) *locally uniformly repulsive* if $\operatorname{graph} x$ is a local uniform repeller.

These concepts will be discussed in the following exercise.

EXERCISE 3.53. Consider the process ϕ generated by the nonautonomous ordinary differential equation

$$\dot{x} = a(t)x + b(t)x^3$$

with continuous functions $a : \mathbb{R} \to \mathbb{R}$ and $b : \mathbb{R} \to (\gamma, \infty)$ for some $\gamma > 0$. Prove that the trivial solution is a

(i) locally forward attractive if $\liminf_{t\to\infty} -a(t)/b(t) > 0$,
(ii) locally pullback attractive if $\liminf_{t\to-\infty} -a(t)/b(t) > 0$,
(iii) locally uniformly attractive if $\inf_{t\in\mathbb{R}} -a(t)/b(t) > 0$,
(iv) locally forward repulsive if $\liminf_{t\to-\infty} a(t) \ge 0$,
(v) locally pullback repulsive if $\liminf_{t\to\infty} a(t) \ge 0$.
(vi) locally uniformly repulsive if $a(t) \ge 0$ for all $t \in \mathbb{R}$.

In particular, this example demonstrates the relationships of the different notions to the time domains, since the conditions which have to be imposed on the equation indicate the relevant time region.

Attractivity and repulsivity of solutions can be obtained by an exponential condition on the linearization along the solution. Because of the concept of the equation of perturbed motion (see Subsection 1.3 of Chapter 2), it suffices to consider the trivial solution

THEOREM 3.54 (Linearized attractivity and repulsivity). *Consider an unbounded interval* $\mathbb{I}$ *of the form* $\mathbb{R}_0^-$, $\mathbb{R}_0^+$ *or* $\mathbb{R}$, *respectively, and let*

$$\dot{x} = B(t)x + F(t,x) \tag{3.33}$$

be a nonautonomous differential equation with continuous functions $B : \mathbb{I} \to \mathbb{R}^{d\times d}$ *and* $F : \mathbb{I} \times U \to \mathbb{R}^d$, $U \subset \mathbb{R}^d$ *a neighborhood of* 0, *such that* $F(t,0) = 0$ *for all* $t \in \mathbb{I}$. *Let* ϕ *denote the process induced by* (3.33) *and* $\Phi : \mathbb{I} \times \mathbb{I} \to \mathbb{R}^{d\times d}$ *denote the transition operator of the linearized equation* $\dot{x} = B(t)x$. *Then the following statements are fulfilled:*

(i) *If there exist* $\beta < 0$, $K \geq 1$ *and* $\delta > 0$ *such that*

$$\|\Phi(t,s)\| \leq K e^{\beta(t-s)} \quad \textit{for all } t \geq s$$

and

$$\|F(t,x)\| \leq \frac{-\beta}{2K}\|x\| \quad \textit{for all } t \in \mathbb{I} \textit{ and } x \in B_\delta(0)\,, \tag{3.34}$$

then one has

$$\operatorname{dist}\big(\phi(t,t_0,B_{\delta/K}(0)),\{0\}\big) \leq \delta\, e^{\frac{\beta}{2}(t-t_0)} \quad \textit{for all } t_0, t \in \mathbb{I} \textit{ with } t_0 \leq t\,,$$

i.e., the trivial solution of (3.33) *is locally pullback (forward, uniformly, respectively) attractive.*

(ii) *If there exist* $\beta > 0$, $K \geq 1$ *and* $\delta > 0$ *such that*

$$\|\Phi(t,s)\| \leq K e^{\beta(t-s)} \quad \textit{for all } t \leq s$$

and

$$\|F(t,x)\| \leq \frac{\beta}{2K}\|x\| \quad \textit{for all } t \in \mathbb{I} \textit{ and } x \in B_\delta(0)\,,$$

then one has

$$\operatorname{dist}\big(\phi(t,t_0,B_{\delta/K}(0)),\{0\}\big) \leq \delta e^{\frac{\beta}{2}(t-t_0)} \quad \textit{for all } t_0, t \in \mathbb{I} \textit{ with } t \leq t_0\,,$$

i.e., the trivial solution of (3.33) *is locally forward (pullback, uniformly, respectively) repulsive.*

PROOF. It suffices to prove (i), since (ii) can be shown analogously. Given $t_0 \in \mathbb{I}$ and $x_0 \in B_\delta(0)$, an estimate for the process ϕ is proved under the additional assumption

$$\phi(t,t_0,x_0) \in B_\delta(0) \quad \text{for all } t \geq t_0\,. \tag{3.35}$$

The solution $\phi(\cdot,t_0,x_0)$ of (3.33) is also a solution of inhomogeneous linear differential equation

$$\dot{x} = B(t)x + F(t,\phi(t,t_0,x_0))\,.$$

Thus, the variation of constants formula implies that

$$\phi(t,t_0,x_0) = \Phi(t,t_0)x_0 + \int_{t_0}^{t} \Phi(t,s)F(s,\phi(s,t_0,x_0))\,\mathrm{d}s \quad \text{for all } t \geq t_0\,,$$

and hence,

$$\begin{aligned}\|\phi(t,t_0,x_0)\| &\leq \|\Phi(t,t_0)\|\,\|x_0\| + \int_{t_0}^{t} \|\Phi(t,s)\|\,\|F(s,\phi(s,t_0,x_0))\|\,\mathrm{d}s\\ &\overset{(3.34)}{\leq} Ke^{\beta(t-t_0)}\|x_0\| + \int_{t_0}^{t} Ke^{\beta(t-s)}\frac{-\beta}{2K}\|\phi(s,t_0,x_0)\|\,\mathrm{d}s\end{aligned}$$

for all $t \geq t_0$ is fulfilled. This implies

$$e^{-\beta t}\|\phi(t,t_0,x_0)\| \leq Ke^{-\beta t_0}\|x_0\| + \frac{-\beta}{2}\int_{t_0}^{t} e^{-\beta s}\|\phi(s,t_0,x_0)\|\,\mathrm{d}s$$

for all $t \geq t_0$. Hence, Gronwall's inequality (see, e.g., ABRAHAM, MARSDEN & RATIU [**1**, Theorem 4.1.7, p. 242]) yields the estimate

$$\|\phi(t,t_0,x_0)\| \leq Ke^{\frac{\beta}{2}(t-t_0)}\|x_0\| \quad \text{for all } t \geq t_0\,. \tag{3.36}$$

Define $\eta := \frac{\delta}{K}$. Since $\frac{\beta}{2} < 0$, the assumption (3.35) is fulfilled for all $t_0 \in \mathbb{I}$ and $x_0 \in B_\eta(0)$, and thus, (3.36) holds for such t_0 and x_0. This implies that

$$\operatorname{dist}\big(\phi(t,t_0,B_\eta(0)),\{0\}\big) \leq K\eta e^{\frac{\beta}{2}(t-t_0)} \quad \text{for all } t_0,t \in \mathbb{I} \text{ with } t_0 \leq t\,.$$

From this inequality, the required conditions for the local attractivity are easily obtained. □

Repellers can be seen as attractors of the process under time reversal as the following proposition shows.

PROPOSITION 3.55 (Process under time reversal). *Let ϕ be an invertible process on a metric space (X,d), and consider the* process under time reversal *ϕ^{-1} which is defined by*

$$\phi^{-1}(t,t_0,x_0) = \phi(-t,-t_0,x_0) \quad \textit{for all } t,t_0 \in \mathbb{T} \textit{ and } x_0 \in X\,.$$

It follows that ϕ^{-1} is also an invertible process, and if one defines

$$\mathcal{M}^{-1} := \{(t,x) \in \mathbb{T} \times X : (-t,x) \in \mathcal{M}\}$$

for a given nonautonomous $\mathcal{M}$, then the following statements are fulfilled:

(i) *$\mathcal{M}$ is a local forward attractor of ϕ if and only if $\mathcal{M}^{-1}$ is a local forward repeller of ϕ^{-1}.*
(ii) *$\mathcal{M}$ is a local pullback attractor of ϕ if and only if $\mathcal{M}^{-1}$ is a local pullback repeller of ϕ^{-1}.*
(iii) *$\mathcal{M}$ is a local uniform attractor of ϕ if and only if $\mathcal{M}^{-1}$ is a local uniform repeller of ϕ^{-1}.*

PROOF. To show that ϕ^{-1} is an invertible process, first note the initial value property

$$\phi^{-1}(t_0,t_0,x_0) = \phi(-t_0,-t_0,x_0) = x_0 \quad \text{for all } t_0 \in \mathbb{T} \text{ and } x_0 \in X\,,$$

and the evolution property is proved by

$$\begin{aligned}\phi^{-1}(t_2,t_0,x_0) &= \phi(-t_2,-t_0,x_0)\\ &= \phi\big(-t_2,-t_1,\phi(-t_1,-t_0,x_0)\big)\\ &= \phi^{-1}\big(t_2,t_1,\phi^{-1}(t_1,t_0,x_0)\big)\end{aligned}$$

for all $t_0,t_1,t_2\in\mathbb{T}$ and $x_0\in X$. Now let $\mathcal{M}$ be a local forward attractor of ϕ, i.e., there exists an $\eta>0$ such that

$$\lim_{t\to\infty}\operatorname{dist}\big(\phi(t,t_0,B_\eta(M_{t_0})),M_t\big)=0 \quad \text{for all } t_0\geq 0\,.$$

This is equivalent to

$$\lim_{t\to\infty}\operatorname{dist}\big(\phi^{-1}\big(-t,-t_0,B_\eta(M_{-(-t_0)})\big),M_{-(-t)}\big)\big)=0 \quad \text{for all } t_0\geq 0\,,$$

which also reads as

$$\lim_{t\to-\infty}\operatorname{dist}\big(\phi^{-1}\big(t,t_0,B_\eta(M_{-t_0})\big),M_{-t}\big)=0 \quad \text{for all } t_0\leq 0\,.$$

This means that $\mathcal{M}^{-1}$ is a local forward repeller of ϕ^{-1}. The other assertions can be shown analogously. □

This theorem shows that in contrast to the autonomous case (see Proposition 1.19), the notions of a local pullback attractor and local forward repeller are not dual. The same holds for local forward attractors and pullback repellers, however, duality is fulfilled for local uniform attractors and repellers by (iii). The lack of duality in case of pullback attraction and repulsion has far reaching consequences which will be discussed in the following chapters. One important consequence is that both concepts do not have the same uniqueness properties.

EXERCISE 3.56. Show that local pullback attractors are locally unique and local pullback repellers are intrinsically non-unique.

Endnotes. For review articles, see Caraballo, Langa & Kloeden [**28**] and Kloeden [**116**]. Nonautonomous sets have a long history in the literature and under that name in, e.g., Aulbach, Rasmussen & Siegmund [**11**] and Rasmussen [**194**]. The definition of a pullback attractor was motivated by that of a random attractor, see the references under Chapter 14 below. In some earlier papers such as Kloeden & Schmalfuß [**140, 142, 141**] and Kloeden & Stonier [**145**], they were called cocycle attractors and the name pullback attractor was later introduced, apparently first in Kloeden [**116**], to distinguish them from forward attractors. See Crauel & Flandoli [**61**] and Pötzsche [**188**] for nonautonomous ω-limit sets. Theorem 3.20 on the existence of a pullback attractor has appeared in many versions in the literature with the original proofs being based on those for the existence of random attractors. Attraction universes were introduced in Schmalfuß [**214**] for random dynamical systems. Similar existence theorems to Theorem 3.27 can be found in Flandoli & Schmalfuß [**78**], Kloeden & Schmalfuß [**140, 142**] and Schmalfuß [**213**], see also Crauel, Debussche & Flandoli [**59**]. Section 4 on the relationship between the different types of nonautonomous attractors is based on Cheban, Kloeden & Schmalfuß [**41**]. The upper semi-continuous dependence of pullback attractors on parameters in Section 5 has been considered in many papers, including explicitly Caraballo & Langa [**32**] and Kloeden [**122**]. Li Desheng & Kloeden [**68, 69**] considered the equi-attraction and the continuous dependence of pullback attractors on parameters. Section 6 on parametrically inflated pullback attractors is based on Wang Yejuan, Li Desheng & Kloeden [**237**] and Section 7 on pullback attractors with continuous fibers on Cheban, Kloeden & Schmalfuß [**41**] and Wang Yejuan, Li Desheng & Kloeden [**237**]. Section 8 on local attractors and repellers is from Rasmussen [**194**]. The figures in this chapter were made by van Geene [**231**] and Storck [**224**].

CHAPTER 4

Morse decompositions

The notion of a (global) pullback attractor was introduced in the previous chapter in order to describe the global dynamical behavior of nonautonomous dynamical systems. A global attractor, however, can be very large and complicated, and the definition of a global attractor does not give information about the dynamical behavior inside the attractor. To study this, the local concepts of attractivity and repulsivity of the previous chapter are needed, which then lead to nonautonomous Morse decompositions. A Morse decomposition is a decomposition of the global attractor into Morse sets and connecting orbits. For autonomous systems, Morse decompositions were first studied in CONLEY **[56]**.

In this chapter, both discrete ($\mathbb{T} = \mathbb{Z}$) and continuous time ($\mathbb{T} = \mathbb{R}$) invertible processes $(t, t_0, x_0) \mapsto \phi(t, t_0, x_0)$ are treated, and it is assumed that $\mathcal{X}$ is a compact and invariant nonautonomous set, which might be given by a pullback attractor of the process. Henceforth, attention is concentrated to the process restricted to this invariant set, i.e., $\phi(\cdot, t_0, x_0)$ is only defined for $(t_0, x_0) \in \mathcal{X}$. One easily verifies that the global pullback attractor of this process is then given by $\mathcal{X}$, the entire extended phase space. Note that this also holds when $\mathcal{X}$ is not the global pullback attractor of the original process. This means in turn that one needs local notions of attractivity (and repulsivity) to proceed with the analysis of the situation.

1. Attractor-repeller pairs

Firstly, the interplay of local attractors and repellers will be discussed. It will be shown that it is quite typical for attractors and repellers to appear together in the case of a compact nonautonomous phase space $\mathcal{X}$, although the situation is not the same as in the autonomous case (which was discussed in Chapter 1). The next theorem on the existence of attractor-repeller pairs is crucial for the whole chapter.

THEOREM 4.1 (Existence of past attractor-repeller pairs). *Let ϕ be an invertible process defined on a compact nonautonomous set $\mathcal{X}$ and let $\mathcal{R}$ be a local forward repeller. Then there exists a uniquely determined local pullback attractor $\mathcal{R}^*$, which is the maximal local pullback attractor outside $\mathcal{R}$ in the sense that $\mathcal{R} \cap \mathcal{R}^* = \emptyset$ and any local pullback attractor $\mathcal{A} \supsetneq \mathcal{R}^*$ has nonempty intersection with $\mathcal{R}$. The pair $(\mathcal{R}^*, \mathcal{R})$ is called a* past attractor-repeller pair.

PROOF. Since $\mathcal{R}$ is a local forward repeller, there exists an $\eta > 0$ such that

$$\lim_{t \to -\infty} \operatorname{dist}\left(\phi(t, t_0, B_\eta(R_{t_0})), R_t\right) = 0 \quad \text{for all } t_0 \leq 0\,. \tag{4.1}$$

Consider the attraction universe $\mathfrak{D}$ containing the nonautonomous sets $\mathcal{D}^\zeta$ for $\zeta \in (0, \eta]$, which are defined by their fibers

$$D_t^\zeta := \begin{cases} X_t \setminus B_\zeta(R_t) & : \ t \le 0\,, \\ X_t & : \ t > 0\,. \end{cases}$$

To show that $\mathcal{D}^\eta$ is pullback absorbing with respect to $\mathfrak{D}$ choose $\zeta \in (0, \eta]$ and $t_0 \in \mathbb{T}$ arbitrarily. If $D_{t_0}^\eta = X_{t_0}$, then obviously

$$\phi(t_0, t, D_t^\zeta) \subset D_{t_0}^\eta \quad \text{for all } t \le 0\,.$$

Otherwise, (4.1) implies that there is a $T > 0$ with

$$\operatorname{dist}\big(\phi(t, t_0, B_\eta(R_{t_0})), R_t\big) < \frac{\zeta}{2} \quad \text{for all } t \le t_0 - T\,,$$

which means that $\phi(t, t_0, B_\eta(R_{t_0})) \subset B_{\zeta/2}(R_t)$ for all $t \le t_0 - T$. Thus, one obtains

$$\phi(t, t_0, D_{t_0}^\eta) = X_t \setminus \phi(t, t_0, B_\eta(R_{t_0})) \supset D_t^\zeta \quad \text{for all } t \le t_0 - T\,.$$

This implies the desired relation $\phi(t_0, t, D_t^\zeta) \subset D_{t_0}^\eta$ for all $t \le t_0 - T$, and Theorem 3.27 from Chapter 3 guarantees the existence of a pullback attractor $\mathcal{R}^*$ with respect to $\mathfrak{D}$ such that $\mathcal{R}^* \subset \mathcal{D}^\eta$. $\mathcal{R}^*$ is also a local pullback attractor, since $\mathcal{D}^{\eta/2} \in \mathfrak{D}$ and $B_{\eta/2}(R_t^*) \subset D_t^{\eta/2}$ for all $t \in \mathbb{T}$.
Let $\mathcal{A} \supsetneq \mathcal{R}^*$ be another local pullback attractor. Then there exists a $t_0 \in \mathbb{T}$ with $A_{t_0} \supsetneq R_{t_0}^*$. Choose an $x \in A_{t_0} \setminus R_{t_0}^*$. Since $\mathcal{A}$ is a local pullback attractor and $x \in A_{t_0}$, there exists an $\tilde{\eta} > 0$ with

$$\lim_{t \to -\infty} \operatorname{dist}\big(\phi(t_0, t, B_{\tilde{\eta}}(\phi(t, t_0, x))), A_{t_0}\big) = 0\,.$$

Due to $\lim_{t \to -\infty} \operatorname{dist}\big(\phi(t, t_0, x), R_t\big) = 0$ (this will be shown in Theorem 4.4 (ii)), there exists a sequence $(y_n)_{n \in \mathbb{N}}$ in R_{t_0} with

$$\lim_{n \to \infty} \operatorname{dist}(y_n, A_{t_0}) = 0\,.$$

Since R_{t_0} and A_{t_0} are compact, this implies that their intersection is nonempty, which is a contradiction. □

The attractor-repeller pair of Theorem 4.1 consists of a local pullback attractor and a local forward repeller. It is called a *past* attractor-repeller pair, since the two components of the pair are the suitable attractivity and repulsivity notions for the past of the system. Via time-reversal (see Proposition 3.55), it is possible to obtain a corresponding theorem about the existence of a *future* attractor-repeller pair consisting of a forward attractor and pullback repeller. The situation of an *all-time* attractor-repeller pair is more subtle, see Exercise 4.3 below.

time domain	*local attractor*	*local repeller*
past	local pullback attractor	local forward repeller
future	local forward attractor	local pullback repeller
all-time	local uniform attractor	local uniform repeller

TABLE 1. Attractors and repellers for the different time domains.

It is natural to ask if the existence of a local pullback attractor implies that of a local forward repeller. The following example shows that this does not hold.

EXAMPLE 4.2. The nonautonomous differential equation

$$\dot{x} = f(t,x)$$

with the function $f : \mathbb{R} \times [0,1] \to \mathbb{R}$ defined by

$$f(t,x) := \begin{cases} |2x-1|-1 & : \ t \geq 0 \text{ and } x \in [0,1]\,, \\ |2x-1|-1 & : \ t < 0 \text{ and } x \in \left[0,\frac{1}{2}\right], \\ (2-2t)\left(x-\frac{1}{2}\right)-1 & : \ t < 0 \text{ and } x \in \left[\frac{1}{2},\frac{2-t}{2-2t}\right], \\ 0 & : \ t < 0 \text{ and } x \in \left(\frac{2-t}{2-2t},1\right], \end{cases}$$

generates a process $\phi : \mathbb{R} \times \mathbb{R} \times X \to X$, where $X = [0,1]$. Then $\mathcal{A} := \mathbb{R} \times \{0\}$ is a local pullback (as well as a local forward) attractor. Suppose that there exists a local forward repeller $\mathcal{A}^*$. By the invariance of $\mathcal{A}^*$, there exist $\gamma > \frac{1}{2}$ and $\tau_1 < 0$ with

$$A_t^* \subset (\gamma, 1] \quad \text{for all } t \leq \tau_1\,.$$

Hence, there exists a $\tau_2 < \tau_1$ with

$$A_s^* = A_t^* \quad \text{and} \quad f\big(t,[\gamma,1]\big) = \{0\} \quad \text{for all } t,s \leq \tau_2\,,$$

which contradicts the assumption that $\mathcal{A}^*$ is a local forward repeller.

EXERCISE 4.3. Show by using Example 4.2 that, in general, there does not exist a uniform attractor-repeller pair, i.e., a local uniform attractor may exist without a corresponding local uniform repeller and the existence of a local uniform repeller need not imply the existence of a local uniform attractor. Why is it sufficient to show only one of these statements?

In this chapter, attention is restricted to the past. Corresponding results for the future are easily obtained via time reversal, whereas general results for all time cannot be expected due to the lack of an attractor-repeller pair. However, one can prove an existence result for all-time attractor-repeller pairs in the context of linear systems. This will be discussed in Subsection 4.2 of Chapter 5.

Some basic properties of past attractor-repeller pairs are stated in the following theorem, including convergence results in both time directions. Note that for the past of the system, convergence in forward time is given by the pullback convergence.

THEOREM 4.4 (Dynamics of past attractor-repeller pairs). *Let ϕ be an invertible process defined on a compact nonautonomous set $\mathcal{X}$ and let $(\mathcal{R}^*, \mathcal{R})$ be a past attractor-repeller pair. Then the following statements hold:*

(i) Isolation in the past. *There exists a $\beta > 0$ such that*

$$B_\beta(R_t^*) \cap B_\beta(R_t) = \emptyset \quad \textit{for all } t \leq 0\,.$$

(ii) Convergence in backward time. *Let $t_0 \in \mathbb{T}$ and $C \subset X_{t_0} \setminus R_{t_0}^*$ be a compact set. Then*

$$\lim_{t\to-\infty} \operatorname{dist}\big(\phi(t,t_0,C), R_t\big) = 0\,.$$

(iii) Convergence in forward time. *Let $(\gamma_t)_{t\in\mathbb{T}}$ be a family of points with $\gamma_t \in X_t$ and $\liminf_{t\to-\infty} \operatorname{dist}(\gamma_t, R_t) > 0$. Then*

$$\lim_{t\to-\infty} \operatorname{dist}\big(\phi(t_0,t,\gamma_t), R_{t_0}^*\big) = 0 \quad \textit{for all } t_0 \in \mathbb{T}\,.$$

PROOF. Let η and $\mathfrak{D} = \{\mathcal{D}^\zeta : \zeta \in (0,\eta]\}$ be defined for the past attractor-repeller pair $(\mathcal{R}^*, \mathcal{R})$ as in the proof of Theorem 4.1.
(i) It was shown in the proof of Theorem 4.1 that $\mathcal{R}^* \subset \mathcal{D}^\eta$. The assertion follows by choosing $\beta := \frac{\eta}{2}$.
(ii) Let $t_0 \in \mathbb{T}$, $C \subset X_{t_0} \setminus R^*_{t_0}$ be a compact set and $\varepsilon \in (0,\eta)$. Since $\mathcal{R}^*$ is a pullback attractor with attraction universe $\mathfrak{D}$ (see proof of Theorem 4.1) and $\mathcal{D}^\varepsilon \in \mathfrak{D}$, there exists a $T > 0$ such that

$$C \cap \phi(t_0, t, D_t^\varepsilon) = \emptyset \quad \text{for all } t \le t_0 - T\,.$$

Hence, $\phi(t, t_0, C) \cap D_t^\varepsilon = \emptyset$ for all $t \le t_0 - T$, which means that $\operatorname{dist}\big(\phi(t,t_0,C), R_t\big) \le \varepsilon$ for all $t \le t_0 - T$.
(iii) Define $\zeta := \frac{1}{2}\min\{\eta, \liminf_{t\to-\infty} \operatorname{dist}(\gamma_t, R_t)\}$ and note that there exists a $t_0 \in \mathbb{T}$ with

$$\gamma_t \in D_t^\zeta \quad \text{for all } t \le t_0\,.$$

This finishes the proof, since $\mathcal{D}^\zeta \in \mathfrak{D}$ and $\mathcal{R}^*$ is pullback attractor with attraction universe $\mathfrak{D}$. □

Theorem 4.1 implies that, given a local forward repeller $\mathcal{R}$, the set $\mathcal{R}^*$ is the uniquely determined local pullback attractor outside of $\mathcal{R}$ with the property of pullback convergence as described in Theorem 4.4 (iii). It is easy to see that such a uniqueness result is not valid for local forward repellers, i.e., it is possible that $(\mathcal{A}, \mathcal{R})$ and $(\mathcal{A}, \tilde{\mathcal{R}})$ are past attractor-repeller pairs with $\mathcal{R} \ne \tilde{\mathcal{R}}$. In this case, however, the fibers of $\mathcal{R}$ and $\tilde{\mathcal{R}}$ converge to each other as time tends to the past.

PROPOSITION 4.5 (Nonuniqueness of past attractor-repeller pairs). *Let $\mathcal{R}$ and $\tilde{\mathcal{R}}$ be local forward repellers of an invertible process ϕ on a compact nonautonomous set $\mathcal{X}$ such that $\mathcal{R}^* = \tilde{\mathcal{R}}^*$. Then*

$$\lim_{t\to-\infty} h\big(R_t, \tilde{R}_t\big) = 0\,.$$

PROOF. Suppose that there exists a sequence $(t_n, \gamma_n)_{n\in\mathbb{N}}$ in $\mathcal{R}$ such that both $\lim_{n\to\infty} t_n = -\infty$ and

$$\liminf_{n\to\infty} \operatorname{dist}\big(\gamma_n, \tilde{R}_{t_n}\big) > 0\,.$$

Then Theorem 4.4 (iii) applied to the attractor-repeller pair $(\tilde{\mathcal{R}}^*, \tilde{\mathcal{R}})$ implies that

$$\lim_{n\to\infty} \operatorname{dist}\big(\phi(0, t_n, \gamma_n), \tilde{R}_0^*\big) = 0\,.$$

Since $\phi(0, t_n, \gamma_n) \in R_0$ and both $\mathcal{R}$ and $\mathcal{R}^* = \tilde{\mathcal{R}}^*$ are compact nonautonomous sets, it follows that $\mathcal{R}_0 \cap \mathcal{R}_0^* \ne \emptyset$. This is a contradiction. □

2. Morse decompositions

The notion of a past attractor-repeller pair is now generalized using Morse decompositions. The definition of a Morse decomposition via finite attractor-repeller sequences is basically the same as in the autonomous case (see Definition 1.43). An alternative definition as proposed in Theorem 1.45 is, however, not possible in the nonautonomous context for reasons which will be explained below.

DEFINITION 4.6 (Past Morse decomposition). Let ϕ be an invertible process defined on a compact nonautonomous set $\mathcal{X}$. A finite family $\{\mathcal{M}^1, \mathcal{M}^2, \dots, \mathcal{M}^n\}$ of nonautonomous sets, the so-called *Morse sets*, is called a *past Morse decomposition* if the representation

$$\mathcal{M}^i = \mathcal{R}^{i,*} \cap \mathcal{R}^{i-1} \quad \text{for all } i \in \{1, \dots, n\}$$

holds with local forward repellers

$$\mathcal{X} = \mathcal{R}^0 \supsetneq \mathcal{R}^1 \supsetneq \cdots \supsetneq \mathcal{R}^n = \emptyset$$

satisfying $\mathcal{R}^{0,*} \subsetneq \mathcal{R}^{1,*} \subsetneq \cdots \subsetneq \mathcal{R}^{n,*}$.

Note that, contrary to the autonomous case, one has to impose the condition $\mathcal{R}^{0,*} \subsetneq \mathcal{R}^{1,*} \subsetneq \cdots \subsetneq \mathcal{R}^{n,*}$ on the local pullback attractors, since Proposition 4.5 indicates that local pullback attractors of the attractor sequence can coincide.

The definition of a past Morse decomposition is a generalization of a past attractor-repeller pair. More precisely, if $(\mathcal{A}, \mathcal{R})$ is a past attractor-repeller pair such that $\emptyset \subsetneq \mathcal{A} \subsetneq \mathcal{X}$, then $\{\mathcal{A}, \mathcal{R}\}$ is a past Morse decomposition.

Basic properties of the Morse sets of a past Morse decomposition are presented in the following proposition.

PROPOSITION 4.7. *Let ϕ be an invertible process defined on a compact nonautonomous set $\mathcal{X}$. Then the Morse sets of a past Morse decomposition $\{\mathcal{M}^1, \dots, \mathcal{M}^n\}$ are nonempty, invariant, pairwise disjoint and* isolated in the past, *i.e., there exists a $\beta > 0$ such that*

$$B_\beta(M_t^i) \cap B_\beta(M_t^j) = \emptyset \quad \textit{for all } t \le 0\,.$$

PROOF. Choose an arbitrary Morse set $\mathcal{M}^i = \mathcal{R}^{i,*} \cap \mathcal{R}^{i-1}$. Because of $\mathcal{R}^{i-1,*} \subsetneq \mathcal{R}^{i,*}$, there is an $x \in \mathcal{R}_0^{i,*} \setminus \mathcal{R}_0^{i-1,*}$ and, since $\mathcal{R}^{i,*} \ni (0,x)$ is a local pullback attractor, there exists an $\eta > 0$ with

$$\lim_{t \to -\infty} \operatorname{dist}\big(\phi(0, t, B_\eta(\phi(t, 0, x))), R_0^{i,*}\big) = 0\,.$$

Due to $\lim_{t \to -\infty} \operatorname{dist}\big(\phi(t, 0, x), R_t^{i-1}\big) = 0$ (see Theorem 4.4 (ii)), there exists a sequence $(y_n)_{n \in \mathbb{N}}$ in R_0^{i-1} with

$$\lim_{n \to \infty} \operatorname{dist}\big(y_n, R_0^{i,*}\big) = 0\,.$$

Since R_0^{i-1} and $R_0^{i,*}$ are compact, this implies that $\mathcal{M}^i = \mathcal{R}^{i,*} \cap \mathcal{R}^{i-1} \neq \emptyset$. Furthermore, $\mathcal{M}^i$ is the intersection of two invariant nonautonomous sets and thus invariant. Choose another Morse set $\mathcal{M}^j = \mathcal{R}^{j,*} \cap \mathcal{R}^{j-1}$. It may be assumed that $j > i$. Then the relation

$$\begin{aligned}\mathcal{M}^i \cap \mathcal{M}^j &= \mathcal{R}^{i,*} \cap \mathcal{R}^{i-1} \cap \mathcal{R}^{j,*} \cap \mathcal{R}^{j-1} \\ &= \mathcal{R}^{i,*} \cap \mathcal{R}^{j-1} \subset \mathcal{R}^{j-1,*} \cap \mathcal{R}^{j-1} = \emptyset\end{aligned}$$

holds. Finally, the fact that the Morse sets are isolated in the past is an easy consequence of Theorem 4.4 (i). □

As in the autonomous case, past Morse decompositions are not uniquely determined.

DEFINITION 4.8 (Finer past Morse decompositions). A past Morse decomposition $\{\mathcal{M}^1, \dots, \mathcal{M}^n\}$ is said to be *finer* than a past Morse decomposition $\{\bar{\mathcal{M}}^1, \dots, \bar{\mathcal{M}}^m\}$ if

$$\lim_{t\to-\infty} \operatorname{dist}\left(\bigcup_{i=1}^n M_t^i, \bigcup_{i=1}^m \bar{M}_t^i\right) = 0\,.$$

This definition generalizes the notion of a finer (autonomous) Morse decomposition as introduced in Definition 1.48. The additional limit in the nonautonomous context is motivated by Proposition 4.5, which says essentially that two past attractor-repeller pairs converge to each other when the local pullback attractors are equal. Examples show that such a (weak) form of nonuniqueness is not valid for arbitrary Morse decompositions, i.e., those consisting of more than two sets. However, it will be seen below that in the special cases of one-dimensional and linear systems, similar results to those in Proposition 4.5 are obtained.

The following theorem shows that the Morse sets are important for the asymptotic behavior of a nonautonomous dynamical system.

THEOREM 4.9 (Forward convergence). *Let ϕ be an invertible process defined on a compact nonautonomous set $\mathcal{X}$ and let $\{\mathcal{M}^1, \dots, \mathcal{M}^n\}$ be a past Morse decomposition obtained by the sequence of local forward repellers $\mathcal{R}^0 \supset \dots \supset \mathcal{R}^n$. Then all families $(\gamma_t)_{t\in\mathbb{T}}$ for which $\gamma_t \in X_t$ and $\liminf_{t\to-\infty} \operatorname{dist}\left(\gamma_t, \bigcup_{j=1,\dots,n} \partial R_t^j\right) > 0$ satisfy*

$$\lim_{t\to-\infty} \operatorname{dist}\left(\phi(t_0, t, \gamma_t), \bigcup_{j=1,\dots,n} M_{t_0}^j\right) = 0 \quad \textit{for all } t_0 \in \mathbb{T}\,.$$

PROOF. Let $t_0 \in \mathbb{T}$, and assume without loss of generality that there exists an $i \in \{1, \dots, n\}$ with

$$\gamma_t \in R_t^{i-1} \quad \text{and} \quad \gamma_t \notin R_t^i \quad \text{for all } t < 0\,.$$

Then, $\liminf_{t\to-\infty} \operatorname{dist}\left(\gamma_t, \partial R_t^i\right) > 0$ yields $\liminf_{t\to-\infty} \operatorname{dist}\left(\gamma_t, R_t^i\right) > 0$. Theorem 4.4 then implies that

$$\lim_{t\to-\infty} \operatorname{dist}\left(\phi(t_0, t, \gamma_t), R_{t_0}^{i,*}\right) = 0\,. \tag{4.2}$$

Assume that there exist an $\varepsilon > 0$ and a sequence $t_n \to -\infty$ with

$$\operatorname{dist}\left(\phi(t_0, t_n, \gamma_{t_n}), M_{t_0}^i\right) \geq \varepsilon \quad \text{for all } n \in \mathbb{N}\,. \tag{4.3}$$

Since $R_{t_0}^{i-1}$ is compact, it can be assumed that the sequence $\left(\phi(t_0, t_n, \gamma_{t_n})\right)_{n\in\mathbb{N}}$ in $R_{t_0}^{i-1}$ is convergent with limit $x_0 \in R_{t_0}^{i-1}$. Moreover, $x_0 \in R_{t_0}^{i,*}$, since (4.2) holds and $R_{t_0}^{i,*}$ is compact. Thus, $x_0 \in M_{t_0}^i = R_{t_0}^{i,*} \cap R_{t_0}^{i-1}$, which contradicts (4.3). □

In contrast to past attractor-repeller pairs, the backward convergence condition described in Theorem 4.4 does not hold for arbitrary past Morse decompositions.

EXERCISE 4.10. Show by means of a counterexample that a backward convergence condition as described in Theorem 4.4 does not hold for arbitrary past Morse decompositions.

However, as will be shown below and in Chapter 5, in the special cases of one-dimensional and linear systems, one obtains similar results to those in Theorem 4.4.

On the other hand, if the backward convergence holds, then the following uniqueness result concerning the local pullback attractors holds.

PROPOSITION 4.11. *Let ϕ be an invertible process defined on a compact nonautonomous set $\mathcal{X}$ and let $\{\mathcal{M}^1, \dots, \mathcal{M}^n\}$ be a past Morse decomposition obtained by the finite sequence of local forward repellers $\mathcal{R}^0 \supset \cdots \supset \mathcal{R}^n$. Moreover, assume that* backward convergence *holds, i.e., for all $(t_0, x_0) \in \mathcal{X}$, there exists an $i \in \{1, \dots, n\}$ with*

$$\lim_{t \to -\infty} \operatorname{dist}\left(\phi(t, t_0, x_0), M_t^i\right) = 0 .$$

Then the representation

$$\mathcal{R}^{i,*} = \left\{(t_0, x_0) \in \mathcal{X} : \lim_{t \to -\infty} \operatorname{dist}\left(\phi(t, t_0, x_0), \bigcup_{j=1,\dots,i} M_t^j\right) = 0\right\}$$

holds for all $i \in \{1, \dots, n\}$, i.e., the local pullback attractors of a past Morse decomposition are uniquely determined.

PROOF. ($\subseteq$) Let $(t_0, x_0) \in \mathcal{R}^{i,*}$. Choose $j \in \{1, \dots, n\}$ such that

$$0 = \lim_{t \to -\infty} \operatorname{dist}\left(\phi(t, t_0, x_0), M_t^j\right) = \lim_{t \to -\infty} \operatorname{dist}\left(\phi(t, t_0, x_0), R_t^{j-1}\right) .$$

If $j > i$, then $\lim_{t \to -\infty} \operatorname{dist}\left(\phi(t, t_0, x_0), R_t^i\right) = 0$, which contradicts Theorem 4.4 (i), since $\phi(t, t_0, x_0) \in R_t^{i,*}$ for all $t \in \mathbb{T}$.
($\supseteq$) Let $(t_0, x_0) \in \mathcal{X} \setminus \mathcal{R}^{i,*}$. Then Theorem 4.4 (ii) implies

$$\lim_{t \to -\infty} \operatorname{dist}\left(\phi(t, t_0, x_0), R_t^i\right) = 0 . \tag{4.4}$$

If $\lim_{t \to -\infty} \operatorname{dist}\left(\phi(t, t_0, x_0), \bigcup_{j=1,\dots,i} M_t^j\right) = 0$, then

$$\lim_{t \to -\infty} \operatorname{dist}\left(\phi(t, t_0, x_0), R_t^{i,*}\right) = 0 ,$$

since $\mathcal{M}^j \subset \mathcal{R}^{i,*}$ for $j \in \{1, \dots, i\}$, which contradicts (4.4) in view of Theorem 4.4 (i). □

3. The one-dimensional case

Stronger results on the convergence behavior of the system and the nonuniqueness of the Morse sets hold for Morse decompositions of one-dimensional processes. Consider an invertible one-dimensional process ϕ defined on a compact nonautonomous set $\mathcal{X}$ whose fibers X_t, $t \in \mathbb{T}$, are compact intervals in $\mathbb{R}$.

In addition to convergence in forward time (i.e., pullback convergence), backward convergence also holds for one-dimensional processes.

THEOREM 4.12. *Let ϕ be an invertible process defined on $\mathcal{X}$ and let $\{\mathcal{M}^1, \dots, \mathcal{M}^n\}$ be a past Morse decomposition obtained by the finite sequence of local forward repellers $\mathcal{R}^0 \supset \cdots \supset \mathcal{R}^n$. Then the following statements hold:*

(i) Convergence in forward time. *All families $(\gamma_t)_{t \in \mathbb{T}}$ for which $\gamma_t \in X_t$ and $\liminf_{t \to -\infty} \operatorname{dist}\left(\gamma_t, \bigcup_{j=1,\dots,n} \partial R_t^j\right) > 0$ satisfy*

$$\lim_{t \to -\infty} \operatorname{dist}\left(\phi(t_0, t, \gamma_t), \bigcup_{j=1,\dots,n} M_{t_0}^j\right) = 0 \quad \textit{for all } t_0 \in \mathbb{T} .$$

(ii) Convergence in backward time. *For all* $(t_0, x_0) \in \mathcal{X}$, *there exists an* $i \in \{1, \dots, n\}$ *with*

$$\lim_{t \to -\infty} \operatorname{dist}\big(\phi(t, t_0, x_0), M_t^i\big) = 0 .$$

PROOF. Since (i) was shown in Theorem 4.9 without restriction on the dimension, it remains to prove (ii). Choose $(t_0, x_0) \in \mathcal{X}$ arbitrarily. Then there exists an $i \in \{1, \dots, n\}$ such that

$$x_0 \in R_{t_0}^{i,*} \quad \text{and} \quad x_0 \notin R_{t_0}^{i-1,*} .$$

If $x_0 \in R_{t_0}^{i-1}$, then the asserted limit relation follows, since then $x_0 \in M_{t_0}^i$ and $\mathcal{M}^i$ is invariant. Assume henceforth that $x_0 \notin R_{t_0}^{i-1}$. In view of the topology on the fibers of $\mathcal{X}$ and the fact that $\lim_{t \to -\infty} \operatorname{dist}\big(\phi(t, t_0, x_0), R_t^{i-1}\big) = 0$ (see Theorem 4.4 (ii)), there exists an $x \in R_{t_0}^{i-1}$ such that

$$\lim_{t \to -\infty} |\phi(t, t_0, x_0) - \phi(t, t_0, x)| = 0 . \tag{4.5}$$

Moreover, since $\mathcal{R}^{i,*}$ is a local pullback attractor, there exists an $\eta > 0$ such that

$$R_{t_0}^{i,*} = \bigcap_{\tau \geq 0} \bigcup_{t \geq \tau} \phi\big(t_0, t_0 - t, B_\eta\big(R_{t_0 - t}^{i,*}\big)\big) =: \limsup_{t \to \infty} \phi\big(t_0, t_0 - t, B_\eta\big(R_{t_0 - t}^{i,*}\big)\big)$$

(see also Exercise 3.49; the closure is not needed here), which implies that

$$\limsup_{t \to \infty} \phi\big(t_0, t_0 - t, B_\eta(\phi(t_0 - t, t_0, x_0))\big) \subset R_{t_0}^{i,*} .$$

Then, $x \in R_{t_0}^{i,*}$ by (4.5). Hence, $x \in M_{t_0}^{i,*}$. □

Proposition 4.5 can also be generalized for one-dimensional systems.

PROPOSITION 4.13. *Let* ϕ *be an invertible process defined on* $\mathcal{X}$, *and let* $\{\mathcal{M}^1, \dots, \mathcal{M}^n\}$ *and* $\{\tilde{\mathcal{M}}_1, \dots, \tilde{\mathcal{M}}_n\}$ *be past Morse decompositions obtained by the finite sequences of local forward repellers* $\mathcal{R}^0 \supset \cdots \supset \mathcal{R}^n$ *and* $\tilde{\mathcal{R}}_0 \supset \cdots \supset \tilde{\mathcal{R}}_n$. *Assume that*

$$\mathcal{R}^{i,*} = \tilde{\mathcal{R}}_i^* \quad \textit{for all } i \in \{1, \dots, n-1\} .$$

Then,

$$\lim_{t \to -\infty} h\big(M_t^i, \tilde{M}_t^i\big) = 0 \quad \textit{for all } i \in \{1, \dots, n\}.$$

PROOF. Choose $i \in \{1, \dots, n\}$. It suffices to show that

$$\lim_{t \to -\infty} \operatorname{dist}\big(M_t^i, \tilde{M}_t^i\big) = 0 .$$

The proof is divided into three parts.
Part 1. It will be shown here that there exists a local forward repeller $\hat{\mathcal{R}}^{i-1} \supset \mathcal{R}^{i-1}$ with $\hat{\mathcal{R}}^{i-1,*} = \mathcal{R}^{i-1,*}$ such that

$$\hat{M}_t^i := R_t^{i,*} \cap \hat{R}_t^{i-1} \quad \text{for all } t \in \mathbb{T}$$

has only finitely many connected components.
Since $\mathcal{R}^{i,*}$ is a local pullback attractor, there exists an $\eta > 0$ such that

$$R_0^{i,*} = \limsup_{t \to -\infty} \phi\big(0, t, B_\eta\big(R_t^{i,*}\big)\big)$$

(see the proof of Theorem 4.12). Moreover, since ϕ is continuous and $B_\eta(R_0^{i,*})$ has only finitely many connected components, it follows that $R_0^{i,*}$ has only finitely many connected components. Now $\mathcal{R}^{i-1}$ is a local forward repeller, so there exists a $\beta > 0$ such that

$$\lim_{t\to-\infty} \operatorname{dist}\big(\phi(t,0,B_\beta(R_0^{i-1})), R_t^{i-1}\big) = 0 .$$

Hence, the nonautonomous set $\hat{\mathcal{R}}^{i-1}$ with fibers defined by

$$\hat{R}_t^{i-1} := \overline{\phi\big(t,0,B_{\beta/2}(R_0^{i-1})\big)} \quad \text{for all } t \in \mathbb{T},$$

is also a local forward repeller satisfying $\hat{\mathcal{R}}^{i-1,*} = \mathcal{R}^{i-1,*}$. Moreover, the set $\hat{R}_t^{i-1}$ has only finitely many connected components for each $t \in \mathbb{T}$, since ϕ is continuous and $B_{\beta/2}(R_0^{i-1})$ has only finitely many connected components.

Part 2. It will be shown here that

$$\lim_{t\to-\infty} \operatorname{dist}\big(\phi(t,0,C), \tilde{M}_t^i\big) = 0$$

for all connected components C of $\hat{M}_0^i$.

Let $C = [c_1, c_2]$ be a connected component of $\hat{M}_0^i$, and choose $\varepsilon > 0$ arbitrarily. By Theorem 4.12 (ii), there exists a $t_1 \le 0$ such that

$$\operatorname{dist}\big(\phi(t,0,c_j), \tilde{M}_t^i\big) \le \frac{\varepsilon}{2} \quad \text{for all } t \le t_1 \text{ and } j \in \{1,2\} . \tag{4.6}$$

Then by Proposition 4.5, there exists a $t_2 \le t_1$ with

$$h\big(\tilde{R}_t^{i-1}, \hat{R}_t^{i-1}\big) \le \frac{\varepsilon}{2} \quad \text{for all } t \le t_2 . \tag{4.7}$$

Let $t \le t_2$ and $x \in \phi(t,0,C)$. If $\min\big\{|x-\phi(t,0,c_1)|, |x-\phi(t,0,c_2)|\big\} \le \frac{\varepsilon}{2}$, then (4.6) implies that $\operatorname{dist}\big(x, \tilde{M}_t^i\big) \le \varepsilon$. Otherwise, since $x \in \hat{R}_t^{i-1}$, by (4.7) there exists a $y \in \tilde{R}_t^{i-1}$ with $|x-y| \le \frac{\varepsilon}{2}$. Obviously,

$$y \in \phi(t,0,C) \subset \phi\big(t,0,\hat{M}_0^i\big) \subset R_t^{i,*} .$$

Hence, $y \in \tilde{M}_t^i$ and thus, $\operatorname{dist}\big(x, \tilde{M}_t^i\big) \le \varepsilon$.

Part 3. Since $\hat{M}_0^i$ has only finitely many connected components, it follows from Step 2 and $\hat{M}^i \supset M^i$ that

$$\lim_{t\to-\infty} \operatorname{dist}\big(M_t^i, \tilde{M}_t^i\big) = 0 .$$

This completes the proof of this theorem. □

Endnotes. Morse decompositions were introduced by Conley [**56**] in order to describe the global asymptotic behavior of (autonomous) dynamical systems on compact metric spaces. The fundamental theory of Morse decompositions can be found in various textbooks and monographs, e.g., in Colonius & Kliemann [**53**, Appendix B2], Rybakowski [**206**, Chapter 3], Robinson [**200**] and Akin [**4**]. Morse decompositions for delay-differential equations are due to Mallet-Paret [**164**]. This chapter is based on Rasmussen [**195**] and contains the fundamental theory of nonautonomous Morse decompositions. Recently, also two different versions of Morse decompositions for random dynamical systems have been introduced by both Ochs [**170**] and Liu [**158**] (see also Crauel, Duc & Siegmund [**60**]).

CHAPTER 5

Linear systems

A comprehensive analysis of nonlinear systems via perturbation techniques requires linear theory. This is due to the fact that in many cases, stability properties of solutions can be derived from the linearization along the solution, the so-called variational equation.

In the linear theory of nonautonomous dynamical systems, the concept of an exponential dichotomy is essential as it extends the idea of hyperbolicity to explicitly time-dependent systems. It was introduced by PERRON [**177, 178**] in the late 1920s, and there have been extensive studies showing the significance of exponential dichotomies both in theory and applications. Based on the notion of an exponential dichotomy, SACKER & SELL [**209**] established a spectral theory for linear skew product flows with compact base flow in the 1970s, the so-called *Sacker–Sell spectral theory*.

Another possibility to approach linear nonautonomous dynamical systems is to consider the induced nonautonomous dynamical system on the projective space. This approach is also fairly classical; for instance, SELGRADE [**215**] found conditions for the existence of a finest Morse decomposition of the skew product flow on the projective space in 1975.

In this chapter, these topics will be treated in the context of processes; although the fundamental concepts for processes and skew product flows are the same, the linear theories for both processes and skew product flows are independent of each other.

Throughout this chapter, let $\mathbb{I}$ be an interval of the form $(-\infty, 0]$, $[0, \infty)$ or $\mathbb{R}$, respectively, and consider a nonautonomous linear differential equation

$$\dot{x} = B(t)x \tag{5.1}$$

with a continuous function $B : \mathbb{I} \to \mathbb{R}^{d\times d}$. Let $\phi : \mathbb{I} \times \mathbb{I} \times \mathbb{R}^d \to \mathbb{R}^d$ be the process induced by (5.1), and let $\Phi : \mathbb{I} \times \mathbb{I} \to \mathbb{R}^{d\times d}$ be the transition operator of (5.1), i.e., $\Phi(t, t_0)x = \phi(t, t_0, x)$ for all $t, t_0 \in \mathbb{I}$ and $x \in \mathbb{R}^d$.

1. Exponential dichotomy

In contrast to the case of an autonomous linear system, it is not sufficient to look at the eigenvalues of the matrices $B(t)$, $t \in \mathbb{I}$, in order to characterize (uniform) asymptotic stability of the nonautonomous system (5.1). The appropriate tool to study this in the nonautonomous case is given by an exponential dichotomy, which will be introduced in this section with respect to three different time domains.

An invariant nonautonomous set $\mathcal{M} \subset \mathbb{I} \times \mathbb{R}^d$ is called *linear invariant manifold* of (5.1) if for each $t \in \mathbb{I}$, the set M_t is a linear subspace of $\mathbb{R}^d$. Given linear invariant manifolds $\mathcal{M}^1$ and $\mathcal{M}^2$ of (5.1), the sets

$$\mathcal{M}^1 \cap \mathcal{M}^2 := \{(t,x) \in \mathbb{I} \times \mathbb{R}^d : x \in M_t^1 \cap M_t^2\} \quad \text{and}$$
$$\mathcal{M}^1 + \mathcal{M}^2 := \{(t,x) \in \mathbb{I} \times \mathbb{R}^d : x \in M_t^1 + M_t^2\}$$

are also linear invariant manifolds of (5.1). A finite sum $\mathcal{M}^1 + \cdots + \mathcal{M}^n$ of linear invariant manifolds is called a *Whitney sum* $\mathcal{M}^1 \oplus \cdots \oplus \mathcal{M}^n$ if the sum $M_t^1 + \cdots + M_t^n$ is direct for all $t \in \mathbb{I}$. An *invariant projector* of (5.1) is a function $P : \mathbb{I} \to \mathbb{R}^{d \times d}$ with

$$P(t) = P(t)^2 \quad \text{and} \quad P(t)\Phi(t,t_0) = \Phi(t,t_0)P(t_0) \quad \text{for all } t, t_0 \in \mathbb{I}\,.$$

The *range*

$$\mathfrak{R}(P) := \{(t,x) \in \mathbb{I} \times \mathbb{R}^d : x \in \mathfrak{R}(P(t))\}$$

and the *null space*

$$\mathfrak{N}(P) := \{(t,x) \in \mathbb{I} \times \mathbb{R}^d : x \in \mathfrak{N}(P(t))\}$$

of an invariant projector P are linear invariant manifolds of (5.1) such that $\mathfrak{R}(P) \oplus \mathfrak{N}(P) = \mathbb{I} \times \mathbb{R}^d$. Since the fibres of $\mathfrak{R}(P)$ and $\mathfrak{N}(P)$ have the same dimension, the rank of P can be defined via

$$\operatorname{rk} P := \dim \mathfrak{R}(P) := \dim \mathfrak{R}(P(t)) \quad \text{for all } t \in \mathbb{R}\,,$$

and one sets

$$\dim \mathfrak{N}(P) := \dim \mathfrak{N}(P(t)) \quad \text{for all } t \in \mathbb{R}\,.$$

The following notion of an exponential dichotomy describes exponential splittings of the linear system.

DEFINITION 5.1 (Exponential dichotomy). Let $\gamma \in \mathbb{R}$ and $P_\gamma : \mathbb{I} \to \mathbb{R}^{d \times d}$ be an invariant projector of (5.1).

(i) If $\mathbb{I} = \mathbb{R}$, then (5.1) is said to admit an *exponential dichotomy (on $\mathbb{R}$)* with growth rate $\gamma \in \mathbb{R}$, constants $\alpha > 0$, $K \geq 1$ and projector P_γ if

$$\|\Phi(t,t_0)P_\gamma(t_0)\| \leq Ke^{(\gamma-\alpha)(t-t_0)} \quad \text{for all } t \geq t_0\,,$$
$$\|\Phi(t,t_0)(\mathbb{1} - P_\gamma(t_0))\| \leq Ke^{(\gamma+\alpha)(t-t_0)} \quad \text{for all } t \leq t_0\,.$$

(ii) If $\mathbb{I}$ is unbounded below, then (5.1) is said to admit an *exponential dichotomy on $\mathbb{R}_0^-$* with growth rate $\gamma \in \mathbb{R}$, constants $\alpha > 0, K \geq 1$ and projector P_γ if

$$\|\Phi(t,t_0)P_\gamma(t_0)\| \leq Ke^{(\gamma-\alpha)(t-t_0)} \quad \text{for all } t_0 \leq t \leq 0\,,$$
$$\|\Phi(t,t_0)(\mathbb{1} - P_\gamma(t_0))\| \leq Ke^{(\gamma+\alpha)(t-t_0)} \quad \text{for all } t \leq t_0 \leq 0\,.$$

(iii) If $\mathbb{I}$ is unbounded above, then (5.1) is said to admit an *exponential dichotomy on $\mathbb{R}_0^+$* with growth rate $\gamma \in \mathbb{R}$, constants $\alpha > 0, K \geq 1$ and projector P_γ if

$$\|\Phi(t,t_0)P_\gamma(t_0)\| \leq Ke^{(\gamma-\alpha)(t-t_0)} \quad \text{for all } 0 \leq t_0 \leq t\,,$$
$$\|\Phi(t,t_0)(\mathbb{1} - P_\gamma(t_0))\| \leq Ke^{(\gamma+\alpha)(t-t_0)} \quad \text{for all } 0 \leq t \leq t_0\,.$$

If (5.1) is almost periodic, then all notions of the above definition coincide (see COPPEL [**57**, Proposition 3, p. 70]).

It is easy to see that if (5.1) admits an exponential dichotomy on $\mathbb{R}$ with growth rate γ, then it also admits an exponential dichotomy on both $\mathbb{R}_0^-$ and $\mathbb{R}_0^+$ with growth rate γ.

EXERCISE 5.2. Show that if (5.1) admits an exponential dichotomy on both $\mathbb{R}_0^-$ and $\mathbb{R}_0^+$ with growth rate γ and the same invariant projector P, then it also admits an exponential dichotomy on $\mathbb{R}$ with growth rate γ.

REMARK 5.3. A big advantage of exponential dichotomies above other types of nonautonomous hyperbolicity is their robustness under a large class of perturbations. Various results can be found in COPPEL [**57**], PALMER [**176**], SAKAMOTO [**210**], PLISS & SELL [**181**] and PÖTZSCHE [**184**]. Moreover, an exponential dichotomy is precisely the assumption required that equilibria persist as bounded entire solutions under temporal perturbations (see PÖTZSCHE [**190**]).

The monotonicity of the exponential function implies the following basic criteria for the existence of exponential dichotomies.

LEMMA 5.4. *Suppose that* (5.1) *admits an exponential dichotomy (on* $\mathbb{R}$, $\mathbb{R}_0^-$, $\mathbb{R}_0^+$, *respectively) with growth rate* γ *and projector* P_γ. *Then the following statements are fulfilled:*

(i) *If* $P_\gamma \equiv \mathbb{1}$, *then* (5.1) *admits an exponential dichotomy (on* $\mathbb{R}$, $\mathbb{R}_0^-$, $\mathbb{R}_0^+$, *respectively) with growth rate* ζ *and invariant projector* $P_\zeta \equiv \mathbb{1}$ *for all* $\zeta > \gamma$.
(ii) *If* $P_\gamma \equiv 0$, *then* (5.1) *admits an exponential dichotomy (on* $\mathbb{R}$, $\mathbb{R}_0^-$, $\mathbb{R}_0^+$, *respectively) with growth rate* ζ *and invariant projector* $P_\zeta \equiv 0$ *for all* $\zeta < \gamma$.

Let $\gamma \in \mathbb{R}$. A function $g : \mathbb{I} \to \mathbb{R}^d$ is called γ^+*-exponentially bounded* if $\mathbb{I}$ is unbounded above and $\sup_{t\in\mathbb{I}\cap[0,\infty)} \|g(t)\|e^{-\gamma t} < \infty$. Accordingly, one says that a function $g : \mathbb{I} \to \mathbb{R}^d$ is γ^-*-exponentially bounded* if $\mathbb{I}$ is unbounded below and $\sup_{t\in\mathbb{I}\cap(-\infty,0]} \|g(t)\|e^{-\gamma t} < \infty$.

If $\mathbb{I}$ is unbounded above, then define for all $\gamma \in \mathbb{R}$

$$\mathcal{S}^\gamma := \left\{(t_0, x) \in \mathbb{I} \times \mathbb{R}^d : \Phi(\cdot, t_0)x \text{ is } \gamma^+\text{-exponentially bounded}\right\},$$

and if $\mathbb{I}$ is unbounded below, then set for all $\gamma \in \mathbb{R}$

$$\mathcal{U}^\gamma := \left\{(t_0, x) \in \mathbb{I} \times \mathbb{R}^d : \Phi(\cdot, t_0)x \text{ is } \gamma^-\text{-exponentially bounded}\right\}.$$

It is obvious that $\mathcal{S}^\gamma$ and $\mathcal{U}^\gamma$ are linear invariant manifolds of (5.1), and given $\gamma \le \zeta$, the relations $\mathcal{S}^\gamma \subset \mathcal{S}^\zeta$ and $\mathcal{U}^\gamma \supset \mathcal{U}^\zeta$ are fulfilled.

The relationship between the projectors of exponential dichotomies with growth rate γ and the sets $\mathcal{S}^\gamma$ and $\mathcal{U}^\gamma$ will now be discussed.

PROPOSITION 5.5. *If* (5.1) *admits an exponential dichotomy on* $\mathbb{R}_0^-$ *with growth rate* γ, *constants* α, K *and projector* P_γ, *then* $\mathfrak{N}(P_\gamma) = \mathcal{U}^\gamma$ *and*

$$\|\Phi(t, t_0)x\| \le K\|x\|e^{\gamma(t-t_0)} \quad \textit{for all } t_0 \le t \le 0 \textit{ and } x \in \mathfrak{R}(P_\gamma(t_0)). \tag{5.2}$$

If (5.1) *admits an exponential dichotomy on* $\mathbb{R}_0^+$ *with growth rate* γ*, constants* α*,* K *and projector* P_γ*, then* $\mathfrak{R}(P_\gamma) = \mathcal{S}^\gamma$ *and*

$$\|\Phi(t,t_0)x\| \le K\|x\|e^{\gamma(t-t_0)} \quad \text{for all } 0 \le t \le t_0 \text{ and } x \in \mathfrak{N}(P_\gamma(t_0))\,.$$

If (5.1) *admits an exponential dichotomy on* $\mathbb{R}$ *with growth rate* γ *and projector* P_γ*, then* $\mathfrak{N}(P_\gamma) = \mathcal{U}^\gamma$ *and* $\mathfrak{R}(P_\gamma) = \mathcal{S}^\gamma$ *are fulfilled.*

PROOF. Suppose that (5.1) admits an exponential dichotomy on $\mathbb{R}_0^-$ with growth rate γ, constants α, K and projector P_γ. This yields

$$\begin{aligned}\|\Phi(t,t_0)P_\gamma(t_0)\| &\le Ke^{(\gamma-\alpha)(t-t_0)} \quad \text{for all } t_0 \le t \le 0\,,\\ \|\Phi(t,t_0)(\mathbb{1} - P_\gamma(t_0))\| &\le Ke^{(\gamma+\alpha)(t-t_0)} \quad \text{for all } t \le t_0 \le 0\,.\end{aligned}$$

The first inequality implies (5.2). Now the relation $\mathfrak{N}(P_\gamma) = \mathcal{U}^\gamma$ will be proved.
($\supseteq$) Choose $(t_0, x) \in \mathcal{U}^\gamma$ arbitrarily. This implies $\|\Phi(t,t_0)x\| \le Ce^{\gamma(t-t_0)}$ for all $t \le t_0$ with some real constant $C > 0$. Write $x = x_1 + x_2$ with $x_1 \in \mathfrak{R}(P_\gamma(t_0))$ and $x_2 \in \mathfrak{N}(P_\gamma(t_0))$. Hence, for all $t \le t_0$,

$$\begin{aligned}\|x_1\| &= \|\Phi(t_0,t)\Phi(t,t_0)P_\gamma(t_0)x\| = \|\Phi(t_0,t)P_\gamma(t)\Phi(t,t_0)x\|\\ &\le Ke^{(\gamma-\alpha)(t_0-t)}\|\Phi(t,t_0)x\| \le CKe^{(\gamma-\alpha)(t_0-t)}e^{\gamma(t-t_0)} = CKe^{-\alpha(t_0-t)}\,.\end{aligned}$$

The right hand side of this inequality converges to zero in the limit $t \to -\infty$. Therefore, $x_1 = 0$, and this means $(t_0, x) \in \mathfrak{N}(P_\gamma)$.
($\subseteq$) Choose $(t_0, x) \in \mathfrak{N}(P_\gamma)$. Thus, for all $t \le t_0 \le 0$, the relation $\|\Phi(t,t_0)x\| \le Ke^{(\gamma+\alpha)(t-t_0)}\|x\|$ is fulfilled. This means that $\Phi(\cdot, t_0)x$ is γ^--exponentially bounded.
The assertions concerning the exponential dichotomy on $\mathbb{R}_0^+$ are treated analogously. If (5.1) admits an exponential dichotomy on $\mathbb{R}$, then (5.1) also admits an exponential dichotomy both on $\mathbb{R}_0^-$ and $\mathbb{R}_0^+$. Hence, $\mathfrak{N}(P_\gamma) = \mathcal{U}^\gamma$ and $\mathfrak{R}(P_\gamma) = \mathcal{S}^\gamma$. □

REMARK 5.6. According to this proposition, an invariant projector is uniquely determined only in case of an exponential dichotomy on $\mathbb{R}$. In addition, the null space of a projector of an exponential dichotomy on $\mathbb{R}_0^-$ and the range of a projector of an exponential dichotomy on $\mathbb{R}_0^+$ are uniquely determined. This implies that the rank of exponential dichotomies with the same growth rate is independent of the choice of the projector. For further information about the kind of nonuniqueness of ranges of projectors of exponential dichotomies on $\mathbb{R}_0^-$ and null spaces of projectors of exponential dichotomies on $\mathbb{R}_0^+$, see Lemma 5.7 in the next section.

2. Dichotomy spectrum

In the previous section, notions of dichotomy were introduced by localizing spaces with different growth behavior. To classify the strength of attractivity and repulsivity of linear systems, the concept of a dichotomy (or Sacker–Sell) spectrum is essential (see SACKER & SELL [**209**]).

Consider again the linear nonautonomous differential equation (5.1). As indicated in Remark 5.6, an invariant projector is uniquely determined only in case of an exponential dichotomy on $\mathbb{R}$. The degree of nonuniqueness of projectors of exponential dichotomies on $\mathbb{R}_0^-$ and $\mathbb{R}_0^+$ is described in the following lemma, which is adapted from AULBACH & SIEGMUND [**12**, Lemma 2.4].

LEMMA 5.7. *Suppose that (5.1) admits an exponential dichotomy on $\mathbb{R}_0^-$ with growth rate γ and projector P, and let $\bar{P}$ be another invariant projector with*

$$\sup_{t\leq 0}\|\bar{P}(t)\|<\infty \quad \textit{and} \quad \mathfrak{N}(P)=\mathfrak{N}(\bar{P})\,.$$

Then (5.1) also admits an exponential dichotomy on $\mathbb{R}_0^-$ with growth rate γ and projector $\bar{P}$. An analogous statement holds for exponential dichotomies on $\mathbb{R}_0^+$ and the ranges of the projectors.

PROOF. Suppose that (5.1) admits an exponential dichotomy on $\mathbb{R}_0^-$ with growth rate γ, constants $\alpha>0$, $K\geq 1$ and projector P, and let $\bar{P}$ be given as above. Observe first that $\sup_{t\leq 0}\|P(t)\|\leq K$, and define $M:=\sup_{t\leq 0}\left\|\bar{P}(t)\right\|$. The relation $\mathfrak{N}(P)=\mathfrak{N}(\bar{P})$ implies the two equations

$$\left(\mathbb{1}-\bar{P}\right)=(\mathbb{1}-P)\left(\mathbb{1}-\bar{P}\right) \quad \text{and} \quad \bar{P}=\left(\mathbb{1}-P+\bar{P}\right)P\,.$$

The first equation yields for all $t_0\leq 0$ and $t\leq t_0$

$$\begin{aligned}\|\Phi(t,t_0)(\mathbb{1}-\bar{P}(t_0))\| &= \|\Phi(t,t_0)(\mathbb{1}-P(t_0))(\mathbb{1}-\bar{P}(t_0))\|\\ &\leq \|\Phi(t,t_0)(\mathbb{1}-P(t_0))\|\,\|\mathbb{1}-\bar{P}(t_0)\|\\ &\leq K(1+M)e^{(\gamma+\alpha)(t-t_0)}\,.\end{aligned}$$

Using the invariance of P and $\bar{P}$, the second equation implies

$$\begin{aligned}\|\Phi(t,t_0)\bar{P}(t_0)\| &= \|\Phi(t,t_0)(\mathbb{1}-P(t_0)+\bar{P}(t_0))P(t_0)\|\\ &\leq \|(\mathbb{1}-P(t)+\bar{P}(t))\|\,\|\Phi(t,t_0)P(t_0)\|\\ &\leq K(1+K+M)e^{(\gamma-\alpha)(t-t_0)}\end{aligned}$$

for all $t\leq 0$ and $t_0\leq t$. □

For the definition of the dichotomy spectra, it is crucial for which growth rates, the linear system (5.1) admits an exponential dichotomy. The growth rates $\gamma=\pm\infty$ are not excluded from our considerations, i.e., one says that (5.1) admits an exponential dichotomy with growth rate ∞ if there exists a $\gamma\in\mathbb{R}$ such that (5.1) admits an exponential dichotomy with growth rate γ and projector $P_\gamma\equiv\mathbb{1}$. Accordingly, one says that (5.1) admits an exponential dichotomy with growth rate $-\infty$ if there exists a $\gamma\in\mathbb{R}$ such that (5.1) admits an exponential dichotomy with growth rate γ and projector $P_\gamma\equiv 0$.

DEFINITION 5.8 (Dichotomy spectrum). Consider the linear system (5.1). Then the *dichotomy spectrum* (5.1) is defined by

$$\begin{aligned}\Sigma^{\leftrightarrow}:=\big\{\gamma\in\overline{\mathbb{R}}:{}&\text{(5.1) does not admit an exponential dichotomy on }\mathbb{R}\\ &\text{with growth rate }\gamma\big\}\,.\end{aligned}$$

The *past dichotomy spectrum* of (5.1) is defined by

$$\begin{aligned}\Sigma^{\leftarrow}:=\big\{\gamma\in\overline{\mathbb{R}}:{}&\text{(5.1) does not admit an exponential dichotomy on }\mathbb{R}_0^-\\ &\text{with growth rate }\gamma\big\}\,.\end{aligned}$$

The *future dichotomy spectrum* of (5.1) is defined by

$$\Sigma^{\rightarrow} := \big\{\gamma \in \overline{\mathbb{R}} : \text{(5.1) does not admit an exponential dichotomy on } \mathbb{R}_0^+ \\ \text{with growth rate } \gamma\big\}\,.$$

The corresponding *resolvent sets* are defined as follows:

$$\rho^{\leftarrow} := \overline{\mathbb{R}} \setminus \Sigma^{\leftarrow}\,, \qquad \rho^{\rightarrow} := \overline{\mathbb{R}} \setminus \Sigma^{\rightarrow} \qquad \text{and} \qquad \rho^{\leftrightarrow} := \overline{\mathbb{R}} \setminus \Sigma^{\leftrightarrow}\,.$$

The dichotomy spectrum is sometimes also called *all-time dichotomy spectrum* to distinguish it from the past and future dichotomy spectrum. It follows directly from the definitions that $\Sigma^{\leftarrow} \subset \Sigma^{\leftrightarrow}$ and $\Sigma^{\rightarrow} \subset \Sigma^{\leftrightarrow}$.

REMARK 5.9. If (5.1) is almost periodic, then the spectra for the past, future and entire time are identical, and they coincide with the Sacker–Sell spectrum from [**209**].

The aim of the following lemma is to analyze the topological structure of the resolvent sets.

LEMMA 5.10. *Suppose that $\rho := \rho^{\leftarrow}, \rho^{\rightarrow}, \rho^{\leftrightarrow}$, respectively. Then $\rho \cap \mathbb{R}$ is open. More precisely, for all $\gamma \in \rho \cap \mathbb{R}$, there exists an $\varepsilon > 0$ such that $B_\varepsilon(\gamma) \subset \rho$. Furthermore, the relation $\operatorname{rk} P_\zeta = \operatorname{rk} P_\gamma$ is fulfilled for all $\zeta \in B_\varepsilon(\gamma)$ and every invariant projector P_γ and P_ζ of the exponential dichotomies of* (5.1) *with growth rates γ and ζ, respectively.*

PROOF. Choose $\gamma \in \rho$ arbitrarily, and let $\mathbb{I} = \mathbb{R}_0^-$ ($\mathbb{I} = \mathbb{R}_0^+$, $\mathbb{I} = \mathbb{R}$, respectively) be the interval corresponding to the time domain. Since (5.1) admits an exponential dichotomy (on $\mathbb{R}$, $\mathbb{R}_0^-$, $\mathbb{R}_0^+$, respectively) with growth rate γ, there exists an invariant projector P_γ and constants $\alpha > 0$, $K \geq 1$ such that

$$\begin{aligned} \|\Phi(t,t_0)P_\gamma(t_0)\| &\leq Ke^{(\gamma-\alpha)(t-t_0)} \quad \text{for all } t \geq t_0 \text{ and } t \in \mathbb{I}\,,\\ \|\Phi(t,t_0)(\mathbb{1} - P_\gamma(t_0))\| &\leq Ke^{(\gamma+\alpha)(t-t_0)} \quad \text{for all } t \leq t_0 \text{ and } t_0 \in \mathbb{I}\,. \end{aligned}$$

Set $\varepsilon := \frac{1}{2}\alpha$, and choose $\zeta \in B_\varepsilon(\gamma)$. It follows that

$$\begin{aligned} \|\Phi(t,t_0)P_\gamma(t_0)\| &\leq Ke^{(\zeta-\frac{\alpha}{2})(t-t_0)} \quad \text{for all } t \geq t_0 \text{ and } t \in \mathbb{I}\,,\\ \|\Phi(t,t_0)(\mathbb{1} - P_\gamma(t_0))\| &\leq Ke^{(\zeta+\frac{\alpha}{2})(t-t_0)} \quad \text{for all } t \leq t_0 \text{ and } t_0 \in \mathbb{I}\,. \end{aligned}$$

This yields $\zeta \in \rho$. Since the ranks of the projectors of exponential dichotomies with the same growth rate are equal (see Remark 5.6), it follows that $\operatorname{rk} P_\zeta = \operatorname{rk} P_\gamma$ for any projector P_ζ of the exponential dichotomy with growth rate ζ. □

LEMMA 5.11. *Assume that $\rho := \rho^{\leftarrow}, \rho^{\rightarrow}, \rho^{\leftrightarrow}$, respectively, let $\gamma_1, \gamma_2 \in \rho \cap \mathbb{R}$ such that $\gamma_1 < \gamma_2$, and choose invariant projectors P_{γ_1} and P_{γ_2} for the corresponding exponential dichotomies with growth rates γ_1 and γ_2. Then the relation $\operatorname{rk} P_{\gamma_1} \leq \operatorname{rk} P_{\gamma_2}$ holds. Moreover, $[\gamma_1, \gamma_2] \subset \rho$ is fulfilled if and only if $\operatorname{rk} P_{\gamma_1} = \operatorname{rk} P_{\gamma_2}$.*

PROOF. The relation $\operatorname{rk} P_{\gamma_1} \leq \operatorname{rk} P_{\gamma_2}$ is a direct consequence of Proposition 5.5, since $\mathcal{S}^{\gamma_1} \subset \mathcal{S}^{\gamma_2}$ and $\mathcal{U}^{\gamma_1} \supset \mathcal{U}^{\gamma_2}$. Assume now that $[\gamma_1, \gamma_2] \subset \rho$. Arguing contrapositively, suppose that $\operatorname{rk} P_{\gamma_1} \neq \operatorname{rk} P_{\gamma_2}$, and choose invariant projectors P_γ for the dichotomies of (5.1) with growth rate γ for all $\gamma \in (\gamma_1, \gamma_2)$. Define

$$\zeta_0 := \sup\big\{\zeta \in [\gamma_1, \gamma_2] : \operatorname{rk} P_\zeta \neq \operatorname{rk} P_{\gamma_2}\big\}\,.$$

Due to Lemma 5.10, there exists an $\varepsilon > 0$ such that $\operatorname{rk} P_{\zeta_0} = \operatorname{rk} P_\zeta$ for all $\zeta \in B_\varepsilon(\zeta_0)$. This is a contradiction to the definition of ζ_0. Conversely, let $\operatorname{rk} P_{\gamma_1} = \operatorname{rk} P_{\gamma_2}$, and consider first the case $\rho = \rho^{\leftarrow}$. Because of $\operatorname{rk} P_{\gamma_1} = \operatorname{rk} P_{\gamma_2}$, Proposition 5.5 yields $\mathfrak{N}(P_{\gamma_1}) = \mathfrak{N}(P_{\gamma_2})$. Due to Lemma 5.7, P_{γ_2} is an invariant projector of the exponential dichotomy on $\mathbb{R}_0^-$ with growth rate γ_1. Thus, one obtains

$$\|\Phi(t,t_0)P_{\gamma_2}(t_0)\| \le K_1 e^{(\gamma_1-\alpha_1)(t-t_0)} \quad \text{for all } t_0 \le t \le 0$$

for some $K_1 \ge 1$ and $\alpha_1 > 0$. P_{γ_2} is also projector of the exponential dichotomy on $\mathbb{R}_0^-$ with growth rate γ_2. Hence,

$$\big\|\Phi(t,t_0)(\mathbb{1} - P_{\gamma_2}(t_0))\big\| \le K_2 e^{(\gamma_2+\alpha_2)(t-t_0)} \quad \text{for all } t \le t_0 \le 0$$

is fulfilled for some $K_2 \ge 1$ and $\alpha_2 > 0$. For all $\gamma \in [\gamma_1, \gamma_2]$, these two inequalities imply by setting $K := \max\{K_1, K_2\}$ and $\alpha := \min\{\alpha_1, \alpha_2\}$ that

$$\|\Phi(t,t_0)P_{\gamma_2}(t_0)\| \le K e^{(\gamma-\alpha)(t-t_0)} \quad \text{for all } t_0 \le t \le 0\,,$$
$$\big\|\Phi(t,t_0)(\mathbb{1} - P_{\gamma_2}(t_0))\big\| \le K e^{(\gamma+\alpha)(t-t_0)} \quad \text{for all } t \le t_0 \le 0\,.$$

This means that $\gamma \in \rho$, and thus, $[\gamma_1, \gamma_2] \subset \rho$. The cases $\rho = \rho^{\rightarrow}, \rho^{\leftrightarrow}$ are treated analogously. □

For an arbitrarily chosen $a \in \mathbb{R}$, define

$$[-\infty, a] := (-\infty, a] \cup \{-\infty\}\,, \qquad [a, \infty] := [a, \infty) \cup \{\infty\}$$

and

$$[-\infty, -\infty] := \{-\infty\}, \qquad [\infty, \infty] := \{\infty\}, \qquad [-\infty, \infty] := \overline{\mathbb{R}}\,.$$

The main theorem of this section, the following *Spectral Theorem*, describes that the spectra consist of at least one and at most d closed intervals.

THEOREM 5.12 (Spectral Theorem). *Given $\Sigma := \Sigma^{\leftarrow}, \Sigma^{\rightarrow}$ or $\Sigma^{\leftrightarrow}$, respectively. Then there exists an $n \in \{1, \dots, d\}$ such that*

$$\Sigma = [a_1, b_1] \cup \cdots \cup [a_n, b_n]$$

with $-\infty \le a_1 \le b_1 < a_2 \le b_2 < \cdots < a_n \le b_n \le \infty$.

PROOF. Due to Lemma 5.10, the resolvent set $\rho \cap \mathbb{R}$ is open. Thus, $\Sigma \cap \mathbb{R}$ is the disjoint union of closed intervals. The relation $(-\infty, b_1] \subset \Sigma$ implies $[-\infty, b_1] \subset \Sigma$, because the assumption of the existence of a $\gamma \in \mathbb{R}$ such that (5.1) admits an exponential dichotomy with growth rate γ and projector $P_\gamma \equiv 0$ leads to $(-\infty, \gamma] \subset \rho$ using Lemma 5.4, and this is a contradiction. Analogously, it follows from $[a_n, \infty) \subset \Sigma$ that $[a_n, \infty] \subset \Sigma$. To show the relation $n \le d$, assume to the contrary that $n \ge d+1$. Thus, there exist

$$\zeta_1 < \zeta_2 < \cdots < \zeta_d \in \rho$$

such that the $d+1$ intervals $(-\infty, \zeta_1)\,, (\zeta_1, \zeta_2)\,, \dots, (\zeta_d, \infty)$ have nonempty intersection with the spectrum Σ. It follows from Lemma 5.11 that

$$0 \le \operatorname{rk} P_{\zeta_1} < \operatorname{rk} P_{\zeta_2} < \cdots < \operatorname{rk} P_{\zeta_d} \le d$$

is fulfilled for invariant projectors P_{ζ_i} of the exponential dichotomy with growth rate ζ_i, $i \in \{1, \dots, n\}$. This implies either $\operatorname{rk} P_{\zeta_1} = 0$ or $\operatorname{rk} P_{\zeta_d} = d$. Thus, either

$$[-\infty, \zeta_1] \cap \Sigma = \emptyset \quad \text{or} \quad [\zeta_d, \infty] \cap \Sigma = \emptyset$$

is fulfilled, and this is a contradiction. To show $n \geq 1$, assume that $\Sigma = \emptyset$. This implies $\{-\infty, \infty\} \subset \rho$. Thus, there exist $\zeta_1, \zeta_2 \in \mathbb{R}$ such that (5.1) admits an exponential dichotomy with growth rate ζ_1 and projector $P_{\zeta_1} \equiv 0$ and an exponential dichotomy with growth rate ζ_2 and projector $P_{\zeta_2} \equiv \mathbb{1}$. Applying Lemma 5.11, one gets $(\zeta_1, \zeta_2) \cap \Sigma \neq \emptyset$. This contradiction yields $n \geq 1$ and finishes the proof of the theorem. □

The Spectral Theorem says that the dichotomy spectra consist of exactly one closed interval in the one-dimensional case. The following exercise shows that there are several possibilities.

EXERCISE 5.13. Consider a scalar linear differential equation of the form

$$\dot{x} = a(t)x\,,$$

where $a : \mathbb{R} \to \mathbb{R}$ is a continuous function. Derive the dichotomy spectra of the following examples.

(i) $\Sigma^{\leftarrow} = \Sigma^{\rightarrow} = \Sigma^{\leftrightarrow} = \{\infty\}$ for $a(t) := |t|$ for all $t \in \mathbb{R}$.
(ii) $\Sigma^{\leftarrow} = \{-\infty\}$, $\Sigma^{\rightarrow} = \{\infty\}$ and $\Sigma^{\leftrightarrow} = \overline{\mathbb{R}}$ for $a(t) := t$ for all $t \in \mathbb{R}$.
(iii) $\Sigma^{\leftarrow} = [-\infty, \beta]$, $\Sigma^{\rightarrow} = \{\beta\}$ and $\Sigma^{\leftrightarrow} = [-\infty, \beta]$ for

$$a(t) := \begin{cases} \beta & : \ t \geq -1\,, \\ \beta - n + n\big(t + 2^{2n} + 1\big) & : \ t \in \big[-2^{2n} - 1, -2^{2n}\big]\,, n \in \mathbb{N}_0\,, \\ \beta - n & : \ t \in \big[-2^{2n+1}, -2^{2n} - 1\big]\,, n \in \mathbb{N}_0\,, \\ \beta - n\big(t + 2^{2n+1} + 1\big) & : \ t \in \big[-2^{2n+1} - 1, -2^{2n+1}\big]\,, n \in \mathbb{N}_0\,, \\ \beta & : \ t \in \big[-2^{2(n+1)}, -2^{2n+1} - 1\big]\,, n \in \mathbb{N}_0\,. \end{cases}$$

(iv) $\Sigma^{\leftarrow} = \{\beta\}$, $\Sigma^{\rightarrow} = [\beta, \infty]$ and $\Sigma^{\leftrightarrow} = [\beta, \infty]$ for

$$a(t) := \begin{cases} \beta & : \ t \leq 1\,, \\ \beta + n\big(t - 2^{2n}\big) & : \ t \in \big[2^{2n}, 2^{2n} + 1\big]\,, n \in \mathbb{N}_0\,, \\ \beta + n & : \ t \in \big[2^{2n} + 1, 2^{2n+1}\big]\,, n \in \mathbb{N}_0\,, \\ \beta + n - n\big(t - 2^{2n+1}\big) & : \ t \in \big[2^{2n+1}, 2^{2n+1} + 1\big]\,, n \in \mathbb{N}_0\,, \\ \beta & : \ t \in \big[2^{2n+1} + 1, 2^{2(n+1)}\big]\,, n \in \mathbb{N}_0\,. \end{cases}$$

(v) $\Sigma^{\leftarrow} = \{\beta\}$, $\Sigma^{\rightarrow} = [\beta, \delta]$ and $\Sigma^{\leftrightarrow} = [\beta, \delta]$ for

$$a(t) := \begin{cases} \beta & : \ t \leq 1\,, \\ \beta + \big(t - 2^{2n}\big)(\delta - \beta) & : \ t \in \big[2^{2n}, 2^{2n} + 1\big]\,, n \in \mathbb{N}_0\,, \\ \delta & : \ t \in \big[2^{2n} + 1, 2^{2n+1}\big]\,, n \in \mathbb{N}_0\,, \\ \delta + \big(t - 2^{2n+1}\big)(\beta - \delta) & : \ t \in \big[2^{2n+1}, 2^{2n+1} + 1\big]\,, n \in \mathbb{N}_0\,, \\ \beta & : \ t \in \big[2^{2n+1} + 1, 2^{2(n+1)}\big]\,, n \in \mathbb{N}_0\,. \end{cases}$$

The following theorem says that each interval of the spectrum corresponds to a linear invariant manifold.

THEOREM 5.14 (Spectral manifolds). *Let*

$$\Sigma := \Sigma^{\leftarrow}, \Sigma^{\rightarrow}, \Sigma^{\leftrightarrow} = [a_1, b_1] \cup \cdots \cup [a_n, b_n]\,,$$

respectively, define the invariant projectors $P_{\gamma_0} := 0$, $P_{\gamma_n} := \mathbb{1}$, *and for* $i \in \{1, \ldots, n-1\}$, *choose* $\gamma_i \in (b_i, a_{i+1})$ *and projectors* P_{γ_i} *of the nonhyperbolic exponential dichotomy of* (5.1) *with growth rate* γ_i. *Then the sets*

$$\mathcal{W}_i := \mathfrak{R}(P_{\gamma_i}) \cap \mathfrak{N}(P_{\gamma_{i-1}}) \quad \textit{for all } i \in \{1, \ldots, n\}$$

are linear invariant manifolds, the so-called spectral manifolds, *such that*

$$\mathcal{W}_1 \oplus \cdots \oplus \mathcal{W}_n = \mathbb{R} \times \mathbb{R}^d$$

and $\mathcal{W}_i \neq \mathbb{R} \times \{0\}$ *for* $i \in \{1, \dots, n\}$.

PROOF. The sets $\mathcal{W}_1, \dots, \mathcal{W}_n$ are obviously linear invariant manifolds. Suppose that there exists an $i \in \{1, \dots, n\}$ with $\mathcal{W}_i = \mathbb{R} \times \{0\}$. In case $i = 1$ or $i = n$, Lemma 5.4 implies $[-\infty, \gamma_1] \cap \Sigma = \emptyset$ or $[\gamma_{n-1}, \infty] \cap \Sigma = \emptyset$, and this is a contradiction. In case $1 < i < n$, due to Lemma 5.11, one obtains

$$\begin{aligned}\dim \mathcal{W}_i &= \dim \big(\mathfrak{R}(P_{\gamma_i}) \cap \mathfrak{N}(P_{\gamma_{i-1}})\big) \\ &= \operatorname{rk} P_{\gamma_i} + d - \operatorname{rk} P_{\gamma_{i-1}} - \dim \big(\mathfrak{R}(P_{\gamma_i}) + \mathfrak{N}(P_{\gamma_{i-1}})\big) \geq 1,\end{aligned}$$

and this is also a contradiction. Now the relation $\mathcal{W}_1 \oplus \cdots \oplus \mathcal{W}_n = \mathbb{R} \times \mathbb{R}^d$ is proven, and without loss of generality, assume that $\Sigma = \Sigma^{\leftarrow}, \Sigma^{\leftrightarrow}$. For $1 \leq i < j \leq n$, due to 5.5, the relations $\mathcal{W}_i \subset \mathfrak{R}(P_{\gamma_i})$ and $\mathcal{W}_j \subset \mathfrak{N}(P_{\gamma_{j-1}}) \subset \mathfrak{N}(P_{\gamma_i})$ are fulfilled. This yields

$$\mathcal{W}_i \cap \mathcal{W}_j \subset \mathfrak{R}(P_{\gamma_i}) \cap \mathfrak{N}(P_{\gamma_i}) = \mathbb{R} \times \{0\},$$

and one obtains

$$\begin{aligned}\mathbb{R} \times \mathbb{R}^d &= \mathcal{W}_1 + \mathfrak{N}(P_{\gamma_1}) = \mathcal{W}_1 + \mathfrak{N}(P_{\gamma_1}) \cap \big(\mathfrak{R}(P_{\gamma_2}) + \mathfrak{N}(P_{\gamma_2})\big) \\ &= \mathcal{W}_1 + \mathfrak{N}(P_{\gamma_1}) \cap \mathfrak{R}(P_{\gamma_2}) + \mathfrak{N}(P_{\gamma_2}) = \mathcal{W}_1 + \mathcal{W}_2 + \mathfrak{N}(P_{\gamma_2}).\end{aligned}$$

Here, the fact that linear subspaces $E, F, G \subset \mathbb{R}^d$ with $E \supset G$ fulfill $E \cap (F + G) = (E \cap F) + G$ was used. It follows inductively that

$$\mathbb{R} \times \mathbb{R}^d = \mathcal{W}_1 + \cdots + \mathcal{W}_n + \mathfrak{N}(P_{\gamma_n}) = \mathcal{W}_1 + \cdots + \mathcal{W}_n.$$

This finishes the proof of this theorem. □

3. Lyapunov spectrum

Lyapunov exponents have been introduced in 1892 by LYAPUNOV [**160**] in his doctoral dissertation, which leads to a corresponding Lyapunov spectral concept. It was shown by PERRON [**178**] in 1930 by means of the example

$$\dot{x} = (\sin(\ln t) + \cos(\ln t))x$$

that negativity of the Lyapunov spectrum does not imply stability. Perron's results impressed the experts in stability theory, and several additional conditions were proposed to guarantee stability of a system with negative Lyapunov exponents. It will turn out in this section that the Lyapunov spectrum is a subset of the dichotomy spectrum. In the case of Perron's example, it is a proper subset, because a dichotomy spectrum contained in the negative half-line implies (exponential) attractivity of the system. For a discussion of Perron's example (also in relation to the so-called Bohl spectrum), see Daleckiĭ & Kreĭn [**63**, p. 123].

DEFINITION 5.15. Consider the linear system (5.1) with transition operator $\Phi : \mathbb{I} \times \mathbb{I} \to \mathbb{R}^{d \times d}$. If $\mathbb{I}$ is unbounded below, for nonzero $\xi \in \mathbb{R}^d$, the numbers

$$\lambda_+^{\leftarrow}(\xi) = \limsup_{t \to \infty} \frac{1}{t} \ln \|\Phi(-t, 0)\xi\| \quad \text{and} \quad \lambda_-^{\leftarrow}(\xi) = \liminf_{t \to \infty} \frac{1}{t} \ln \|\Phi(-t, 0)\xi\|$$

are called *upper and lower Lyapunov exponent for* $t \to -\infty$. If $\mathbb{I}$ is unbounded above, the numbers

$$\lambda_+^{\to}(\xi) = \limsup_{t\to\infty} \frac{1}{t} \ln \|\Phi(t,0)\xi\| \quad \text{and} \quad \lambda_-^{\to}(\xi) = \liminf_{t\to\infty} \frac{1}{t} \ln \|\Phi(t,0)\xi\|$$

are called *upper and lower Lyapunov exponent for* $t \to \infty$. The *Lyapunov spectrum for* $t \to -\infty$ is defined by

$$\sigma^{\leftarrow} := \bigcup_{0\neq\xi\in\mathbb{R}^d} \left[\lambda_-^{\leftarrow}(\xi), \lambda_+^{\leftarrow}(\xi)\right],$$

and the *Lyapunov spectrum for* $t \to \infty$ is defined by

$$\sigma^{\to} := \bigcup_{0\neq\xi\in\mathbb{R}^d} \left[\lambda_-^{\to}(\xi), \lambda_+^{\to}(\xi)\right].$$

There exist $n^-, n^+ \in \{1,\dots,d\}$ and $\xi_1^-,\dots,\xi_{n^-}^-, \xi_1^+,\dots,\xi_{n^+}^+ \in \mathbb{R}^d$ such that

$$\sigma^{\leftarrow} = \left[\lambda_-^{\leftarrow}(\xi_1), \lambda_+^{\leftarrow}(\xi_1)\right] \cup \dots \cup \left[\lambda_-^{\leftarrow}(\xi_{n^-}), \lambda_+^{\leftarrow}(\xi_{n^-})\right]$$

and

$$\sigma^{\to} = \left[\lambda_-^{\to}(\xi_1), \lambda_+^{\to}(\xi_1)\right] \cup \dots \cup \left[\lambda_-^{\to}(\xi_{n^+}), \lambda_+^{\to}(\xi_{n^+})\right].$$

In the following, the relationship between the dichotomy spectra and the Lyapunov spectra is discussed.

THEOREM 5.16. *The relations* $-\sigma^{\leftarrow} \subset \Sigma^{\leftarrow}$ *and* $\sigma^{\to} \subset \Sigma^{\to}$ *hold.*

PROOF. Given $\lambda \in -\sigma^{\leftarrow}$, there exists a $\xi \in \mathbb{R}^d$ with $-\lambda \in \left[\lambda_-^{\leftarrow}(\xi), \lambda_+^{\leftarrow}(\xi)\right]$. Initially, suppose that $\lambda \in \mathbb{R}$, and arguing contrapositively, assume that (5.1) admits an exponential dichotomy on $\mathbb{R}_0^-$ with growth rate $\gamma := \lambda$, constants $K \geq 1, \alpha > 0$ and invariant projector P_γ, i.e.,

$$\begin{aligned} \|\Phi(t,t_0)P_\gamma(t_0)\| &\leq Ke^{(\gamma-\alpha)(t-t_0)} \quad \text{for all } t_0 \leq t \leq 0, \\ \left\|\Phi(t,t_0)(\mathbb{1} - P_\gamma(t_0))\right\| &\leq Ke^{(\gamma+\alpha)(t-t_0)} \quad \text{for all } t \leq t_0 \leq 0. \end{aligned}$$

Write $\xi = \xi_1 + \xi_2$ with $\xi_1 \in \mathfrak{R}(P_\gamma(0))$ and $\xi_2 \in \mathfrak{N}(P_\gamma(0))$. If $\xi_1 = 0$, then one obtains

$$\begin{aligned} \lambda_+^{\leftarrow}(\xi) &= \limsup_{t\to\infty} \frac{1}{t} \ln \|\Phi(-t,0)\xi\| \leq \limsup_{t\to\infty} \frac{1}{t} \ln \left(Ke^{-(\gamma+\alpha)t}\right) \\ &= -\gamma - \alpha = -\lambda - \alpha \leq \lambda_+^{\leftarrow}(\xi) - \alpha. \end{aligned}$$

This is a contradiction. Otherwise ($\xi_1 \neq 0$), observe that for all $t \geq 0$,

$$\begin{aligned} \|\xi_1\| = \|P_\gamma(0)\xi\| &= \left\|\Phi(0,-t)\Phi(-t,0)P_\gamma(0)\xi\right\| \\ &= \left\|\Phi(0,-t)P_\gamma(-t)\Phi(-t,0)\xi\right\| \leq Ke^{(\gamma-\alpha)t}\|\Phi(-t,0)\xi\| \end{aligned}$$

is fulfilled. Thus, $\|\Phi(-t,0)\xi\| \geq K^{-1}e^{-(\gamma-\alpha)t}\|\xi_1\|$ for all $t \geq 0$, and therefore,

$$\begin{aligned} \lambda_-^{\leftarrow}(\xi) &= \liminf_{t\to\infty} \frac{1}{t} \ln \|\Phi(-t,0)\xi\| \geq \liminf_{t\to\infty} \frac{1}{t} \ln \left(K^{-1}e^{-(\gamma-\alpha)t}\|\xi_1\|\right) \\ &= \alpha - \gamma = \alpha - \lambda \geq \alpha + \lambda_-^{\leftarrow}(\xi). \end{aligned}$$

This is also a contradiction, and hence, $\lambda \in \Sigma^{\leftarrow}$. The case $\lambda \notin \mathbb{R}$ is now treated, and without loss of generality, assume that $\lambda = \infty$. Moreover, suppose that $-\infty \notin \Sigma^{\leftarrow}$. Thus, there exist $\gamma \in \mathbb{R}$, $K \geq 1$ and $\alpha > 0$ with

$$\|\Phi(-t,0)\| \leq Ke^{-(\gamma+\alpha)t} \quad \text{for all } t \geq 0\,,$$

and this relation implies the contradiction

$$\limsup_{t\to\infty} \frac{1}{t} \ln\big(\|\Phi(-t,0)\|\big) \leq -\gamma - \alpha < \infty\,.$$

Hence, $\lambda \in \Sigma^{\leftarrow}$. The remaining assertion $\sigma^{\leftarrow} \subset \Sigma^{\leftarrow}$ can be proved analogously. □

The following exercise shows that the Lyapunov spectra do not coincide with the dichotomy spectra.

EXERCISE 5.17. Show that the Lyapunov spectra of

$$\dot{x} = \big(t\sin(t) - \cos(t)\big)x$$

do not coincide with the dichotomy spectra.

4. Morse decompositions

In this section, linear nonautonomous differential equations systems are studied in the projective space (see the Appendix), leading to attractor-repeller and Morse decompositions.

Consider the linear nonautonomous differential equation (5.1) with transition operator $\Phi : \mathbb{I} \times \mathbb{I} \to \mathbb{R}^{d\times d}$. Then a process $\mathbb{P}\Phi : \mathbb{I} \times \mathbb{I} \times \mathbb{P}^{d-1} \to \mathbb{P}^{d-1}$ is induced by the definition

$$\mathbb{P}\Phi(t,t_0)\mathbb{P}x := \mathbb{P}(\Phi(t,t_0)x) \quad \text{for all } t,t_0 \in \mathbb{I} \text{ and } x \in \mathbb{R}^d$$

(see COLONIUS & KLIEMANN [**53**, Lemma 5.2.1, p. 149]).

In first instance, only the past (or equivalently, the future) of the linear system is considered.

4.1. Past Morse decompositions in projective space.

It is assumed throughout this subsection that $\mathbb{I} = \mathbb{R}_0^-$. The main observation of the following proposition is that local pullback attractors in $\mathbb{P}^{d-1}$ give rise to linear subspaces in $\mathbb{R}^d$.

PROPOSITION 5.18. *Let $\mathcal{A}$ be a local pullback attractor of $\mathbb{P}\Phi$. Then for all $t_0 \in \mathbb{I}$ and all compact sets $C \subset \mathbb{S}^{d-1} \setminus \mathbb{P}^{-1}A_{t_0}$, one has*

$$\lim_{t\to-\infty} \frac{\sup_{v\in\mathbb{S}^{d-1}\cap\mathbb{P}^{-1}A_{t_0}} \|\Phi(t,t_0)v\|}{\inf_{w\in C}\|\Phi(t,t_0)w\|} = 0\,.$$

Moreover, for all $t_0 \in \mathbb{I}$, the set $\mathbb{P}^{-1}A_{t_0}$ is a linear subspace of $\mathbb{R}^d$.

PROOF. Let $\mathcal{A}$ be a local pullback attractor of $\mathbb{P}\Phi$, and choose $t_0 \in \mathbb{I}$ and a compact set $C \subset \mathbb{S}^{d-1} \setminus \mathbb{P}^{-1}A_{t_0}$ arbitrarily. Firstly, define for any nonzero $v \in \mathbb{P}^{-1}A_{t_0}$ and $w \in C$ the two-dimensional linear subspace $L_{v,w} \subset \mathbb{R}^d$ by

$$L_{v,w} := \{rv + sw : r,s \in \mathbb{R}\}\,.$$

The proof of this proposition is divided into five parts.
Part 1. It is shown that for all nonzero $v \in \mathbb{P}^{-1}A_{t_0}$ and $w \in C$ such that $\mathbb{P}v$ is a boundary point of $A_{t_0} \cap \mathbb{P}L_{v,w}$ relative to $\mathbb{P}L_{v,w}$, one has

$$\lim_{t\to-\infty} \frac{\|\Phi(t,t_0)v\|}{\|\Phi(t,t_0)w\|} = 0\,.$$

Since $\mathcal{A}$ is a local pullback attractor, there exists an $\eta > 0$ such that

$$\lim_{t\to-\infty} \operatorname{dist}_{\mathbb{P}}\left(\mathbb{P}\Phi(t_0,t)B_{2\eta}(A_t), A_{t_0}\right) = 0\,. \tag{5.3}$$

Due to Lemma A.1, there exists a $\delta \in (0,1)$ such that $d_{\mathbb{P}}(\mathbb{P}u_1, \mathbb{P}u_2) \le \eta$ holds for all $u_1, u_2 \in \mathbb{R}^d \setminus \{0\}$ with

$$\frac{\langle u_1, u_2\rangle^2}{\|u_1\|^2\|u_2\|^2} \ge 1-\delta\,.$$

Suppose to the contrary that there exist a $\gamma > 0$ and a sequence $\{t_n\}_{n\in\mathbb{N}}$ with $\lim_{n\to\infty} t_n = -\infty$ such that

$$\frac{\|\Phi(t_n,t_0)w\|}{\|\Phi(t_n,t_0)v\|} \le \gamma \quad \text{for all } n \in \mathbb{N}\,.$$

For nonzero $c \in \mathbb{R}$ with $|c|$ sufficiently small, this implies that for all $n \in \mathbb{N}$,

$$\begin{aligned}
&\frac{\langle \Phi(t_n,t_0)(cw+v), \Phi(t_n,t_0)v\rangle^2}{\|\Phi(t_n,t_0)(cw+v)\|^2\|\Phi(t_n,t_0)v\|^2}\\
&= \frac{c^2\langle \Phi(t_n,t_0)w,\Phi(t_n,t_0)v\rangle^2 + 2c\|\Phi(t_n,t_0)v\|^2\langle \Phi(t_n,t_0)w,\Phi(t_n,t_0)v\rangle + \|\Phi(t_n,t_0)v\|^4}{c^2\|\Phi(t_n,t_0)w\|^2\|\Phi(t_n,t_0)v\|^2 + 2c\|\Phi(t_n,t_0)v\|^2\langle \Phi(t_n,t_0)w,\Phi(t_n,t_0)v\rangle + \|\Phi(t_n,t_0)v\|^4}\\
&\ge 1-\delta
\end{aligned}$$

holds. Hence, for $|c| > 0$ sufficiently small, one has

$$\operatorname{dist}_{\mathbb{P}}(\mathbb{P}\Phi(t_n,t_0)\mathbb{P}(cw+v), A_{t_n}) \le \eta \quad \text{for all } n \in \mathbb{N}\,.$$

This implies

$$\begin{aligned}
\operatorname{dist}_{\mathbb{P}}(\mathbb{P}(cw+v), A_{t_0}) &= \lim_{n\to\infty} \operatorname{dist}_{\mathbb{P}}(\mathbb{P}(cw+v), A_{t_0})\\
&= \lim_{n\to\infty} \operatorname{dist}_{\mathbb{P}}\big(\mathbb{P}\Phi(t_0,t_n)\underbrace{\mathbb{P}\Phi(t_n,t_0)\mathbb{P}(cw+v)}_{\in B_{2\eta}(A_{t_n})}, A_{t_0}\big)\\
&\overset{(5.3)}{=} 0\,.
\end{aligned}$$

This is a contradiction, since $\mathbb{P}v$ is assumed to be a boundary point of $A_{t_0} \cap \mathbb{P}L_{v,w}$ in $\mathbb{P}L_{v,w}$, and thus, the first part of this proof is finished.
Part 2. It is shown that for all nonzero $v \in \mathbb{P}^{-1}A_{t_0}$ and $w \in C$, the intersection $A_{t_0} \cap \mathbb{P}L_{v,w}$ is a singleton. For this purpose, note that any point in $\mathbb{P}L_{v,w} \setminus \{\mathbb{P}v\}$ is given by $\mathbb{P}(w+cv)$ for some $c \in \mathbb{R}$. It follows from Part 1 that

$$\begin{aligned}
&\lim_{t\to-\infty} \frac{\langle \Phi(t,t_0)(w+cv), \Phi(t,t_0)w\rangle^2}{\|\Phi(t,t_0)(w+cv)\|^2\|\Phi(t,t_0)w\|^2}\\
&= \lim_{t\to-\infty} \frac{\|\Phi(t,t_0)w\|^4 + 2c\|\Phi(t,t_0)w\|^2\langle \Phi(t,t_0)v,\Phi(t,t_0)w\rangle + c^2\langle \Phi(t,t_0)v,\Phi(t,t_0)w\rangle^2}{\|\Phi(t,t_0)w\|^4 + 2c\|\Phi(t,t_0)w\|^2\langle \Phi(t,t_0)v,\Phi(t,t_0)w\rangle + c^2\|\Phi(t,t_0)v\|^2\|\Phi(t,t_0)w\|^2} = 1
\end{aligned}$$

if $\mathbb{P}v$ is a boundary point of $A_{t_0} \cap \mathbb{P}L_{v,w}$ relative to $\mathbb{P}L_{v,w}$. This implies with Lemma A.1 that

$$\lim_{t\to-\infty} d_{\mathbb{P}}(\mathbb{P}\Phi(t,t_0)\mathbb{P}(w+cv), \mathbb{P}\Phi(t,t_0)\mathbb{P}w) = 0\,,$$

and hence, $\mathbb{P}(w + cv) \notin A_{t_0}$. Therefore, $A_{t_0} \cap \mathbb{P}L_{v,w}$ consists of a single point.
Part 3. It follows directly from Part 1 and Part 2 that for all nonzero $v \in \mathbb{P}^{-1}A_{t_0}$ and $w \in C$, one has

$$\lim_{t\to\infty} \frac{\|\Phi(-t, t_0)v\|}{\|\Phi(-t, t_0)w\|} = 0\,.$$

Part 4. It is now shown that $\mathbb{P}^{-1}A_{t_0}$ is a linear subspace of $\mathbb{R}^d$. Already proved was that for any two-dimensional subspace $L_{v,w}$, the set $A_{t_0} \cap \mathbb{P}L_{v,w}$ is either empty, equals $\mathbb{P}L_{v,w}$ or consists of a single point. This implies that $\mathbb{P}^{-1}A$ intersects each fiber in a linear subspace.
Part 5. Finally, it will be shown that

$$\lim_{t\to-\infty} \frac{\sup_{v\in\mathbb{S}^{d-1}\cap\mathbb{P}^{-1}A_{t_0}} \|\Phi(t, t_0)v\|}{\inf_{w\in C} \|\Phi(t, t_0)w\|} = 0\,.$$

Assume to the contrary that there exist sequences $\{t_n\}_{n\in\mathbb{N}}$ in $\mathbb{R}$, $\{v_n\}_{n\in\mathbb{N}}$ in $\mathbb{S}^{d-1}\cap \mathbb{P}^{-1}A_{t_0}$ and $\{w_n\}_{n\in\mathbb{N}}$ in C such that $\lim_{n\to\infty} t_n = -\infty$ and one may assume that $\lim_{n\to\infty} v_n = v$ and $\lim_{n\to\infty} w_n = w$ for some $v \in \mathbb{P}^{-1}A_{t_0} \cap \mathbb{S}^{d-1}$ and $w \in C$, and the following property is fulfilled: There exists a $\gamma > 0$ such that

$$\frac{\|\Phi(t_n, t_0)w_n\|}{\|\Phi(t_n, t_0)v_n\|} \le \gamma \quad \text{for all } n \in \mathbb{N}\,.$$

Similarly to Part 1, for nonzero $c \in \mathbb{R}$ with $|c|$ sufficiently small, this implies that for all $n \in \mathbb{N}$,

$$\begin{aligned}
&\frac{\langle\Phi(t_n,t_0)(cw_n+v_n),\Phi(t_n,t_0)v_n\rangle^2}{\|\Phi(t_n,t_0)(cw_n+v_n)\|^2\|\Phi(t_n,t_0)v_n\|^2}\\
&= \frac{c^2\langle\Phi(t_n,t_0)w_n,\Phi(t_n,t_0)v_n\rangle^2+2c\|\Phi(t_n,t_0)v_n\|^2\langle\Phi(t_n,t_0)w_n,\Phi(t_n,t_0)v_n\rangle+\|\Phi(t_n,t_0)v_n\|^4}{c^2\|\Phi(t_n,t_0)w_n\|^2\|\Phi(t_n,t_0)v_n\|^2+2c\|\Phi(t_n,t_0)v_n\|^2\langle\Phi(t_n,t_0)w_n,\Phi(t_n,t_0)v_n\rangle+\|\Phi(t_n,t_0)v_n\|^4}\\
&\ge 1-\delta
\end{aligned}$$

holds, with $\delta \in (0, 1)$ chosen as in Part 1. Hence, for $|c| > 0$ sufficiently small, one obtains

$$\operatorname{dist}_{\mathbb{P}}(\mathbb{P}\Phi(t_n, t_0)\mathbb{P}(cw_n + v_n), A_{t_n}) \le \eta \quad \text{for all } n \in \mathbb{N}\,.$$

Since $\mathbb{P}(cw + v) \notin A_{t_0}$ (due to Part 2, $A_{t_0} \cap \mathbb{P}L_{v,w}$ is a singleton), there exist an $n_0 \in \mathbb{N}$ and a $\beta > 0$ such that $cw_n + v_n \notin B_\beta(A_{t_0})$ for all $n \ge n_0$. Similarly to Part 1, using (5.3), this implies a contradiction. □

One cannot expect that the fibers of local forward repellers give rise to linear subspaces, since they are intrinsically nonunique. The following proposition, however, says that for any local pullback attractor, it is easy to find a linear counterpart in form of a local forward repeller.

PROPOSITION 5.19. *Let $\mathcal{A}$ be a local pullback attractor of $\mathbb{P}\Phi$ and $\mathcal{R} \subset \mathbb{I} \times \mathbb{P}^{d-1}$ be an invariant nonautonomous set such that $\mathbb{P}^{-1}R_t$ is a linear subspace of $\mathbb{R}^d$ and*

$$\mathbb{P}^{-1}A_t \oplus \mathbb{P}^{-1}R_t = \mathbb{R}^d \quad \text{for all } t \in \mathbb{I}\,.$$

Then $\mathcal{R}$ is a local forward repeller, and the relation $\mathcal{A} = \mathcal{R}^*$ is fulfilled.

PROOF. The proof of this proposition is divided into five parts.
Part 1. It will be shown that for all $t_0 \in \mathbb{I}$ and compact sets $C \subset \mathbb{P}^{d-1}$ with

$C \cap A_{t_0} = \emptyset$, one has

$$\lim_{t\to-\infty} \inf_{0\neq v\in\mathbb{P}^{-1}C} \frac{\|\Phi(t,t_0)v_r\|}{\|\Phi(t,t_0)v\|} = \lim_{t\to\infty} \sup_{0\neq v\in\mathbb{P}^{-1}C} \frac{\|\Phi(t,t_0)v_r\|}{\|\Phi(t,t_0)v\|} = 1\,,$$

where $v = v_a + v_r$ with $v_a \in \mathbb{P}^{-1}A_{t_0}$ and $v_r \in \mathbb{P}^{-1}R(p)$. The first assertion follows from

$$\begin{aligned}
&\lim_{t\to-\infty} \inf_{0\neq v\in\mathbb{P}^{-1}C} \frac{\|\Phi(t,t_0)v_r\|}{\|\Phi(t,t_0)v\|}\\
&\geq \left(\lim_{t\to-\infty} \sup_{0\neq v\in\mathbb{P}^{-1}C} \frac{\|\Phi(t,t_0)v_a\|}{\|\Phi(t,t_0)v_r\|} + 1\right)^{-1}\\
&= \left(\lim_{t\to-\infty} \sup_{v\in\mathbb{P}^{-1}C,\, v_a\neq 0} \frac{\|v_a\|\|\Phi(t,t_0)\frac{v_a}{\|v_a\|}\|}{\|v_r\|\|\Phi(t,t_0)\frac{v_r}{\|v_r\|}\|} + 1\right)^{-1}\\
&\overset{5.18}{=} 1
\end{aligned}$$

and

$$\begin{aligned}
&\lim_{t\to-\infty} \inf_{0\neq v\in\mathbb{P}^{-1}C} \frac{\|\Phi(t,t_0)v_r\|}{\|\Phi(t,t_0)v\|}\\
&\leq \left(\lim_{t\to-\infty} \sup_{0\neq v\in\mathbb{P}^{-1}C} \left|1 - \frac{\|\Phi(t,t_0)v_a\|}{\|\Phi(t,t_0)v_r\|}\right|\right)^{-1}\\
&= \left(\lim_{t\to-\infty} \sup_{v\in\mathbb{P}^{-1}C,\, v_a\neq 0} \left|1 - \frac{\|v_a\|\|\Phi(t,t_0)\frac{v_a}{\|v_a\|}\|}{\|v_r\|\|\Phi(t,t_0)\frac{v_r}{\|v_r\|}\|}\right|\right)^{-1}\\
&\overset{5.18}{=} 1\,.
\end{aligned}$$

In both relations, 5.18 is applicable, because the set $\{v_a : v \in \mathbb{P}^{-1}C \cap \mathbb{S}^{d-1}\}$ is compact and the set $\{v_r : v \in \mathbb{P}^{-1}C \cap \mathbb{S}^{d-1}\}$ is bounded away from zero. This is due to the fact that the projector $P \in \mathbb{R}^{d\times d}$ with range A_{t_0} and null space R_{t_0} satisfies

$$\begin{aligned}
\{v_a : v \in \mathbb{P}^{-1}C \cap \mathbb{S}^{d-1}\} &= P\big(\mathbb{P}^{-1}C \cap \mathbb{S}^{d-1}\big) \qquad \text{and}\\
\{v_r : v \in \mathbb{P}^{-1}C \cap \mathbb{S}^{d-1}\} &= (\mathbb{1} - P)\big(\mathbb{P}^{-1}C \cap \mathbb{S}^{d-1}\big)\,.
\end{aligned}$$

The assertion

$$\lim_{t\to-\infty} \sup_{0\neq v\in\mathbb{P}^{-1}C} \frac{\|\Phi(t,t_0)v_r\|}{\|\Phi(t,t_0)v\|} = 1$$

follows analogously.

Part 2. For all $t_0 \in \mathbb{I}$ and compact sets $C \subset \mathbb{P}^{d-1}$ with $C \cap A_{t_0} = \emptyset$, one has

$$\lim_{t\to-\infty} \operatorname{dist}_{\mathbb{P}}\big(\mathbb{P}\Phi(t,t_0)C, R_t\big) = 0\,.$$

With v_a and v_r defined as in Part 1, for all $t \geq 0$ and $v \in \mathbb{S}^{d-1} \cap \mathbb{P}^{-1}C$, consider the expression

$$\begin{aligned}
&\frac{\langle \Phi(t,t_0)v, \Phi(t,t_0)v_r\rangle^2}{\|\Phi(t,t_0)v\|^2\|\Phi(t,t_0)v_r\|^2} \\
&= \frac{\big(\langle \Phi(t,t_0)v_a, \Phi(t,t_0)v_r\rangle + \langle \Phi(t,t_0)v_r, \Phi(t,t_0)v_r\rangle\big)^2}{\|\Phi(t,t_0)v\|^2\|\Phi(t,t_0)v_r\|^2} \\
&= \frac{\langle \Phi(t,t_0)v_a, \Phi(t,t_0)v_r\rangle^2}{\|\Phi(t,t_0)v\|^2\|\Phi(t,t_0)v_r\|^2} + \frac{\|\Phi(t,t_0)v_r\|^2}{\|\Phi(t,t_0)v\|^2} + \frac{2\langle \Phi(t,t_0)v_a, \Phi(t,t_0)v_r\rangle}{\|\Phi(t,t_0)v\|^2}\,.
\end{aligned}$$

Using the Cauchy-Schwarz inequality, one obtains the following relations:

$$\begin{aligned}
0 \leq \lim_{t\to-\infty} \sup_{v\in\mathbb{S}^{d-1}\cap\mathbb{P}^{-1}C} \frac{\langle \Phi(t,t_0)v_a, \Phi(t,t_0)v_r\rangle^2}{\|\Phi(t,t_0)v\|^2\|\Phi(t,t_0)v_r\|^2} \leq \\
\lim_{t\to-\infty} \sup_{v\in\mathbb{S}^{d-1}\cap\mathbb{P}^{-1}C} \frac{\|\Phi(t,t_0)v_a\|^2}{\|\Phi(t,t_0)v\|^2} \overset{5.18}{=} 0
\end{aligned}$$

and

$$\begin{aligned}
0 &\leq \lim_{t\to-\infty} \sup_{v\in\mathbb{S}^{d-1}\cap\mathbb{P}^{-1}C} \frac{2\big|\langle \Phi(t,t_0)v_a, \Phi(t,t_0)v_r\rangle\big|}{\|\Phi(t,t_0)v\|^2} \\
&\leq \lim_{t\to-\infty} \sup_{v\in\mathbb{S}^{d-1}\cap\mathbb{P}^{-1}C} 2\frac{\|\Phi(t,t_0)v_a\|}{\|\Phi(t,t_0)v\|}\frac{\|\Phi(t,t_0)v_r\|}{\|\Phi(t,t_0)v\|} \\
&\overset{\text{Part 1}}{=} \lim_{t\to-\infty} \sup_{v\in\mathbb{S}^{d-1}\cap\mathbb{P}^{-1}C} \frac{2\|\Phi(t,t_0)v_a\|}{\|\Phi(t,t_0)v\|} \\
&\overset{5.18}{=} 0\,.
\end{aligned}$$

Hence, it follows that

$$\begin{aligned}
&\lim_{t\to-\infty} \inf_{v\in\mathbb{S}^{d-1}\cap\mathbb{P}^{-1}C} \frac{\langle \Phi(t,t_0)v, \Phi(t,t_0)v_r\rangle^2}{\|\Phi(t,t_0)v\|^2\|\Phi(t,t_0)v_r\|^2} \\
&= \lim_{t\to-\infty} \inf_{v\in\mathbb{S}^{d-1}\cap\mathbb{P}^{-1}C} \Bigg(\frac{\langle \Phi(t,t_0)v_a, \Phi(t,t_0)v_r\rangle^2}{\|\Phi(t,t_0)v\|^2\|\Phi(t,t_0)v_r\|^2} + \frac{\|\Phi(t,t_0)v_r\|^2}{\|\Phi(t,t_0)v\|^2} \\
&\qquad + \frac{2\langle \Phi(t,t_0)v_a, \Phi(t,t_0)v_r\rangle}{\|\Phi(t,t_0)v\|^2}\Bigg) \\
&\overset{\text{Part 1}}{=} 1\,.
\end{aligned}$$

Using Lemma A.1, this implies the assertion.

Part 3. It is shown that $\mathcal{A}$ and $\mathcal{R}$ are isolated, i.e., there exists a $\beta > 0$ such that

$$B_\beta(A_t) \cap B_\beta(R_t) = \emptyset \quad \text{for all } t \in \mathbb{I}\,.$$

This follows from the fact that $\mathcal{A}$ is a local pullback attractor, i.e., there exists an $\eta > 0$ such that

$$\lim_{t\to-\infty} \operatorname{dist}_{\mathbb{P}}\big(\mathbb{P}\Phi(0,t)B_\eta(A_t), A_0\big) = 0\,.$$

Part 4. It is a direct consequence of Part 2 and Part 3 that $\mathcal{R}$ is a local forward repeller.

Part 5. Finally, it is shown that the relation $\mathcal{A} = \mathcal{R}^*$ is fulfilled. Define $\eta > 0$ and

$\mathcal{D}^\zeta$ for $\zeta \in (0,\eta]$ as in the proof of Theorem 4.1 of Chapter 4, and consider also the collection $\mathfrak{D} := \{\mathcal{D}^\zeta : \zeta \in (0,\eta]\}$. It is sufficient to show that $\mathcal{A}$ is a pullback attractor with respect to the attraction universe $\mathfrak{D}$ (cf. also Exercise 3.49). Thereto, fix $\zeta \in (0,\eta]$ and $t_0 \in \mathbb{I}$. Furthermore, choose $\varepsilon > 0$ arbitrarily and consider the compact set $C := \mathbb{P}^{d-1} \setminus B_\varepsilon(A_{t_0})$. Due to Part 2, one has

$$\lim_{t\to-\infty} \operatorname{dist}_{\mathbb{P}}\big(\mathbb{P}\Phi(t,t_0)C, R_t\big) = 0\,.$$

This implies that there exists a $\tau > 0$ such that $\mathbb{P}\Phi(t,t_0)C \cap D_t^\zeta = \emptyset$ for all $t \le \tau$. Thus,

$$\operatorname{dist}_{\mathbb{P}}\big(\mathbb{P}\Phi(t_0,t)D_t^\zeta, A_{t_0}\big) \le \varepsilon \quad \text{for all } t \le \tau\,,$$

i.e., $\mathcal{A}$ is a pullback attractor with respect to the attraction universe $\mathfrak{D}$. This finishes the proof of this proposition. □

LEMMA 5.20. *For $n \in \mathbb{N}$, consider nontrivial linear subspaces $W_n \subsetneq V_n \subsetneq \mathbb{R}^d$. Furthermore, let $\{x_n\}_{n\in\mathbb{N}}$ be a sequence in $\mathbb{R}^d$ such that the following hypotheses are fulfilled:*

(i) *$x_n \notin V_n$ for all $n \in \mathbb{N}$,*
(ii) *$\lim_{n\to\infty} d_{\mathbb{P}}(\mathbb{P}x_n, \mathbb{P}V_n) = 0$,*
(iii) *there exists an $\varepsilon > 0$ such that $d_{\mathbb{P}}(\mathbb{P}x_n, \mathbb{P}W_n) \ge \varepsilon$ for all $n \in \mathbb{N}$.*

For all $n \in \mathbb{N}$, define $C_n := W_n \oplus \{\lambda x_n : \lambda \in \mathbb{R}\}$. Then the limit relation

$$\lim_{n\to\infty} \operatorname{dist}_{\mathbb{P}}(\mathbb{P}C_n, \mathbb{P}V_n) = 0$$

holds.

PROOF. One may assume that that $\|x_n\| = 1$ for all $n \in \mathbb{N}$. Due to hypothesis (ii), there exists a sequence $\{v_n\}_{n\in\mathbb{N}}$ with $v_n \in V_n$ and $\|v_n\| = 1$ for all $n \in \mathbb{N}$ such that $\lim_{n\to\infty} \|x_n - v_n\| = 0$. Since $\mathbb{P}C_n$ is a compact subset of $\mathbb{P}^{d-1}$, there exists a sequence $\{c_n\}_{n\in\mathbb{N}}$ with $c_n \in C_n$ for all $n \in \mathbb{N}$ such that $\operatorname{dist}_{\mathbb{P}}(\mathbb{P}C_n, \mathbb{P}V_n) = \operatorname{dist}_{\mathbb{P}}(\mathbb{P}c_n, \mathbb{P}V_n)$. Without loss of generality, assume that c_n is of the form

$$c_n := x_n + w_n \quad \text{for all } n \in \mathbb{N}\,,$$

where $\{w_n\}_{n\in\mathbb{N}}$ is a sequence with $w_n \in W_n$ for all $n \in \mathbb{N}$, and define

$$r_n := v_n + w_n \quad \text{for all } n \in \mathbb{N}$$

and $\beta_n := \langle x_n, w_n\rangle$, $\delta_n := \langle v_n, w_n\rangle$ and $\gamma_n := \langle x_n, v_n\rangle$ for all $n \in \mathbb{N}$. Then for all $n \in \mathbb{N}$, it follows that

$$\frac{\langle c_n, r_n\rangle^2}{\|c_n\|^2\|r_n\|^2}$$
$$= \frac{\gamma_n^2 + \overbrace{\beta_n^2 + \delta_n^2 + \|w_n\|^4 + 2\gamma_n\|x_n\|^2 + 2\beta_n\delta_n}^{=:\xi_n} + \overbrace{2(\beta_n+\delta_n)(\|w_n\|^2+\gamma_n)}^{=:\eta_n}}{1 + \underbrace{2\|w_n\|^2 + \|w_n\|^4 + 4\beta_n\delta_n}_{=:\tilde{\xi}_n} + \underbrace{2(\beta_n+\delta_n)(\|w_n\|^2+1)}_{\tilde{\eta}_n}}\,,$$

and it is easy to see that $\lim_{n\to\infty}\gamma_n = 1$, $\lim_{n\to\infty}\frac{\xi_n}{\tilde{\xi}_n} = 1$ and $\lim_{n\to\infty}\frac{\eta_n}{\tilde{\eta}_n} = 1$. This implies that

$$\lim_{n\to\infty} \frac{\langle c_n, r_n\rangle^2}{\|c_n\|^2\|r_n\|^2} = 1$$

if $\|c_n\|\|r_n\|$ is bounded away from 0 in the limit $n \to \infty$. To see that this is fulfilled, hypothesis (iii) is needed, which yields that there exists a $\delta \in (0,1)$ with

$$\frac{\langle x_n, w\rangle}{\|w\|} \le \delta \quad \text{for all } n \in \mathbb{N} \text{ and } w \in W_n$$

(see Lemma A.1). This means that for all $n \in \mathbb{N}$ and $w \in W_n$, one has

$$\|x_n - w\|^2 = 1 - 2\langle x_n, w\rangle + \|w\|^2 \ge 1 - 2\delta\|w\| + \|w\|^2 \ge \gamma \quad \text{for some } \gamma > 0\,,$$

and this finishes the proof of this lemma. □

In contrast to nonautonomous Morse decompositions in the general case, convergence in both directions to the Morse sets is fulfilled in the linear case.

THEOREM 5.21. *Let $\{\mathcal{M}^1, \dots, \mathcal{M}^n\}$ be a past Morse decomposition obtained by the finite sequence of local forward repellers $\mathcal{R}^0 \supset \dots \supset \mathcal{R}^n$ such that $\mathbb{P}^{-1}\mathcal{R}^i_t$ is a linear subspace of $\mathbb{R}^d$ for $i \in \{1, \dots, n-1\}$ and $t \in \mathbb{I}$. Then the following statements hold:*

(i) Convergence in forward time. *For all $t_0 \in \mathbb{I}$ and all families $\{\gamma_t\}_{t\in\mathbb{I}}$ in $\mathbb{P}^{d-1}$ with $\liminf_{t\to-\infty} \operatorname{dist}_{\mathbb{P}}\big(\gamma_t, \bigcup_{j=1,\dots,n} \partial R^j_t\big) > 0$,*

$$\lim_{t\to-\infty} \operatorname{dist}_{\mathbb{P}}\big(\mathbb{P}\Phi(t_0,t)\gamma_t, \textstyle\bigcup_{j=1,\dots,n} M^j_{t_0}\big) = 0\,.$$

(ii) Convergence in backward time. *For all $(t_0, x) \in \mathbb{I} \times \mathbb{P}^{d-1}$, there exists an $i \in \{1, \dots, n\}$ with*

$$\lim_{t\to-\infty} \operatorname{dist}_{\mathbb{P}}\big(\mathbb{P}\Phi(t,t_0)x, M^i_t\big) = 0\,.$$

PROOF. The convergence in forward time is valid also for general Morse decompositions and was proved in Theorem 4.9. To prove (ii), choose $(t_0, x) \in \mathbb{I} \times \mathbb{P}^{d-1}$ arbitrarily. Then there exists an $i \in \{1, \dots, n\}$ such that

$$x \in R^{i,*}_{t_0} \quad \text{and} \quad x \notin R^{i-1,*}_{t_0}\,.$$

If $x \in R^{i-1}_{t_0}$, the above limit relation follows, since then $x \in M^i_{t_0}$ and $\mathcal{M}^i$ is invariant. Assume therefore $x \notin R^{i-1}_{t_0}$ from now on. To obtain a contradiction, also assume that there exist an $\varepsilon > 0$ and a sequence $\{t_n\}_{n\in\mathbb{N}}$ in $\mathbb{R}$ with $\lim_{n\to\infty} t_n = -\infty$ such that

$$\operatorname{dist}_{\mathbb{P}}\big(\mathbb{P}\Phi(t_n,t_0)x, M^i_{t_n}\big) \ge \varepsilon \quad \text{for all } n \in \mathbb{N}\,.$$

Define $C := \mathbb{P}^{-1}M^i_{t_0} \oplus \mathbb{P}^{-1}\{x\}$. Since $\lim_{t\to-\infty} \operatorname{dist}_{\mathbb{P}}(\mathbb{P}\Phi(t,t_0)x, R^{i-1}_t) = 0$ (see Proposition 4.4), Lemma 5.20 implies that

$$\lim_{n\to\infty} \operatorname{dist} d_{\mathbb{P}}\big(\mathbb{P}\Phi(t_n,t_0)\mathbb{P}C, R^{i-1}_{t_n}\big) = 0\,. \tag{5.4}$$

Define $\bar{C} := \mathbb{P}^{-1}R^{i-1}_{t_0} \oplus \mathbb{P}^{-1}\{x\}$ and obtain

$$\begin{aligned}\dim\big(\bar{C} \cap \mathbb{P}^{-1}R^{i-1,*}_{t_0}\big) &= \dim \bar{C} + \dim \mathbb{P}^{-1}R^{i-1,*}_{t_0} - \dim\big(\bar{C} + \mathbb{P}^{-1}R^{i-1,*}_{t_0}\big)\\ &= d + 1 - d = 1\,.\end{aligned}$$

Let $y = v + w$ be a nonzero element of $\bar{C} \cap \mathbb{P}^{-1}R^{i-1,*}_{t_0}$ with $v \in \mathbb{P}^{-1}\{x\}$ and $w \in \mathbb{P}^{-1}R^{i-1}_{t_0}$. Since y and v are in $\mathbb{P}^{-1}R^{i,*}_{t_0}$, w is also an element of $\mathbb{P}^{-1}R^{i,*}_{t_0}$. Hence, $w \in \mathbb{P}^{-1}M^i_{t_0}$. This implies $y \in C$, and hence, from (5.4), one obtains

$$\lim_{n\to\infty} d_{\mathbb{P}}\big(\mathbb{P}\Phi(t_n,t_0)\mathbb{P}y, R^{i-1}_{t_n}\big) = 0\,.$$

This is a contradiction, since $\mathbb{P}y \in R_{t_0}^{i-1,*}$ and $\mathcal{R}^{i-1}$ and $\mathcal{R}^{i-1,*}$ are isolated (cf. Proposition 4.4). □

The following exercise is a generalization of Proposition 4.5.

EXERCISE 5.22. Let $\{\mathcal{M}^1, \dots, \mathcal{M}^n\}$ and $\{\hat{\mathcal{M}}^1, \dots, \hat{\mathcal{M}}^n\}$ be past Morse decompositions obtained by the finite sequences of local forward repellers $\mathcal{R}^0 \supset \cdots \supset \mathcal{R}^n$ and $\hat{\mathcal{R}}^0 \supset \cdots \supset \hat{\mathcal{R}}^n$ such that $\mathbb{P}^{-1}R_t^i$ and $\mathbb{P}^{-1}\hat{R}_t^i$ are linear subspaces of $\mathbb{R}^d$ for $i \in \{1, \dots, n-1\}$ and $t \in \mathbb{I}$. Assume that

$$\mathcal{R}^{i,*} = \hat{\mathcal{R}}^{i,*} \quad \text{for all } i \in \{1, \dots, n-1\}.$$

Then the convergence in Hausdorff distance

$$\lim_{t \to -\infty} h_{\mathbb{P}}\big(M_t^i, \hat{M}_t^i\big) = 0 \quad \text{for all } i \in \{1, \dots, n\}$$

is fulfilled.

The following theorem is an analogue to the Theorem of Selgrade (see SELGRADE [**215**, Theorem 9.7] and COLONIUS & KLIEMANN [**53**, Theorem 5.2.5]).

THEOREM 5.23 (Finest past Morse decomposition). *There exists a* finest past Morse *decomposition $\{\mathcal{M}^1, \dots, \mathcal{M}^n\}$ of $\mathbb{P}\Phi$, i.e., for any other past Morse decomposition $\{\tilde{\mathcal{M}}^1, \dots, \tilde{\mathcal{M}}^m\}$, the relation*

$$\lim_{t \to -\infty} \operatorname{dist}_{\mathbb{P}}\big(\textstyle\bigcup_{i=1}^n M_t^i, \bigcup_{i=1}^m \tilde{M}_t^i\big) = 0$$

holds. Moreover, $n \le d$ and the following decomposition in a Whitney sum is fulfilled:

$$\mathbb{P}^{-1}\mathcal{M}^1 \oplus \cdots \oplus \mathbb{P}^{-1}\mathcal{M}^n = \mathbb{I} \times \mathbb{R}^d.$$

PROOF. It will first be shown that for any two local pullback attractors $\mathcal{A}$ and $\bar{\mathcal{A}}$, either the relation

$$\mathcal{A} \subset \bar{\mathcal{A}} \quad \text{or} \quad \mathcal{A} \supset \bar{\mathcal{A}}$$

is fulfilled. Supposing the contrary, there exists a $t_0 \in \mathbb{I}$ and elements

$$x \in \mathbb{S}^{d-1} \cap \big(\mathbb{P}^{-1}A_{t_0} \setminus \mathbb{P}^{-1}\bar{A}_{t_0}\big) \quad \text{and} \quad \bar{x} \in \mathbb{S}^{d-1} \cap \big(\mathbb{P}^{-1}\bar{A}_{t_0} \setminus \mathbb{P}^{-1}A_{t_0}\big).$$

Because of Proposition 5.18, one obtains

$$\lim_{t \to -\infty} \frac{\|\Phi(t, t_0)x\|}{\|\Phi(t, t_0)\bar{x}\|} = 0 \quad \text{and} \quad \lim_{t \to -\infty} \frac{\|\Phi(t, t_0)\bar{x}\|}{\|\Phi(t, t_0)x\|} = 0.$$

This is a contradiction. Proposition 5.18 also implies that the fibers of local pullback attractors correspond to linear subspaces. Thus, there are at most $d+1$ local pullback attractors of $\mathbb{P}\Phi$, namely

$$\emptyset = \mathcal{A}^0 \subsetneq \mathcal{A}^1 \subsetneq \cdots \subsetneq \mathcal{A}^n = \mathbb{I} \times \mathbb{P}^{d-1}$$

with $n \le d$. Due to Proposition 5.19, it is possible to choose a sequence of local forward repellers $\mathbb{I} \times \mathbb{P}^{d-1} = \mathcal{R}^0 \supsetneq \mathcal{R}^1 \supsetneq \cdots \supsetneq \mathcal{R}^n = \emptyset$ such that $\mathcal{R}^{i,*} = \mathcal{A}^i$ for $i \in \{0, \dots, n\}$. Denote by $\{\mathcal{M}^1, \dots, \mathcal{M}^n\}$ the corresponding past Morse decomposition, and let $\{\tilde{\mathcal{M}}^1, \dots, \tilde{\mathcal{M}}^m\}$ be another past Morse decomposition, obtained by the sequence $\mathbb{I} \times \mathbb{P}^{d-1} = \tilde{\mathcal{R}}^0 \supsetneq \tilde{\mathcal{R}}^1 \supsetneq \cdots \supsetneq \tilde{\mathcal{R}}^m = \emptyset$ of local forward repellers. Then for each $i \in \{0, \dots, m\}$, there exists an $n_i \in \{0, \dots, n\}$ such that $\tilde{\mathcal{R}}^{i,*} = \mathcal{A}^{n_i}$. Consider now the past Morse decomposition $\{\bar{\mathcal{M}}^1, \dots, \bar{\mathcal{M}}^n\}$, which is obtained by the local forward repellers $\mathcal{R}^{n_0}, \dots, \mathcal{R}^{n_m}$. Due to Exercise 5.22, one has

$$\lim_{t \to -\infty} h_{\mathbb{P}}\big(\bigcup_{i=1}^m \bar{M}_t^i, \bigcup_{i=1}^m \tilde{M}_t^i\big) = 0 \,.$$

Moreover, it is easy to see that $\bigcup_{i=1}^n \mathcal{M}^i \subset \bigcup_{i=1}^m \bar{\mathcal{M}}^i$ holds. This finishes the proof of the first assertion of this theorem. To show

$$\mathbb{P}^{-1}\mathcal{M}^1 \oplus \dots \oplus \mathbb{P}^{-1}\mathcal{M}^n = \mathbb{I} \times \mathbb{R}^d \,,$$

first note that for $1 \le i < j \le n$, one has $\mathbb{P}^{-1}\mathcal{M}^i \cap \mathbb{P}^{-1}\mathcal{M}^j = \mathbb{I} \times \{0\}$ (see Proposition 4.7). Furthermore, Proposition 5.19 implies

$$\begin{aligned}
\mathbb{I} \times \mathbb{R}^d &= \mathbb{P}^{-1}\mathcal{R}^{1,*} + \mathbb{P}^{-1}\mathcal{R}^1 \\
&= \mathbb{P}^{-1}\mathcal{M}^1 + \big(\mathbb{P}^{-1}\mathcal{R}^1 \cap (\mathbb{P}^{-1}\mathcal{R}^{2,*} + \mathbb{P}^{-1}\mathcal{R}^2)\big) \\
&= \mathbb{P}^{-1}\mathcal{M}^1 + \big(\mathbb{P}^{-1}\mathcal{R}^1 \cap \mathbb{P}^{-1}\mathcal{R}^{2,*}\big) + \mathbb{P}^{-1}\mathcal{R}^2 \\
&= \mathbb{P}^{-1}\mathcal{M}^1 + \mathbb{P}^{-1}\mathcal{M}^2 + \mathbb{P}^{-1}\mathcal{R}^2 \,.
\end{aligned}$$

Here, the fact that linear subspaces $E, F, G \subset \mathbb{R}^d$ with $E \supset G$ satisfy

$$E \cap (F + G) = (E \cap F) + G$$

was used. It follows inductively that

$$\mathbb{I} \times \mathbb{R}^d = \mathbb{P}^{-1}\mathcal{M}^1 + \dots + \mathbb{P}^{-1}\mathcal{M}^n + \mathbb{P}^{-1}\mathcal{R}^n = \mathbb{P}^{-1}\mathcal{M}^1 + \dots + \mathbb{P}^{-1}\mathcal{M}^n \,.$$

This finishes the proof of this theorem. □

REMARK 5.24. A finest past Morse decomposition $\{\mathcal{M}^1, \dots, \mathcal{M}^n\}$ is not uniquely determined, but it follows directly from the above theorem that any other finest Morse decomposition $\{\tilde{\mathcal{M}}^1, \dots, \tilde{\mathcal{M}}^m\}$ satisfies

$$\lim_{t \to -\infty} h_{\mathbb{P}}\big(\bigcup_{i=1}^n \mathcal{M}_t^i, \bigcup_{i=1}^m \tilde{M}_t^i\big) = 0 \,.$$

Moreover, the relation $n = m$ holds.

4.2. All-time Morse decompositions in projective space. The construction of past attractor-repeller pairs is possible only in one direction, i.e., a local forward repeller implies a local pullback attractor (see Theorem 4.1 of Chapter 4), but Example 4.2 showed that in general, a local forward repeller cannot be constructed from a local pullback attractor. Note that the local pullback attractor from Example 4.2 is also a local uniform attractor and no corresponding local uniform repeller exists, since this would be also a local forward repeller. Similarly, one can show that, in general, there is no method to construct a local uniform attractor from a local uniform repeller.

In this subsection, it is shown that in the linear setting, it is possible to obtain an local uniform repeller from a local uniform attractor and vice versa. A step towards this result is the following proposition, which says, among other things, that both local uniform attractors and repellers in $\mathbb{P}^{d-1}$ give rise to linear subspaces in $\mathbb{R}^d$. It is assumed throughout this subsection that $\mathbb{I} = \mathbb{R}$.

PROPOSITION 5.25. *The following statements hold:*

(i) *Let* $\mathcal{A} \notin \{\emptyset, \mathbb{I} \times \mathbb{P}^{d-1}\}$ *be a local uniform attractor of* $\mathbb{P}\Phi$. *Then, for all* $\beta \in (0, \sqrt{2})$, *one has*

$$\lim_{t\to\infty} \sup_{t_0\in\mathbb{R}} \frac{\sup_{v\in\mathbb{S}^{d-1}\cap\mathbb{P}^{-1}A_{t_0}} \|\Phi(t_0-t,t_0)v\|}{\inf_{w\in\mathbb{S}^{d-1}\cap\mathbb{P}^{-1}(\mathbb{P}^{d-1}\setminus B_\beta(A_{t_0}))} \|\Phi(t_0-t,t_0)w\|} = 0\,. \tag{5.5}$$

Moreover, for all $t_0 \in \mathbb{R}$, *the set* $\mathbb{P}^{-1}A_{t_0}$ *is a linear subspace of* $\mathbb{R}^d$.

(ii) *Let* $\mathcal{R} \notin \{\emptyset, \mathbb{I} \times \mathbb{P}^{d-1}\}$ *be a local uniform repeller of* $\mathbb{P}\Phi$. *Then, for all* $\beta \in (0, \sqrt{2})$, *one has*

$$\lim_{t\to\infty} \sup_{t_0\in\mathbb{R}} \frac{\sup_{v\in\mathbb{S}^{d-1}\cap\mathbb{P}^{-1}R_{t_0}} \|\Phi(t_0+t,t_0)v\|}{\inf_{w\in\mathbb{S}^{d-1}\cap\mathbb{P}^{-1}(\mathbb{P}^{d-1}\setminus B_\beta(R_{t_0}))} \|\Phi(t_0+t,t_0)w\|} = 0\,.$$

Moreover, for all $t_0 \in \mathbb{R}$, *the set* $\mathbb{P}^{-1}R_{t_0}$ *is a linear subspace of* $\mathbb{R}^d$.

PROOF. Note first that β is supposed to be less than $\sqrt{2}$ only to guarantee that the infimum in the denominator of (5.5) is taken over a nonempty set. $\mathcal{A}$ is also a local pullback attractor, and hence, Proposition 5.18 implies that for all $t \in \mathbb{R}$, the set $\mathbb{P}^{-1}A_t$ is a linear subspace. The definition of a local uniform attractor implies the existence of an $\eta > 0$ such that

$$\lim_{t\to\infty} \sup_{t_0\in\mathbb{R}} \operatorname{dist}_{\mathbb{P}}\big(\mathbb{P}\Phi(t_0+t,t_0)B_{2\eta}(A_{t_0}), A_{t_0+t}\big) = 0\,. \tag{5.6}$$

Due to Lemma A.1, there exists a $\delta \in (0,1)$ such that for all nonzero $u_1, u_2 \in \mathbb{R}^d$ with $\langle u_1, u_2\rangle^2/(\|u_1\|^2\|u_2\|^2) \geq 1-\delta$, one has $d_{\mathbb{P}}(\mathbb{P}u_1, \mathbb{P}u_2) \leq \eta$. Assume to the contrary that there exist sequences $(t_n)_{n\in\mathbb{N}}$, $(s_n)_{n\in\mathbb{N}}$ in $\mathbb{R}$, $(v_n)_{n\in\mathbb{N}}$ in $\mathbb{S}^{d-1} \cap \mathbb{P}^{-1}A_{s_n}$ and $(w_n)_{n\in\mathbb{N}}$ in $\in \mathbb{S}^{d-1} \cap \mathbb{P}^{-1}\big(\mathbb{P}^{d-1} \setminus B_\beta(A_{t_0})\big)$ such that $\lim_{n\to\infty}(s_n - t_n) = -\infty$ and the following property is fulfilled: There exists a $\gamma > 0$ with $\|\Phi(t_n,s_n)w_n\|/\|\Phi(t_n,s_n)v_n\| \leq \gamma$ for all $n \in \mathbb{N}$. Using the abbreviation $\Phi_n := \Phi(t_n,s_n)$, for nonzero $c \in \mathbb{R}$ with $|c|$ sufficiently small, this implies that for all $n \in \mathbb{N}$,

$$\begin{aligned}
&\frac{\big\langle \Phi_n(cw_n+v_n), \Phi_n v_n\big\rangle^2}{\|\Phi_n(cw_n+v_n)\|^2\|\Phi_n v_n\|^2} \\
&= \frac{c^2\big\langle \Phi_n w_n, \Phi_n v_n\big\rangle^2 + 2c\|\Phi_n v_n\|^2\big\langle \Phi_n w_n, \Phi_n v_n\big\rangle + \|\Phi_n v_n\|^4}{c^2\|\Phi_n w_n\|^2\|\Phi_n v_n\|^2 + 2c\|\Phi_n v_n\|^2\big\langle \Phi_n w_n, \Phi_n v_n\big\rangle + \|\Phi_n v_n\|^4} \geq 1-\delta
\end{aligned}$$

holds. Thus, for such a fixed c, one obtains

$$\operatorname{dist}_{\mathbb{P}}\big(\mathbb{P}\Phi(t_n,s_n)\mathbb{P}(cw_n+v_n), A_{t_n}\big) \leq \eta \quad \text{for all } n \in \mathbb{N}\,,$$

and this means that

$$\begin{aligned}
&\lim_{n\to\infty} \operatorname{dist}_{\mathbb{P}}\big(\mathbb{P}(cw_n+v_n), A_{s_n}\big) \\
&= \lim_{n\to\infty} \operatorname{dist}_{\mathbb{P}}\big(\mathbb{P}\Phi(s_n,t_n)\underbrace{\mathbb{P}\Phi(t_n,s_n)\mathbb{P}(cw_n+v_n)}_{\in B_{2\eta}(A_{t_n})}, A_{s_n}\big) \overset{(5.6)}{=} 0\,.
\end{aligned}$$

This is a contradiction, because $\mathbb{P}w_n \notin B_\beta(A_{s_n})$ implies that there exists an $\alpha > 0$ with $\mathbb{P}(cw_n+v_n) \notin B_\alpha(A_{s_n})$ for all $n \in \mathbb{N}$. □

A local uniform attractor is also a local forward attractor, and hence, the formalism of constructing attractor-repeller pairs leads to a corresponding local pullback repeller, which is given by (5.8) below. In the linear case, the rate of attraction of the local forward attractor equals the rate of repulsion of the local pullback repeller, and therefore, the local pullback repeller is also a local uniform repeller, since the local forward attractor is also a local uniform attractor. This fact is the main idea for the proof of the following theorem.

THEOREM 5.26. *The following statements hold:*

(i) *Given a local uniform attractor $\mathcal{A}$ of $\mathbb{P}\Phi$, i.e., there exists an $\eta > 0$ with*

$$\lim_{t\to\infty} \sup_{t_0\in\mathbb{R}} \operatorname{dist}_{\mathbb{P}} \big(\mathbb{P}\Phi(t_0+t,t_0)B_\eta(A_{t_0}), A_{t_0+t}\big) = 0\,. \tag{5.7}$$

Then the nonautonomous set $\mathcal{A}^*$, defined by

$$A_t^* := \bigcap_{\tau\geq 0} \overline{\bigcup_{s\geq\tau} \mathbb{P}\Phi(t,s)\big(\mathbb{P}^{d-1}\setminus B_\eta(A_s)\big)} \quad \textit{for all } t\in\mathbb{R}\,, \tag{5.8}$$

is a local uniform repeller such that $B_\eta(A_t)\cap A_t^* = \emptyset$ for all $t\in\mathbb{R}$, and $(\mathcal{A},\mathcal{A}^*)$ is called an all-time attractor-repeller pair. *Moreover, the following decomposition in a Whitney sum holds:*

$$\mathbb{P}^{-1}\mathcal{A}\oplus\mathbb{P}^{-1}\mathcal{A}^* = \mathbb{R}\times\mathbb{R}^d\,. \tag{5.9}$$

(ii) *Given a local uniform repeller $\mathcal{R}$ of $\mathbb{P}\Phi$, i.e., there exists an $\eta > 0$ with*

$$\lim_{t\to\infty} \sup_{t_0\in\mathbb{R}} \operatorname{dist}_{\mathbb{P}} \big(\mathbb{P}\Phi(t_0-t,t_0)B_\eta(R_{t_0}), R_{t_0-t}\big) = 0\,.$$

Then the nonautonomous set $\mathcal{R}^*$, defined by

$$R_t^* := \bigcap_{\tau\geq 0} \overline{\bigcup_{s\geq\tau} \mathbb{P}\Phi(t,s)\big(\mathbb{P}^{d-1}\setminus B_\eta(R_s)\big)} \quad \textit{for all } t\in\mathbb{R}\,,$$

is a local uniform attractor such that $R_t^*\cap B_\eta(R_t) = \emptyset$ for all $t\in\mathbb{R}$, and $(\mathcal{R}^*,\mathcal{R})$ is called an all-time attractor-repeller pair. *Moreover, the following decomposition in a Whitney sum holds: $\mathbb{P}^{-1}\mathcal{R}^*\oplus\mathbb{P}^{-1}\mathcal{R} = \mathbb{R}\times\mathbb{R}^d$.*

(iii) *One has $(\mathcal{A}^*)^* = \mathcal{A}$ and $(\mathcal{R}^*)^* = \mathcal{R}$ for local uniform attractors $\mathcal{A}$ and local uniform repellers $\mathcal{R}$.*

PROOF. (i) Firstly note that $A_t^*\cap B_\eta(A_t) = \emptyset$ for all $t\in\mathbb{R}$. Otherwise, since $\mathcal{A}$ and $\mathcal{A}^*$ are compact nonautonomous sets, (5.7) leads to $A_t^*\cap A_t\neq\emptyset$ for some $t\in\mathbb{R}$, but this is a contradiction due to Theorem 4.1 (it was proved there that past and future attractor-repeller pairs are disjoint, and (5.8) is the formula for the corresponding local pullback repeller for a given local forward attractor $\mathcal{A}$). Hence, due to Proposition 5.25, for fixed $\gamma\in(0,\eta)$, one has

$$\lim_{t\to\infty} \sup_{t_0\in\mathbb{R}} \frac{\sup_{v\in\mathbb{S}^{d-1}\cap\mathbb{P}^{-1}A_{t_0}} \|\Phi(t_0-t,t_0)v\|}{\inf_{w\in\mathbb{S}^{d-1}\cap\mathbb{P}^{-1}B_\gamma(A_{t_0}^*)} \|\Phi(t_0-t,t_0)w\|} = 0\,. \tag{5.10}$$

The relation (5.9) is a consequence of Proposition 5.19. The remaining proof of (i) is divided into two parts.

Part 1. It will be shown that

$$1 = \lim_{t\to\infty} \inf_{t_0\in\mathbb{R}} \inf_{0\neq v\in\mathbb{P}^{-1}B_\gamma(A^*_{t_0})} \frac{\|\Phi(t_0-t,t_0)v_r^{t_0}\|}{\|\Phi(t_0-t,t_0)v\|}$$
$$= \lim_{t\to\infty} \sup_{t_0\in\mathbb{R}} \sup_{0\neq v\in\mathbb{P}^{-1}B_\gamma(A^*_{t_0})} \frac{\|\Phi(t_0-t,t_0)v_r^{t_0}\|}{\|\Phi(t_0-t,t_0)v\|},$$

where $v = v_a^{t_0} + v_r^{t_0}$ with $v_a^{t_0} \in \mathbb{P}^{-1}A_{t_0}$ and $v_r^{t_0} \in \mathbb{P}^{-1}A^*_{t_0}$. The first part of the equality (concerning the infimum) follows from

$$\lim_{t\to\infty} \inf_{t_0\in\mathbb{R}} \inf_{0\neq v\in\mathbb{P}^{-1}B_\gamma(A^*_{t_0})} \frac{\|\Phi(t_0-t,t_0)v_r^{t_0}\|}{\|\Phi(t_0-t,t_0)v\|}$$
$$\geq \left(\lim_{t\to\infty} \sup_{t_0\in\mathbb{R}} \sup_{0\neq v\in\mathbb{P}^{-1}B_\gamma(A^*_{t_0})} \frac{\|\Phi(t_0-t,t_0)v_a^{t_0}\|}{\|\Phi(t_0-t,t_0)v_r^{t_0}\|} + 1\right)^{-1}$$
$$= \left(\lim_{t\to\infty} \sup_{t_0\in\mathbb{R}} \sup_{v\in\mathbb{P}^{-1}B_\gamma(A^*_{t_0}),v_a^{t_0}\neq 0} \frac{\|v_a^{t_0}\| \left\|\Phi(t_0-t,t_0)\frac{v_a^{t_0}}{\|v_a^{t_0}\|}\right\|}{\|v_r^{t_0}\| \left\|\Phi(t_0-t,t_0)\frac{v_r^{t_0}}{\|v_r^{t_0}\|}\right\|} + 1\right)^{-1}$$
$$\overset{(5.10)}{=} 1$$

and

$$\lim_{t\to\infty} \inf_{t_0\in\mathbb{R}} \inf_{0\neq v\in\mathbb{P}^{-1}B_\gamma(A^*_{t_0})} \frac{\|\Phi(t_0-t,t_0)v_r^{t_0}\|}{\|\Phi(t_0-t,t_0)v\|}$$
$$\leq \left(\lim_{t\to\infty} \sup_{t_0\in\mathbb{R}} \sup_{0\neq v\in\mathbb{P}^{-1}B_\gamma(A^*_{t_0})} \left|1 - \frac{\|\Phi(t_0-t,t_0)v_a^{t_0}\|}{\|\Phi(t_0-t,t_0)v_r^{t_0}\|}\right|\right)^{-1}$$
$$= \left(\lim_{t\to\infty} \sup_{t_0\in\mathbb{R}} \sup_{v\in\mathbb{P}^{-1}B_\gamma(A^*_{t_0}),v_a^{t_0}\neq 0} \left|1 - \frac{\|v_a^{t_0}\| \left\|\Phi(t_0-t,t_0)\frac{v_a^{t_0}}{\|v_a^{t_0}\|}\right\|}{\|v_r^{t_0}\| \left\|\Phi(t_0-t,t_0)\frac{v_r^{t_0}}{\|v_r^{t_0}\|}\right\|}\right|\right)^{-1}$$
$$\overset{(5.10)}{=} 1\,.$$

In both relations, the last equality holds, because the set $V_a := \{v_a^{t_0} : t_0 \in \mathbb{R}, v \in \mathbb{S}^{d-1} \cap \mathbb{P}^{-1}B_\gamma(A^*_{t_0})\}$ is bounded and the set $V_r := \{v_r^{t_0} : t_0 \in \mathbb{R}, v \in \mathbb{S}^{-1} \cap \mathbb{P}^{-1}B_\gamma(A^*_{t_0})\}$ is bounded away from zero. The second assertion concerning the supremum instead of the infimum follows analogously.

Part 2. It will be proved that

$$\lim_{t\to\infty} \sup_{t_0\in\mathbb{R}} \operatorname{dist}_{\mathbb{P}}\left(\mathbb{P}\Phi(t_0-t,t_0)B_\gamma(A^*_{t_0}), A^*_{t_0-t})\right) = 0\,, \tag{5.11}$$

i.e., $\mathcal{A}^*$ is a local uniform repeller. With $v_r^{t_0}$ and $v_a^{t_0}$ defined as in Part 1, for all $t_0 \in \mathbb{R}$, $t \geq 0$ and $v \in \mathbb{S}^{d-1} \cap \mathbb{P}^{-1} B_\gamma(A^*_{t_0})$, the relation

$$\begin{aligned}
&\frac{\langle \Phi(t_0-t,t_0)v, \Phi(t_0-t,t_0)v_r^{t_0}\rangle^2}{\|\Phi(t_0-t,t_0)v\|^2 \|\Phi(t_0-t,t_0)v_r^{t_0}\|^2} \\
&= \frac{\langle \Phi(t_0-t,t_0)v_a^{t_0}, \Phi(t_0-t,t_0)v_r^{t_0}\rangle^2}{\|\Phi(t_0-t,t_0)v\|^2 \|\Phi(t_0-t,t_0)v_r^{t_0}\|^2} + \frac{\|\Phi(t_0-t,t_0)v_r^{t_0}\|^2}{\|\Phi(t_0-t,t_0)v\|^2} \\
&\quad + \frac{2\langle \Phi(t_0-t,t_0)v_a^{t_0}, \Phi(t_0-t,t_0)v_r^{t_0}\rangle}{\|\Phi(t_0-t,t_0)v\|^2}
\end{aligned}$$

holds. Using the Cauchy-Schwarz inequality, one obtains the following relations:

$$\begin{aligned}
0 \leq \lim_{t\to\infty} \sup_{t_0\in\mathbb{R}} \sup_{v\in\mathbb{S}^{d-1}\cap\mathbb{P}^{-1}B_\gamma(A^*_{t_0})} \frac{\langle \Phi(t_0-t,t_0)v_a^{t_0}, \Phi(t_0-t,t_0)v_r^{t_0}\rangle^2}{\|\Phi(t_0-t,t_0)v\|^2 \|\Phi(t_0-t,t_0)v_r^{t_0}\|^2} \leq \\
\lim_{t\to\infty} \sup_{t_0\in\mathbb{R}} \sup_{v\in\mathbb{S}^{d-1}\cap\mathbb{P}^{-1}B_\gamma(A^*_{t_0})} \frac{\|\Phi(t_0-t,t_0)v_a^{t_0}\|^2}{\|\Phi(t_0-t,t_0)v\|^2} \overset{(5.10)}{=} 0
\end{aligned}$$

and

$$\begin{aligned}
0 &\leq \lim_{t\to\infty} \sup_{t_0\in\mathbb{R}} \sup_{v\in\mathbb{S}^{d-1}\cap\mathbb{P}^{-1}B_\gamma(A^*_{t_0})} \frac{2\left|\langle \Phi(t_0-t,t_0)v_a^{t_0}, \Phi(t_0-t,t_0)v_r^{t_0}\rangle\right|}{\|\Phi(t_0-t,t_0)v\|^2} \\
&\leq \lim_{t\to\infty} \sup_{t_0\in\mathbb{R}} \sup_{v\in\mathbb{S}^{d-1}\cap\mathbb{P}^{-1}B_\gamma(A^*_{t_0})} 2\,\frac{\|\Phi(t_0-t,t_0)v_a^{t_0}\|}{\|\Phi(t_0-t,t_0)v\|}\,\frac{\|\Phi(t_0-t,t_0)v_r^{t_0}\|}{\|\Phi(t_0-t,t_0)v\|} \\
&\overset{\text{Part 1}}{=} \lim_{t\to\infty} \sup_{t_0\in\mathbb{R}} \sup_{v\in\mathbb{S}^{d-1}\cap\mathbb{P}^{-1}B_\gamma(A^*_{t_0})} \frac{2\|\Phi(t_0-t,t_0)v_a^{t_0}\|}{\|\Phi(t_0-t,t_0)v\|} \overset{(5.10)}{=} 0
\end{aligned}$$

(the fact from Part 1 that V_a is bounded was used). Hence, due to Part 1, it follows that

$$\begin{aligned}
&\lim_{t\to\infty} \inf_{t_0\in\mathbb{R}} \inf_{v\in\mathbb{S}^{d-1}\cap\mathbb{P}^{-1}B_\gamma(A^*_{t_0})} \frac{\langle \Phi(t_0-t,t_0)v, \Phi(t_0-t,t_0)v_r^{t_0}\rangle^2}{\|\Phi(t_0-t,t_0)v\|^2 \|\Phi(t_0-t,t_0)v_r^{t_0}\|^2} = \\
&= \lim_{t\to\infty} \inf_{t_0\in\mathbb{R}} \inf_{v\in\mathbb{S}^{d-1}\cap\mathbb{P}^{-1}B_\gamma(A^*_{t_0})} \left(\frac{\langle \Phi(t_0-t,t_0)v_a^{t_0}, \Phi(t_0-t,t_0)v_r^{t_0}\rangle^2}{\|\Phi(t_0-t,t_0)v\|^2 \|\Phi(t_0-t,t_0)v_r^{t_0}\|^2} \right. \\
&\quad \left. + \frac{\|\Phi(t_0-t,t_0)v_r^{t_0}\|^2}{\|\Phi(t_0-t,t_0)v\|^2} + \frac{2\langle \Phi(t_0-t,t_0)v_a^{t_0}, \Phi(t_0-t,t_0)v_r^{t_0}\rangle}{\|\Phi(t_0-t,t_0)v\|^2} \right) = 1\,.
\end{aligned}$$

Using Lemma A.1, this implies (5.11).
(ii) can be proved analogously to (i).
(iii) Let $\mathcal{A}$ be a local uniform attractor. Then $\mathcal{A}$ is also a local forward attractor, and due to Theorem 4.1 from Chapter 4, $\mathcal{A}^*$ given by (5.8) is the maximal local pullback repeller outside of $\mathcal{A}$ (with respect to set inclusion). Reapplication of Theorem 4.1 shows that $(\mathcal{A}^*)^*$ is the maximal local pullback attractor outside of $\mathcal{A}^*$. This proves $\mathcal{A} \subset (\mathcal{A}^*)^*$. Assume to the contrary that there exist $t_0 \in \mathbb{R}$ and $x \in (A^*)^*_{t_0} \setminus A_{t_0}$. Then there exists a $\tau \geq 0$ such that $\mathbb{P}\Phi(t_0-t,t_0)x \notin B_\eta(A_{t_0-t})$ for all $t \geq \tau$, because otherwise, (5.7) would lead to $x \in A_{t_0}$ (note that $\mathcal{A}$ is a compact nonautonomous set). Hence, (5.8) implies that $x \in A^*_{t_0}$, and this is a contradiction, since $\mathcal{A}^*$ and $(\mathcal{A}^*)^*$ are disjoint due to (i). □

EXERCISE 5.27. In PALMER & SIEGMUND [**174**], so-called *generalized attractor-repeller pairs* are introduced. These consist of two invariant and linear nonautonomous sets $\mathcal{A}$ and $\mathcal{R}$ of Φ fulfilling the following three conditions:

(a) $A_{t_0} \oplus R_{t_0} = \mathbb{R}^d$ for all $t_0 \in \mathbb{R}$,
(b) given $t_0 \in \mathbb{R}$, $0 \neq \xi \in A_{t_0}$ and $0 \neq \eta \in R_{t_0}$,

$$\frac{\|\Phi(t,t_0)\eta\|}{\|\Phi(t,t_0)\xi\|} \to 0 \text{ as } t \to \infty \quad \text{and} \quad \frac{\|\Phi(t,t_0)\xi\|}{\|\Phi(t,t_0)\eta\|} \to 0 \text{ as } t \to -\infty\,,$$

(c) the angle between A_{t_0} and R_{t_0} is bounded below by a positive number.

It is easy to see that local uniform attractor repeller-pairs are also generalized attractor-repeller pairs. Show by a counterexample that the reversal of this statement is not true.

Analogously to the case of the past and future considered in Chapter 4, all-time Morse decompositions are obtained as intersections of attractors and repellers.

DEFINITION 5.28. A family $\{\mathcal{M}^1, \mathcal{M}^2, \dots, \mathcal{M}^n\}$ of nonautonomous sets, the so-called *Morse sets*, is called *all-time Morse decomposition* if the representation $\mathcal{M}^i = \mathcal{A}^i \cap \mathcal{A}^{i-1,*}$ for all $i \in \{1, \dots, n\}$ holds with local uniform attractors $\emptyset = \mathcal{A}_0 \subsetneq \mathcal{A}_1 \subsetneq \cdots \subsetneq \mathcal{A}_n = \mathbb{R} \times \mathbb{P}^{d-1}$.

All-time Morse decompositions have the following basic properties.

PROPOSITION 5.29. *The Morse sets of an all-time Morse decomposition* $\{\mathcal{M}^1, \dots, M_n\}$ *are nonempty, invariant, pairwise disjoint and* isolated, *i.e., there exists a* $\beta > 0$ *such that for all* $1 \le i < j \le n$*, one has* $B_\beta(\mathcal{M}^i_t) \cap B_\beta(\mathcal{M}^j_t) = \emptyset$ *for all* $t \in \mathbb{R}$*. Moreover, the sets* $\mathbb{P}^{-1}\mathcal{M}^i_t$*,* $i \in \{1, \dots, n\}$*, are linear subspaces of* $\mathbb{R}^d$ *for all* $t \in \mathbb{R}$*, and one has the following decomposition in a Whitney sum:* $\mathbb{P}^{-1}\mathcal{M}^1 \oplus \cdots \oplus \mathbb{P}^{-1}\mathcal{M}^n = \mathbb{R} \times \mathbb{R}^d$.

PROOF. Due to the fact that every all-time Morse decomposition is a past Morse decomposition, Proposition 4.7 yields that the Morse sets are nonempty, invariant and pairwise disjoint. Moreover, the Morse sets are isolated, since this is an obvious consequence of the fact that attractor-repeller pairs are isolated (cf. Theorem 5.26). To show the decomposition in a Whitney sum, first note that for $1 \le i < j \le n$, one has $\mathbb{P}^{-1}\mathcal{M}^i \cap \mathbb{P}^{-1}\mathcal{M}^j = \mathbb{R} \times \{0\}$. Furthermore, (5.9) implies

$$\begin{aligned} \mathbb{R} \times \mathbb{R}^d &= \mathbb{P}^{-1}\mathcal{A}^1 + \mathbb{P}^{-1}\mathcal{A}^{1,*} = \mathbb{P}^{-1}\mathcal{M}^1 + \big(\mathbb{P}^{-1}\mathcal{A}^{1,*} \cap \big(\mathbb{P}^{-1}\mathcal{A}^2 + \mathbb{P}^{-1}\mathcal{A}^{2,*}\big)\big) \\ &= \mathbb{P}^{-1}\mathcal{M}^1 + \big(\mathbb{P}^{-1}\mathcal{A}^{1,*} \cap \mathbb{P}^{-1}\mathcal{A}^2\big) + \mathbb{P}^{-1}\mathcal{A}^{2,*} \\ &= \mathbb{P}^{-1}\mathcal{M}^1 + \mathbb{P}^{-1}\mathcal{M}^2 + \mathbb{P}^{-1}\mathcal{A}^{2,*}\,. \end{aligned}$$

It was used here that linear subspaces $E, F, G \subset \mathbb{R}^d$ with $E \supset G$ fulfill $E \cap (F+G) = (E \cap F) + G$. Finally, it follows inductively that

$$\mathbb{R} \times \mathbb{R}^d = \mathbb{P}^{-1}\mathcal{M}^1 + \cdots + \mathbb{P}^{-1}\mathcal{M}^n + \mathbb{P}^{-1}R_n = \mathbb{P}^{-1}\mathcal{M}^1 + \cdots + \mathbb{P}^{-1}\mathcal{M}^n\,.$$

This finishes the proof of this theorem. □

The following theorem shows that the Morse sets are crucial for the asymptotic behavior of the system.

THEOREM 5.30. *For an all-time Morse decomposition* $\{\mathcal{M}^1, \dots, \mathcal{M}^n\}$, *the following statements are fulfilled:*

(i) Convergence in forward time. *For all* $(t_0, x) \in \mathbb{R} \times \mathbb{P}^{d-1}$, *there exists an* $i \in \{1, \dots, n\}$ *with* $\lim_{t\to\infty} \operatorname{dist}_{\mathbb{P}} \big(\mathbb{P}\Phi(t, t_0)x, M_t^i\big) = 0$.

(ii) Convergence in backward time. *For all* $(t_0, x) \in \mathbb{R} \times \mathbb{P}^{d-1}$, *there exists an* $i \in \{1, \dots, n\}$ *with* $\lim_{t\to-\infty} \operatorname{dist}_{\mathbb{P}} \big(\mathbb{P}\Phi(t, t_0)x, M_t^i\big) = 0$.

PROOF. This follows from Theorem 5.21, since an all-time Morse decomposition is both a past and a future Morse decomposition. □

The proof of the existence of a finest all-time Morse decomposition is left as an exercise to the reader (see also Theorem 5.23).

THEOREM 5.31. *There exists a* finest all-time Morse decomposition $\{\mathcal{M}^1, \dots, \mathcal{M}^n\}$, *i.e., given another all-time Morse decomposition* $\{\bar{\mathcal{M}}^1, \dots, \bar{\mathcal{M}}^m\}$, *then for all* $i \in \{1, \dots, m\}$, *there exists* $j \in \{1, \dots, n\}$ *with* $\mathcal{M}^j \subset \bar{\mathcal{M}}^i$. *Moreover,* $n \leq d$ *holds and the following decomposition in a Whitney sum is fulfilled:*

$$\mathbb{P}^{-1}\mathcal{M}^1 \oplus \cdots \oplus \mathbb{P}^{-1}\mathcal{M}^n = \mathbb{R} \times \mathbb{R}^d .$$

EXERCISE 5.32. Prove Theorem 5.31.

This chapter is concluded by the observation that the spectral manifolds of a linear system give rise to a Morse decomposition in the projective space. This result was proved in RASMUSSEN [**199**].

THEOREM 5.33. *Let*

$$\Sigma = \Sigma^{\leftarrow}, \Sigma^{\rightarrow}, \Sigma^{\leftrightarrow} = [a_1, b_1] \cup \cdots \cup [a_n, b_n]$$

be the dichotomy spectra of (5.1), *respectively, define the invariant projectors* $P_{\gamma_0} := 0$, $P_{\gamma_n} := \mathbb{1}$, *and for* $i \in \{1, \dots, n-1\}$, *choose* $\gamma_i \in (b_i, a_{i+1})$ *and projectors* P_{γ_i} *of the nonhyperbolic exponential dichotomy of* (5.1) *with growth rate* γ_i. *Then the sets*

$$\mathcal{M}^i := \mathbb{P}\big(\mathfrak{R}(P_{\gamma_i}) \cap \mathfrak{N}(P_{\gamma_{i-1}})\big) \quad \textit{for all } i \in \{1, \dots, n\}$$

are the Morse sets of a past (future, all-time, respectively) Morse decomposition of $\mathbb{P}\Phi$.

Note that it is possible that the above Morse decomposition defined by the spectral intervals is coarser than the finest Morse decomposition of Theorem 5.23 and Theorem 5.31 (see COLONIUS, KLOEDEN & RASMUSSEN [**54**]).

Endnotes. The concept of an exponential dichotomy is due to Perron [**177, 178**], and the literature in this field is abundant, see Barreira & Valls [**17**], Chicone & Latushkin [**45**], Coppel [**57**], Daleckiĭ & Kreĭn [**63**], Massera & Schäffer [**165**], Palmer [**171, 173**] and Sacker & Sell [**208**] for important books and articles in this field (see also Coffman & Schäffer [**50**] and Henry [**93**] for the discrete case). The dichotomy spectrum as discussed in Section 2 is based on the Sacker–Sell spectrum [**209**], which was extended to arbitrary invertible systems in Ben-Artzi & Gohberg [**20**], to noninvertible systems in Aulbach & Siegmund [**12**] and to different time domains in Rasmussen [**199**]. Apart from the Sacker–Sell spectrum, there are other approaches to characterize spectral properties of linear systems, and the relationship of the Sacker–Sell spectrum to other spectral notions has been worked out by several authors, see Johnson, Palmer & Sell [**102**] and Colonius & Kliemann [**52, 51**]. The Lyapunov spectrum from Section 3 is based on Lyapunov exponents which have been introduced in [**160**]. Daleckiĭ & Kreĭn [**63**] is a classical reference for the Bohl spectrum, which was introduced already in 1913. A spectral notion which is based on operator semigroup theory and an insight of Mather [**166**] was developed in Chicone & Swanson [**46**], and Fredholm properties of evolution operators have been related to exponential dichotomies on half-lines in Palmer [**175**] (see also Ben-Artzi, Gohberg & Kaashoek [**21**] and Latushkin & Pogan [**156**]). These studies have been extremely important for stability theory of traveling waves (see Härterich, Sandstede & Scheel [**92**] and Sandstede [**212**]). Morse decompositions of linear systems with chain recurrent base space have been investigated by Selgrade [**215**] and Salamon & Zehnder [**211**] (see also Colonius & Kliemann [**53**]). The approach presented in Section 4 is from Rasmussen [**195, 197**] and is based on the theory of the nonautonomous Morse decompositions from Chapter 4.

CHAPTER 6

Invariant manifolds

In the study of nonlinear systems, invariant manifolds play a central role. Firstly, they help to understand the *local* dynamical behavior near an equilibrium, a periodic solution or, in the nonautonomous context, an arbitrary solution. For instance, stable and unstable manifolds describe the saddle point structure around hyperbolic equilibria, and center manifolds capture the essential dynamics which makes them to a main object in bifurcation and stability theory. But also for *global* studies, invariant manifolds are important. They serve as separatrix between different domains of attraction, and global attractors consist of unstable manifolds and equilibria.

The construction of stable and unstable invariant manifolds goes back to both POINCARÉ [**182**] and HADAMARD [**88**]. Later, the theory was extended from hyperbolic to nonhyperbolic systems, from finite to infinite-dimensional and from time-independent to time-dependent equations.

1. Global invariant manifolds

Invariant manifolds for a linear system

$$\dot{x} = B(t)x$$

have been treated already in Chapter 5. They are given, for instance, as the spectral manifolds, which are time-dependent linear spaces belonging to some spectral interval and can be characterized dynamically via exponential boundedness.

In this section, a small nonlinear perturbation of this situation will be discussed, in which case the nonautonomous invariant manifolds can be seen as small perturbations of the spectral manifolds. Consider a nonlinear ordinary differential equation

$$\dot{x} = B(t)x + F(t,x) \tag{6.1}$$

with continuous functions $B : \mathbb{R} \to \mathbb{R}^{d\times d}$ and $F : \mathbb{R}\times\mathbb{R}^d \to \mathbb{R}^d$ such that $F(t,0) = 0$ for all $t \in \mathbb{R}$. The process induced by (6.1) will be denoted by $\phi : \mathbb{R}\times\mathbb{R}\times\mathbb{R}^d \to \mathbb{R}^d$ and is globally defined due to Hypothesis 6.2 below. In addition to (6.1), consider the corresponding linear ordinary differential equation, given by

$$\dot{x} = B(t)x\,, \tag{6.2}$$

whose transition operator is denoted by $\Phi : \mathbb{R}\times\mathbb{R} \to \mathbb{R}^{d\times d}$.

Let $P_+ : \mathbb{R} \to \mathbb{R}^{d\times d}$ be an invariant projector of (6.2). Then the function $P_- : \mathbb{R} \to \mathbb{R}^{d\times d}$, given by $P_-(t) := \mathbb{1} - P_+(t)$ for all $t \in \mathbb{R}$, is also an invariant projector, the so-called *complementary projector*. In the following, the symbol $P_\pm$ simultaneously stands for P_+ and P_-, respectively, and similarly with further notation in this chapter.

The notion of a global nonautonomous invariant manifold is now introduced.

DEFINITION 6.1 (Global nonautonomous invariant manifold). Let $P_\pm : \mathbb{R} \to \mathbb{R}^{d\times d}$ be complementary invariant projectors of (6.2), and consider continuous functions $w^\pm : \mathbb{R}\times\mathbb{R}^d \to \mathbb{R}^d$ satisfying

(i) $w^\pm(t,0) = 0$ for all $t \in \mathbb{R}$,
(ii) $w^\pm(t,x) = w^\pm\big(t, P_\pm(t)x\big) \in \mathfrak{R}(P_\mp(t))$ for all $t\in\mathbb{R}$ and $x \in \mathbb{R}^d$.

Then the nonautonomous graphs

$$\mathcal{W}^\pm := \big\{\big(t, x + w^\pm(t,x)\big) \in \mathbb{R}\times\mathbb{R}^d : x \in \mathfrak{R}(P_\pm(t))\big\}$$

are called *global nonautonomous invariant manifolds* of (6.1) if

$$\big(t, \phi(t,t_0,x_0)\big) \in \mathcal{W}^\pm \quad \text{for all } (t_0,x_0)\in\mathcal{W}^\pm \text{ and } t\in\mathbb{R}\,.$$

To prove existence results for global nonautonomous invariant manifolds of (6.1), the following hypotheses on the linear part and the nonlinearity are needed.

HYPOTHESIS 6.2. Suppose that the following two statements hold.

(H1) *Hypothesis on linear part.* The linear system (6.2) admits an exponential dichotomy on $\mathbb{R}$, i.e., there exist complementary invariant projectors $P_\pm : \mathbb{R}\to\mathbb{R}^{d\times d}$ such that the inequalities

$$\|\Phi(t,t_0)P_+(t_0)\| \le Ke^{\alpha_+(t-t_0)} \quad \text{for all } t \ge t_0\,,$$
$$\|\Phi(t,t_0)P_-(t_0)\| \le Ke^{\alpha_-(t-t_0)} \quad \text{for all } t \le t_0$$

hold with real constants $K \ge 1$ and $\alpha_- < \alpha_+$.

(H2) *Hypothesis on nonlinearity.* The nonlinear part is globally Lipschitz continuous such that for some $L < \frac{1}{4K}(\alpha_- - \alpha_+)$, one has

$$\|F(t,x) - F(t,y)\| \le L\|x-y\| \quad \text{for all } t\in\mathbb{R} \text{ and } x,y\in\mathbb{R}^d\,.$$

The following theorem says that under this hypothesis, global invariant manifolds exist, and solutions lying in these manifolds fulfill certain exponential growth conditions (recall that the notion of exponential boundedness was introduced on page 81).

THEOREM 6.3. *Consider the nonautonomous differential equation* (6.1), *and assume that both* (H1) *and* (H2) *from Hypothesis 6.2 hold. Then there exist continuous functions* $w^\pm : \mathbb{R}\times\mathbb{R}^d\to\mathbb{R}^d$ *such that the graphs*

$$\mathcal{W}^\pm := \big\{\big(t, x + w^\pm(t,x)\big) \in \mathbb{R}\times\mathbb{R}^d : x \in \mathfrak{R}(P_\pm(t))\big\}$$

are global nonautonomous invariant manifolds which fulfill the representations

$$\mathcal{W}^\pm := \big\{(t,x)\in\mathbb{R}\times\mathbb{R}^d : \phi(\cdot,t,x) \text{ is } \gamma^\pm\text{-exponentially bounded}\big\} \tag{6.3}$$

for all $\gamma \in (\alpha_+ + 2KL, \alpha_- - 2KL)$. *The invariant manifold* $\mathcal{W}^+$ *is called* pseudo-stable manifold, *whereas* $\mathcal{W}^-$ *is called* pseudo-unstable manifold. *Moreover, the estimates*

$$\big\|w^\pm(t,x) - w^\pm(t,y)\big\| \le \frac{2K^2L}{\alpha_- - \alpha_+ - 4KL}\|x-y\|$$

hold for all $t\in\mathbb{R}$ *and* $x,y\in\mathbb{R}^d$.

The main idea of the proof of this theorem is to characterize the invariant manifolds by the exponential growth rates of the solutions contained in it. Let $\gamma \in \mathbb{R}$ be an exponential growth rate, and consider the sets

$$\mathfrak{X}_\gamma^\pm := \left\{ \psi \in C(\mathbb{R}, \mathbb{R}^d) : \sup_{t \in \mathbb{R}_0^\pm} e^{-\gamma t} \|\psi(t)\| < \infty \right\}$$

of all continuous and $\gamma^\pm$-exponentially bounded functions. It is straightforward to check that $\mathfrak{X}_\gamma^\pm$ are Banach spaces with respect to the norms

$$\|\psi\|_{t_0,\gamma}^+ := \sup_{t \in [t_0,\infty)} e^{-\gamma(t-t_0)} \|\psi(t)\|$$

and

$$\|\psi\|_{t_0,\gamma}^- := \sup_{t \in (-\infty,t_0]} e^{-\gamma(t-t_0)} \|\psi(t)\| ,$$

which depend also on some initial time $t_0 \in \mathbb{R}$. Of key importance for the proof of this theorem are the *Lyapunov–Perron operators* $T_{t_0}^\pm : \mathfrak{X}_{t_0,\gamma}^\pm \times \mathbb{R}^d \to \mathfrak{X}_{t_0,\gamma}^\pm$, defined by

$$T_{t_0}^+(t, \psi, x_0) := \Phi(t, t_0) P_+(t_0) x_0 + \int_{t_0}^\infty G(t, s) F(s, \psi(s))\, \mathrm{d}s ,$$

$$T_{t_0}^-(t, \psi, x_0) := \Phi(t, t_0) P_-(t_0) x_0 + \int_{-\infty}^{t_0} G(t, s) F(s, \psi(s))\, \mathrm{d}s ,$$

where the abbreviation $\mathcal{T}_{t_0}^+(t, \psi, \xi) := \mathcal{T}_{t_0}^+(\psi, \xi)(t)$ has been used, and *Green's function* G is defined by

$$G(t, s) := \begin{cases} -\Phi(t, s) P_-(s) & : \; t < s , \\ \Phi(t, s) P_+(s) & : \; t \geq s . \end{cases}$$

It will be shown below with the contraction mapping theorem that the Lyapunov–Perron operator has a uniquely determined fixed point, leading to a point on the invariant manifold. Before proving the theorem, some preparation is needed.

PROPOSITION 6.4. *Consider the linear inhomogeneous equation*

$$\dot{x} = B(t) x + b(t) , \tag{6.4}$$

where $b : \mathbb{R} \to \mathbb{R}^d$ is continuous, suppose that (H1) *from Hypothesis 6.2 is fulfilled, and let $t_0 \in \mathbb{R}$ and $\gamma \in (\alpha_-, \alpha_+)$. Then the following statements hold:*

(i) *Let $b \in \mathfrak{X}_{t_0,\gamma}^+$ and $x_0 \in \mathbb{R}^d$. Then there exists a uniquely determined γ^+-exponentially bounded solution $\psi : \mathbb{R} \to \mathbb{R}^d$ of* (6.4) *satisfying $P_+(t_0)\psi(t_0) = P_+(t_0) x_0$ and with the representation as a Lyapunov–Perron integral*

$$\psi(t) = \Phi(t, t_0) P_+(t_0) x_0 + \int_{t_0}^\infty G(t, s) b(s)\, \mathrm{d}s \quad \textit{for all } t \in \mathbb{R} .$$

In addition,

$$\|\psi\|_{t_0,\gamma}^+ \leq K \left(\|P_+(t_0) x_0\| + \frac{\|b\|_{t_0,\gamma}^+}{\gamma - \alpha_+} + \frac{\|b\|_{t_0,\gamma}^+}{\alpha_- - \gamma} \right) .$$

(ii) *Let $b \in \mathfrak{X}^-_{t_0,\gamma}$ and $x_0 \in \mathbb{R}^d$. Then there exists a uniquely determined γ^--exponentially bounded solution $\psi : \mathbb{R} \to \mathbb{R}^d$ of (6.4) satisfying $P_-(t_0)\psi(t_0) = P_-(t_0)x_0$ and with the representation as a Lyapunov–Perron integral*

$$\psi(t) = \Phi(t,t_0)P_-(t_0)x_0 + \int_{-\infty}^{t_0} G(t,s)b(s)\,\mathrm{d}s \quad \textit{for all } t \in \mathbb{R}\,.$$

In addition,

$$\|\psi\|^-_{t_0,\gamma} \le K\left(\|P_-(t_0)x_0\| + \frac{\|b\|^-_{t_0,\gamma}}{\gamma - \alpha_+} + \frac{\|b\|^-_{t_0,\gamma}}{\alpha_- - \gamma}\right).$$

PROOF. Since the proofs of (i) and (ii) are similar, only the first assertion will be proved. Let $t_0 \in \mathbb{R}$, and choose $b \in \mathfrak{X}^+_{t_0,\gamma}$ and $x_0 \in \mathbb{R}^d$. The proof that the Lyapunov–Perron integral ψ is a solution to (6.4) is left as Exercise 6.5 below. For $t \ge t_0$, one has

$$\begin{aligned}
&\|P_+(t)\psi(t)\| \\
&\le \|\Phi(t,t_0)P_+(t_0)\|\|P_+(t_0)x_0\| + \int_{t_0}^t \|\Phi(t,s)P_+(s)\|\|b(s)\|\,\mathrm{d}s \\
&\le Ke^{\alpha_+(t-t_0)}\|P_+(t_0)x_0\| + \int_{t_0}^t Ke^{\alpha_+(t-s)}\|b\|^+_{t_0,\gamma}e^{\gamma(s-t_0)}\,\mathrm{d}s \\
&\le Ke^{\alpha_+(t-t_0)}\|P_+(t_0)x_0\| + K\|b\|^+_{t_0,\gamma}e^{\alpha_+t-\gamma t_0}\int_{t_0}^t e^{s(\gamma-\alpha_+)}\,\mathrm{d}s \\
&\le Ke^{\alpha_+(t-t_0)}\|P_+(t_0)x_0\| + K\|b\|^+_{t_0,\gamma}e^{\alpha_+t-\gamma t_0}\frac{e^{t(\gamma-\alpha_+)} - e^{t_0(\gamma-\alpha_+)}}{\gamma-\alpha_+} \\
&\le Ke^{\gamma(t-t_0)}\left(\|P_+(t_0)x_0\| + \|b\|^+_{t_0,\gamma}e^{t(\alpha_+-\gamma)}\frac{e^{t(\gamma-\alpha_+)} - e^{t_0(\gamma-\alpha_+)}}{\gamma-\alpha_+}\right) \\
&\le Ke^{\gamma(t-t_0)}\left(\|P_+(t_0)x_0\| + \|b\|^+_{t_0,\gamma}\frac{1}{\gamma-\alpha_+}\right)
\end{aligned}$$

and

$$\begin{aligned}
&\|P_-(t)\psi(t)\| \\
&\le \int_t^\infty \|\Phi(t,s)P_-(s)\|\|b(s)\|\,\mathrm{d}s \le \int_t^\infty Ke^{\alpha_-(t-s)}\|b\|^+_{t_0,\gamma}e^{\gamma(s-t_0)}\,\mathrm{d}s \\
&\le Ke^{\alpha_-t-\gamma t_0}\|b\|^+_{t_0,\gamma}\int_t^\infty e^{s(\gamma-\alpha_-)}\,\mathrm{d}s \le Ke^{\alpha_-t-\gamma t_0}\|b\|^+_{t_0,\gamma}\frac{1}{\alpha_- - \gamma}e^{-t(\alpha_- - \gamma)} \\
&= Ke^{\gamma(t-t_0)}\|b\|^+_{t_0,\gamma}\frac{1}{\alpha_- - \gamma}\,.
\end{aligned}$$

This implies that

$$\begin{aligned}
e^{-\gamma(t-t_0)}\|\psi(t)\| &\le e^{-\gamma(t-t_0)}\big(\|P_+(t)\psi(t)\| + \|P_-(t)\psi(t)\|\big) \\
&\le K\left(\|P_+(t_0)x_0\| + \frac{\|b\|^+_{t_0,\gamma}}{\gamma-\alpha_+} + \frac{\|b\|^+_{t_0,\gamma}}{\alpha_- - \gamma}\right),
\end{aligned}$$

and this means that ψ is γ^+-exponentially bounded with the estimate as stated in the proposition. Assume that there is another γ^+-exponentially bounded solution $\bar{\psi}$ to (6.4) such that $P_+(t_0)\bar{\psi}(t_0) = P_+(t_0)x_0$. Then $\psi - \bar{\psi}$ is a γ^+-exponentially bounded solution to the homogeneous equation $\dot{x} = B(t)x$ fulfilling $P_+(t_0)\big(\psi(t_0) - \bar{\psi}(t_0)\big) = 0$. The second equation in the exponential dichotomy in (H1) implies that $\psi(t_0) - \bar{\psi}(t_0) = 0$, which means that $\psi = \bar{\psi}$. This finishes the proof of the proposition. □

EXERCISE 6.5. Show that the functions $\psi : \mathbb{R} \to \mathbb{R}^d$ defined in both Proposition 6.4 (i) and (ii) are indeed solutions to (6.4). Why does the integral used for their definition converge?

PROOF OF THEOREM 6.3. Firstly, fix $\gamma \in (\alpha_+ + 2KL, \alpha_- - 2KL)$, and choose $\delta \in \big(2KL, \frac{1}{2}(\alpha_- - \alpha_+)\big)$ such that $\gamma \in [\alpha_+ + \delta, \alpha_- - \delta]$. The proof of this theorem is divided into five parts.

Part 1. To see that the Lyapunov–Perron operator $T^+_{t_0} : \mathfrak{X}^+_{t_0,\gamma} \times \mathbb{R}^d \to \mathfrak{X}^+_{t_0,\gamma}$ for $t_0 \in \mathbb{R}$ is is well-defined, note that, similarly to the proof of Proposition 6.4, one obtains for some $x_0 \in \mathbb{R}^d$, $\psi \in \mathfrak{X}^+_{t_0,\gamma}$ and $t \geq t_0$ the relations

$$\|P_+(t)T^+_{t_0}(t,\psi,x_0)\| \leq Ke^{\gamma(t-t_0)}\left(\|P_+(t_0)x_0\| + L\frac{\|\psi\|^+_{t_0,\gamma}}{\gamma - \alpha_+}\right)$$

and

$$\|P_-(t)T^+_{t_0}(t,\psi,x_0)\| \leq KLe^{\gamma(t-t_0)}\frac{\|\psi\|^+_{t_0,\gamma}}{\alpha_- - \gamma}.$$

This implies that $T^+_{t_0}(\psi,x_0)$ is γ^+-exponentially bounded with

$$\|T^+_{t_0}(\psi,x_0)\|^+_{t_0,\gamma} \leq K\left(\|P_+(t_0)x_0\| + L\|\psi\|^+_{t_0,\gamma}\left(\frac{1}{\gamma - \alpha_+} + \frac{1}{\alpha_- - \gamma}\right)\right).$$

Part 2. It is shown now that the mapping $T^+_{t_0}(\cdot,x_0) : \mathfrak{X}^+_{t_0,\gamma} \to \mathfrak{X}^+_{t_0,\gamma}$ is a contraction for arbitrarily chosen $(t_0,x_0) \in \mathbb{R} \times \mathbb{R}^d$. Similarly to the proof of Proposition 6.4, one obtains for $\psi_1,\psi_2 \in \mathfrak{X}^+_{t_0,\gamma}$ that

$$\big\|P_+(t)\big(T^+_{t_0}(t,\psi_1,x_0) - T^+_{t_0}(t,\psi_2,x_0)\big)\big\| \leq KLe^{\gamma(t-t_0)}\frac{\|\psi_1 - \psi_2\|^+_{t_0,\gamma}}{\gamma - \alpha_+}$$

and

$$\big\|P_-(t)\big(T^+_{t_0}(t,\psi_1,x_0) - T^+_{t_0}(t,\psi_2,x_0)\big)\big\| \leq KLe^{\gamma(t-t_0)}\frac{\|\psi_1 - \psi_2\|^+_{t_0,\gamma}}{\alpha_- - \gamma}$$

for $t \geq t_0$. This implies that

$$\begin{aligned}&\big\|T^+_{t_0}(\psi_1,x_0) - T^+_{t_0}(\psi_2,x_0)\big\|^+_{t_0,\gamma}\\ &\leq KL\|\psi_1 - \psi_2\|^+_{t_0,\gamma}\left(\frac{1}{\gamma - \alpha_+} + \frac{1}{\alpha_- - \gamma}\right) \leq \underbrace{\frac{2KL}{\delta}}_{<1}\|\psi_1 - \psi_2\|^+_{t_0,\gamma}.\end{aligned}$$

Hence, $T^+_{t_0}(\cdot,x_0)$ is a contraction on the Banach space $\mathfrak{X}^+_{t_0,\gamma}$, and the contraction mapping theorem implies that $T^+_{t_0}(\cdot,x_0)$ has a uniquely determined fixed point $\psi^*(t_0,x_0) \in \mathfrak{X}^+_{t_0,\gamma}$. Note that this fixed point does not depend on γ, since $\mathfrak{X}^+_{t_0,\gamma} \subset \mathfrak{X}^+_{t_0,\alpha_- - \delta}$.

Part 3. Let $(t_0, x_0) \in \mathbb{R} \times \mathbb{R}^d$. It is shown now that a function $\psi \in \mathfrak{X}^+_{t_0,\gamma}$ is a solution to (6.1) fulfilling $P_+(t_0)\psi(t_0) = P_+(t_0)x_0$ if and only if ψ is a fixed point of $T^+_{t_0}(\cdot, x_0)$.
($\Rightarrow$) Let $\psi \in \mathfrak{X}^+_{t_0,\gamma}$ be a solution to (6.1) fulfilling $P_+(t_0)\psi(t_0) = P_+(t_0)x_0$. Then ψ also solves the linear inhomogeneous equation

$$\dot{x} = B(t)x + F(t, \psi(t)) .$$

One easily verifies that the inhomogeneity $F(\cdot, \psi(\cdot))$ lies in $\mathfrak{X}^+_{t_0,\gamma}$. Proposition 6.4 (i) now yields that this equation has a uniquely determined γ^+-exponentially bounded solution satisfying the projection condition mentioned in Proposition 6.4 (i). It turns out that ψ is exactly this solution, and Proposition 6.4 (i) also implies the representation as Lyapunov–Perron integral

$$\psi(t) = \Phi(t, t_0)P_+(t_0)x_0 + \int_{t_0}^{\infty} G(t, s)F(t, \psi(t))\, \mathrm{d}s \quad \text{for all } t \in \mathbb{R} .$$

This, however, means that ψ is a fixed point of $T^+_{t_0}(\cdot, x_0)$.
($\Leftarrow$) Conversely, suppose that ψ is a fixed point of $T^+_{t_0}(\cdot, x_0)$. Analogously to Exercise 6.5, one shows that ψ then is a solution to (6.1), and the relation $P_+(t_0)\psi(t_0) = P_+(t_0)x_0$ follows immediately.
In particular, this part showed that $\psi^*(t_0, x_0)$ is the uniquely determined γ^+-exponentially bounded solution of (6.1) fulfilling $P_+(t_0)\psi^*(t_0, x_0)(t_0) = P_+(t_0)x_0$.
Part 4. The above part motivates the definition of

$$w^+(t, x) := P_-(t)\psi^*(t, x)(t) ,$$

which immediately implies that $w^+(t, x) \in \mathfrak{R}P_-(t)$, and one has

$$\begin{aligned} w^+(t, x) &= P_-(t)\psi^*(t, x)(t) = P_-(t)\psi^*(t, P_+(t)x)(t) \\ &= w^+(t, P_+(t)x) . \end{aligned}$$

Moreover, $w^+(t, 0) = 0$, since the trivial solution is γ^+-exponentially bounded. Now the representation (6.3) will be proved.
($\subset$) Let $(t, x + w^+(t, x)) \in \mathcal{W}^+$ with $x \in \mathfrak{R}(P_+(t))$. Then Part 3 implies that $x + w^+(t, x) = P_+(t)x + P_+(t)\psi^*(t, x)(t) = P_+(t)\psi^*(t, x)(t) + P_+(t)\psi^*(t, x)(t) = \psi^*(t, x)(t)$. Hence, it follows that $\phi(\cdot, t, x) = \psi^*(t, x)(\cdot)$, which means that $\phi(\cdot, t, x)$ is γ^+-exponentially bounded.
($\supset$) Let $(t, x) \in \mathbb{R} \times \mathbb{R}^d$ such that $\varphi(\cdot, t, x)$ is γ^+-exponentially bounded. Then Part 3 implies that $\varphi(\cdot, t, x) = \psi^*(t, P_+(t)x)(\cdot)$, in particular $x = \psi^*(t, P_+(t)x)(t) = P_+(t)\psi^*(t, P_+(t)x)(t) + P_-(t)\psi^*(t, P_+(t)x)(t) = P_+(t)x + w^+(t, P_+(t)x)$. This implies that $(t, x) \in \mathcal{W}^+$.
Part 5. Finally, the estimate for w^+ is proved in this last part. Firstly, note that

$$\big\| T^+_{t_0}(\psi, x_1) - T^+_{t_0}(\psi, x_2) \big\|^+_{t_0,\gamma} \le K \big\| x_1 - x_2 \big\|$$

for all $t \ge t_0$, $\psi \in \mathfrak{X}^+_{t_0,\gamma}$ and $x_1, x_2 \in \mathbb{R}^d$, which implies that

$$\begin{aligned}
&\left\|\psi^*(t_0,x_1) - \psi^*(t_0,x_2)\right\|^+_{t_0,\gamma} \\
&= \left\|T^+_{t_0}(\psi^*(t_0,x_1),x_1) - T^+_{t_0}(\psi^*(t_0,x_2),x_2)\right\|^+_{t_0,\gamma} \\
&\le \left\|T^+_{t_0}(\psi^*(t_0,x_1),x_1) - T^+_{t_0}(\psi^*(t_0,x_1),x_2)\right\|^+_{t_0,\gamma} \\
&\quad + \left\|T^+_{t_0}(\psi^*(t_0,x_1),x_2) - T^+_{t_0}(\psi^*(t_0,x_2),x_2)\right\|^+_{t_0,\gamma} \\
&\le K\|x_1 - x_2\| + \frac{2KL}{\delta}\|\psi^*(t_0,x_1) - \psi^*(t_0,x_1)\|^+_{t_0,\gamma}\,.
\end{aligned}$$

This means that

$$\left\|\psi^*(t_0,x_1) - \psi^*(t_0,x_2)\right\|^+_{t_0,\gamma} \le \frac{K\delta}{\delta - 2KL}\|x_1 - x_2\|\,.$$

In addition, one has

$$\left\|P_-(t_0)(T^+_{t_0}(t_0,\psi,x_1) - T^+_{t_0}(t_0,\psi,x_2))\right\| = 0$$

for all $t \ge t_0$, $\psi \in \mathfrak{X}^+_{t_0,\gamma}$ and $x_1, x_2 \in \mathbb{R}^d$, and this implies that

$$\begin{aligned}
&\left\|P_-(t_0)(\psi^*(t_0,t_0,x_1) - \psi^*(t_0,t_0,x_2))\right\| \\
&= \left\|P_-(t_0)(T^+_{t_0}(t_0,\psi^*(t_0,x_1),x_1) - T^+_{t_0}(t_0,\psi^*(t_0,x_2),x_2))\right\| \\
&\le \left\|P_-(t_0)(T^+_{t_0}(t_0,\psi^*(t_0,x_1),x_1) - T^+_{t_0}(t_0,\psi^*(t_0,x_1),x_2))\right\| \\
&\quad + \left\|P_-(t_0)(T^+_{t_0}(t_0,\psi^*(t_0,x_1),x_2) - T^+_{t_0}(t_0,\psi^*(t_0,x_2),x_2))\right\| \\
&\le 0 + KL\frac{\|\psi^*(t_0,x_1) - \psi^*(t_0,x_2)\|^+_{t_0,\gamma}}{\alpha_- - \gamma} \\
&\le \frac{KL}{\delta}\|\psi^*(t_0,x_1) - \psi^*(t_0,x_2)\|^+_{t_0,\gamma}\,.
\end{aligned}$$

Finally, this means that

$$\begin{aligned}
&\left\|w^+(t_0,x_1) - w^+(t_0,x_2)\right\| \\
&= \left\|P_-(t_0)(\psi^*(t_0,t_0,x_1) - \psi^*(t_0,t_0,x_2))\right\| \\
&\le \frac{KL}{\delta}\|\psi^*(t_0,x_1) - \psi^*(t_0,x_2)\|^+_{t_0,\gamma} \\
&\le \frac{KL}{\delta}\frac{K\delta}{\delta - 2KL}\|x_1 - x_2\| = \frac{K^2L}{\delta - 2KL}\|x_1 - x_2\|\,.
\end{aligned}$$

The claimed estimate follows from the limit $\delta \to \frac{1}{2}(\alpha_- - \alpha_+)$. This finishes the proof of the theorem. □

REMARK 6.6 (Differentiability of invariant manifolds). It was proved in this section that the invariant manifolds of the Lipschitz continuous differential equation (6.1) are given as graphs of Lipschitz continuous functions. Even if the function F of the right hand side of (6.1) is a C^m-function, the manifolds $\mathcal{W}^\pm$ are only C^1 in general. If, however, the so-called gap condition

$$m\alpha_+ < \alpha_-$$

holds, then the pseudo-stable manifold $\mathcal{W}^+$ is C^m, and if the gap condition

$$\alpha_+ < m\alpha_-$$

is fulfilled, then the pseudo-unstable manifold $\mathcal{W}^-$ is C^m.

REMARK 6.7 (Inertial manifolds). Two important pillars of this book are *attractors* and *invariant manifolds.* The connection between these two is given by *inertial manifolds*, which are invariant manifolds of finite dimension for evolutionary equations. The importance of inertial manifolds is given by the facts that they allow a reduction to finite-dimensional dynamics and that global attractors are contained in inertial manifolds. The treatment of inertial manifolds is out of the scope of this book; see FOIAS, SELL & TEMAM [**80**] and SELL & YOU [**220**] for the autonomous theory, and CHEPYZHOV & GORITSKY [**42**] and KOKSCH & SIEGMUND [**148**] for the nonautonomous case, and see JONES & STUART [**103**], PÖTZSCHE [**185, 186**] for discrete/discretization theory.

2. Local invariant manifolds

The invariant manifold theorem of the first section of this chapter is directly applicable only in rare cases due to the strong global Lipschitz condition. It will be described in this section how a cut-off function helps to prove the existence of invariant manifolds in the vicinity of an arbitrary solution, if the linearization along it admits an exponential dichotomy. For simplicity, assume that the solution under consideration is given by the trivial solution (which can be reached via the equation of perturbed motion, see Subsection 1.3 of Chapter 2).

Consider a nonlinear ordinary differential equation

$$\dot{x} = B(t)x + F(t,x) \tag{6.5}$$

with a continuous function $B : \mathbb{R} \to \mathbb{R}^{d\times d}$ and a C^1-function $F : \mathbb{R} \times \mathbb{R}^d \to \mathbb{R}^d$ such that $F(t,0) = 0$ for all $t \in \mathbb{R}$. The process induced by (6.5) will be denoted by ϕ. In addition to (6.5), consider the corresponding linear ordinary differential equation, given by

$$\dot{x} = B(t)x\,, \tag{6.6}$$

whose transition operator is denoted by $\Phi : \mathbb{R} \times \mathbb{R} \to \mathbb{R}^{d\times d}$.

DEFINITION 6.8 (Local nonautonomous invariant manifolds). Let $P_\pm : \mathbb{R} \to \mathbb{R}^{d\times d}$ be complementary invariant projectors of (6.6), and consider a neighborhood U of 0 and continuous functions $w^\pm : \mathbb{R} \times U \to \mathbb{R}^d$ satisfying

(i) $w^\pm(t,0) = 0$ for all $t \in \mathbb{R}$,
(ii) $\lim_{x\to 0} \frac{w^\pm(t,x)}{\|x\|} = 0$ uniformly in $t \in \mathbb{R}$,
(iii) $w^\pm(t,x) = w^\pm\big(t, P_\pm(t)x\big) \in \mathfrak{R}(P_\mp(t))$ for all $t \in \mathbb{R}$ and $x \in U$.

Then the graphs

$$\mathcal{W}^\pm := \big\{\big(\tau, \xi + w^\pm(\tau,\xi)\big) \in \mathbb{R} \times \mathbb{R}^d : \xi \in \mathfrak{R}(P_\pm(\tau)) \cap U\big\}$$

are called *local nonautonomous invariant manifolds* of (6.5) if

$$\big(t, \phi(t,\tau,\xi)\big) \in \mathcal{W}^\pm \qquad \begin{array}{l}\text{for all } (\tau,\xi) \in \mathcal{W}^\pm \text{ and } t \in \mathbb{R} \text{ such that}\\ \phi\big(\tau + c(t-\tau), \tau, \xi\big) \in U \text{ for all } c \in [0,1]\,.\end{array}$$

To obtain local nonautonomous invariant manifolds, the following hypothesis is needed. Note that the conditions on the linear part are the same as in Hypothesis 6.2.

HYPOTHESIS 6.9. Suppose that the following two statements hold.

(H1) *Hypothesis on linear part.* The linear system (6.2) admits an exponential dichotomy on $\mathbb{R}$, i.e., there exist complementary invariant projectors $P_\pm : \mathbb{R} \to \mathbb{R}^{d\times d}$ such that the inequalities

$$\|\Phi(t,t_0)P_+(t_0)\| \le Ke^{\alpha_+(t-t_0)} \quad \text{for all } t \ge t_0\,,$$

$$\|\Phi(t,t_0)P_-(t_0)\| \le Ke^{\alpha_-(t-t_0)} \quad \text{for all } t \le t_0$$

hold with real constants $K \ge 1$ and $\alpha_+ < \alpha_-$.

(H2) *Hypothesis on nonlinearity.* The limit relation

$$\lim_{x\to 0} \sup_{t\in\mathbb{R}} \|D_2F(t,x)\| = 0$$

holds.

The following theorem states the existence of local invariant manifolds under this hypothesis.

THEOREM 6.10. *Consider the nonautonomous differential equation* (6.5), *and assume that both* (H1) *and* (H2) *from Hypothesis 6.9 hold. Then there exists a neighborhood U of* 0 *and continuous functions $w^\pm : \mathbb{R}\times U \to \mathbb{R}^d$ such that the graphs*

$$\mathcal{W}^\pm := \big\{(t, x + w^\pm(t,x)) \in \mathbb{R}\times\mathbb{R}^d : x \in U \cap \mathfrak{R}(P_\pm(t))\big\}$$

are local nonautonomous invariant manifolds of (6.5).

PROOF. The proof is an application of the global case which was discussed in Theorem 6.3, and one needs to use a suitable cut-off function such that the local condition (H2) from Hypothesis 6.9 passes over to the global condition (H2) from Hypothesis 6.2. Fix such a smooth cut-off function $\chi : \mathbb{R}^d \to [0,1]$ with

$$\chi(x) = 1 \text{ for all } x \text{ with } \|x\| \le 1 \quad \text{and} \quad \chi(x) = 0 \text{ for all } x \text{ with } \|x\| \ge 2$$

(see, e.g., ABRAHAM, MARSDEN & RATIU [**1**, Lemma 4.2.13]). For any $\sigma > 0$, define the function $F_\sigma : \mathbb{R}\times\mathbb{R}^d \to \mathbb{R}^d$ by

$$F_\sigma(t,x) := \chi\big(\tfrac{x}{\sigma}\big)F(t,x) \quad \text{for all } (t,x) \in \mathbb{R}\times\mathbb{R}^d\,.$$

Due to the mean value inequality, the relation $F(\cdot,0) \equiv 0$ leads to

$$\|F(t,x)\| \le \|x\| \sup_{s\in[0,1]} \|D_2F(t,sx)\| \quad \text{for all } x \in \mathbb{R}^d \text{ and } t \in \mathbb{R}\,.$$

Since $D_2F_\sigma(t,x) = \chi\big(\frac{x}{\sigma}\big)D_2F(t,x) + \frac{1}{\sigma}D\chi\big(\frac{x}{\sigma}\big)F(t,x)$, one has for all $t\in\mathbb{R}$ that

$$\begin{aligned}&\sup_{x\in\mathbb{R}^d} \|D_2F_\sigma(t,x)\|\\ &\le \sup_{\|x\|\le 2\sigma} \|D_2F(t,x)\| + \frac{1}{\sigma} \sup_{\|x\|\le 2} D\chi(x) \sup_{\|x\|\le 2\sigma} \left(\|x\| \sup_{s\in[0,1]} \|D_2F(t,sx)\|\right).\end{aligned}$$

Hence,

$$\sup_{(t,x)\in\mathbb{R}\times\mathbb{R}^d} \|D_2F_\sigma(t,x)\| \le \left(1 + 2\sup_{x\in\mathbb{R}^d} D\chi(x)\right) \sup_{\|x\|\le 2\sigma,\, t\in\mathbb{R}} \|D_2F(t,x)\|\,.$$

By (H2), this implies that

$$\lim_{\sigma\to 0} \sup_{(t,x)\in\mathbb{R}\times\mathbb{R}^d} \|D_2F_\sigma(t,x)\| = 0\,, \tag{6.7}$$

and thus, there exists a $\tilde{\rho} > 0$ such that

$$\dot{y} = B(t)y + F_{\tilde{\rho}}(t, y) \tag{6.8}$$

fulfills the (global) hypotheses of Theorem 6.3. This means that there exist continuous functions $\tilde{w}^{\pm} : \mathbb{R} \times \mathbb{R}^d \to \mathbb{R}^d$ such that the graphs

$$\tilde{\mathcal{W}}^{\pm} := \big\{\big(t, x + \tilde{w}^{\pm}(t, x)\big) \in \mathbb{R} \times \mathbb{R}^d : x \in \mathfrak{R}(P_{\pm}(t))\big\}$$

are global nonautonomous invariant manifolds of (6.8). Since the systems (6.5) and (6.8) coincide for $t \in \mathbb{R}$ and $x \in B_{\tilde{\rho}}(0)$, and the functions $w^{\pm}$ are globally Lipschitz continuous (see Theorem 6.3), there exists a $\rho > 0$ such that with the functions

$$w^{\pm} : \mathbb{R} \times B_\rho(0) \to \mathbb{R}^d, \quad w^{\pm}(t, x) := \tilde{w}^{\pm}(t, x) \quad \text{for all } t \in \mathbb{R} \text{ and } x \in B_\rho(0),$$

the sets

$$\mathcal{W}^{\pm} := \big\{\big(t, x + w^{\pm}(t, x)\big) \in \mathbb{R} \times \mathbb{R}^d : x \in \mathfrak{R}(P_{\pm}(t)) \cap U_\rho(0)\big\}$$

are subsets of $\mathbb{R} \times B_{\tilde{\rho}}(0)$.
It remains to show that $\mathcal{W}^{\pm}$ are local nonautonomous invariant manifolds of (6.5) in the sense of Definition 6.8. The conditions (i) and (iii) of this definition follow directly, since $\mathcal{W}^{\pm}$ was constructed from corresponding global invariant manifolds. (ii) also holds, since it was shown in Theorem 6.3 that the functions $\tilde{w}^{\pm}$ describing the global invariant manifolds are globally Lipschitz continuous with the Lipschitz constant $\frac{2K^2L}{\alpha_- - \alpha_+ - 4KL}$ uniformly for all $t \in \mathbb{R}$, where L is the global Lipschitz constant of the nonlinearity. It is easy to see that this expression converges to 0 in the limit $L \to 0$, and due to (6.7), the constant L can be made arbitrarily small, which implies the assertion. □

The following exercise shows that solutions lying in the local manifolds fulfill certain exponential growth conditions.

EXERCISE 6.11. Assume that the hypotheses of Theorem 6.10 are fulfilled. Show that in case $\alpha_+ < 0$, for all $\gamma > \alpha_+$, there exists a $\delta_+ > 0$ such that

$$\phi(\cdot, t, x) \text{ is } \gamma^+\text{-exponentially bounded for all } (t, x) \in \mathcal{W}^+ \text{ with } \|x\| \le \delta_+ .$$

Analogously, in the case $\alpha_- > 0$, show that for all $\gamma < \alpha_-$ there exists a $\delta_- > 0$ such that

$$\phi(\cdot, t, x) \text{ is } \gamma^-\text{-exponentially bounded for all } (t, x) \in \mathcal{W}^{\pm} \text{ with } \|x\| \le \delta_-$$

holds. Why is a characterization similarly to (6.3) not possible in general for both manifolds?

3. Hierarchies of invariant manifolds

An exponential dichotomy of a linear system is characterized by the growth behavior of solutions lying in two invariant nonautonomous sets, which are given by the range and null space of an invariant projector. In the first section of this chapter, a nonlinear perturbation of this situation was considered, leading to both a pseudo-stable and pseudo-unstable nonautonomous invariant manifold. An exponential dichotomy, however, gives just a coarse spectral description of a linear system, and it was seen in Theorem 5.14 that the whole extended phase space can be obtained as Whitney sum of the spectral manifolds. Under a small nonlinear perturbation, these spectral manifolds persist and yield a hierarchy of invariant manifolds. Firstly, this will be shown locally.

Consider a nonlinear ordinary differential equation

$$\dot{x} = B(t)x + F(t, x) \tag{6.9}$$

with a continuous function $B : \mathbb{R} \to \mathbb{R}^{d\times d}$ and a C^1-function $F : \mathbb{R} \times \mathbb{R}^d \to \mathbb{R}^d$ such that $F(t, 0) = 0$ for all $t \in \mathbb{R}$. The process induced by (6.9) will be denoted by ϕ. In addition to (6.9), consider the corresponding linear ordinary differential equation, given by

$$\dot{x} = B(t)x\,, \tag{6.10}$$

whose transition operator is denoted by $\Phi : \mathbb{R} \times \mathbb{R} \to \mathbb{R}^{d\times d}$. In the following, Hypothesis 6.9 is extended in order to describe the spectral situation of the linearization in more detail.

HYPOTHESIS 6.12. Suppose that the following two statements hold.

(H1) *Hypothesis on linear part.* The dichotomy spectrum of (6.2) consists of at least two spectral intervals, i.e., is given by

$$\Sigma^{\leftrightarrow} = [a_1, b_1] \cup \cdots \cup [a_n, b_n]\,,$$

where $d \geq n \geq 2$.

(H2) *Hypothesis on nonlinearity.* The limit relation

$$\lim_{x\to 0} \sup_{t\in\mathbb{R}} \|D_2F(t, x)\| = 0$$

holds.

The hypothesis on the linear part (H1) implies that there exist real constants $K \geq 1$ and $\alpha_+^i < \alpha_-^i \in [b_i, a_{i+1}]$ for $i \in \{1, \ldots, n-1\}$ and complementary invariant projectors $P_\pm^i : \mathbb{R} \to \mathbb{R}^{d\times d}$ for $i \in \{1, \ldots, n-1\}$ such that the inequalities

$$\|\Phi(t, t_0)P_+^i(t_0)\| \leq Ke^{\alpha_+^i(t-t_0)} \quad \text{for all } t \geq t_0\,,$$
$$\|\Phi(t, t_0)P_-^i(t_0)\| \leq Ke^{\alpha_-^i(t-t_0)} \quad \text{for all } t \leq t_0$$

hold for $i \in \{1, \ldots, n-1\}$. Hence, one can apply Theorem 6.10 to obtain local nonautonomous invariant manifolds $\mathcal{W}_i^\pm$ for $i \in \{1, \ldots, n-1\}$. Motivated by the intersections of the ranges and null spaces of invariant projectors taken in Theorem 5.14, one defines

$$\mathcal{W}_{i,j} := \mathcal{W}_i^+ \cap \mathcal{W}_j^- \quad \text{for all } 1 \leq j < i \leq n-1\,.$$

The invariant sets obtained by this procedure form the hierarchy of nonautonomous invariant manifolds:

$$
\begin{array}{ccccccccc}
\mathcal{W}_1^+ & \subset & \mathcal{W}_2^+ & \subset & \cdots & \subset & \mathcal{W}_{n-1}^+ & \subset & \mathbb{R}\times\mathbb{R}^d \\
 & & \cup & & & & \cup & & \cup \\
 & & \mathcal{W}_{2,1} & \subset & \cdots & \subset & \mathcal{W}_{n-1,1} & \subset & \mathcal{W}_1^- \\
 & & & & & & \cup & & \cup \\
 & & & & \ddots & & \vdots & & \vdots \\
 & & & & & & \mathcal{W}_{n-1,n-2} & \subset & \mathcal{W}_{n-2}^- \\
 & & & & & & & & \cup \\
 & & & & & & & & \mathcal{W}_{n-1}^-
\end{array}
$$

As already defined above, the manifolds $\mathcal{W}_i^+$, $i \in \{1,\dots,n-1\}$, are called pseudo-stable manifolds, and $\mathcal{W}_i^-$, $i \in \{1,\dots,n-1\}$, are the pseudo-unstable manifolds. In the hyperbolic case $0 \notin \Sigma^\pm$, choose i maximal with $b_i < 0$. Then $\mathcal{W}_i^+$ is called *stable manifold*, $\mathcal{W}_i^-$ is called *unstable manifold*, and the sets $\mathcal{W}_j^+$ for $1 \le j < i$ and $\mathcal{W}_j^-$ for $i < j \le n-1$ are called *strongly stable* and *strongly unstable manifolds*, respectively. In the nonhyperbolic case $0 \in \Sigma^\pm$, choose i minimal with $b_i > 0$. Then $\mathcal{W}_{i,i-1}$ is called *center manifold*, $\mathcal{W}_i^+$ is called *center-stable manifold*, and $\mathcal{W}_{i-1}^-$ is called *center-unstable manifold*.

The same construction can of course be done in the global case, where (H2) of Hypothesis 6.12 is replaced by

(H2') *Hypothesis on nonlinearity.* The nonlinear part is globally Lipschitz continuous such that for some $L < \frac{1}{4K}\min_{i\in\{1,\dots,n\}}(\alpha_-^i - \alpha_+^i)$, one has

$$\|F(t,x) - F(t,y)\| \le L\|x-y\| \quad \text{for all } t \in \mathbb{R} \text{ and } x,y \in \mathbb{R}^d\,.$$

Then the invariant manifolds of the hierarchy can be characterized by exponential growth rates as the following exercise shows.

EXERCISE 6.13. Suppose that (H1) and (H2') is fulfilled. Show that the (global) invariant manifolds of the hierarchy fulfill the following dynamical characterizations:

(i) For all $i \in \{1,\dots,n-1\}$, one has

$$\mathcal{W}_i^+ = \big\{(t,x) \in \mathbb{R}\times\mathbb{R}^d : \phi(\cdot,t,x) \text{ is } \tfrac{1}{2}(\alpha_+^i + \alpha_-^i)^+\text{-exponentially bounded}\big\}\,,$$
$$\mathcal{W}_i^- = \big\{(t,x) \in \mathbb{R}\times\mathbb{R}^d : \phi(\cdot,t,x) \text{ is } \tfrac{1}{2}(\alpha_+^i + \alpha_-^i)^-\text{-exponentially bounded}\big\}\,.$$

(ii) For all $1 \le j < i \le n-1$, one has

$$\mathcal{W}_{i,j} = \big\{(t,x) : \phi(\cdot,t,\xi) \text{ is } \tfrac{1}{2}(\alpha_+^i + \alpha_-^i)^+\text{- and } \tfrac{1}{2}(\alpha_+^j + \alpha_-^j)^-\text{-exponentially bounded}\big\}\,.$$

4. Taylor approximation

The explicit computation of invariant manifolds is possible only in rare cases. Nevertheless, in many situations, it suffices to know only their Taylor approximation up to a certain order, like, e.g., in bifurcation theory or to apply the center manifold reduction.

Consider a nonlinear ordinary differential equation

$$\dot{x} = B(t)x + F(t,x) \tag{6.11}$$

with a continuous function $B : \mathbb{R} \to \mathbb{R}^{d\times d}$ and a C^1-function $F : \mathbb{R} \times \mathbb{R}^d \to \mathbb{R}^d$ such that $F(t,0) = 0$ for all $t \in \mathbb{R}$. The process induced by (6.11) will be denoted by ϕ. In addition to (6.11), consider the corresponding linear ordinary differential equation, given by

$$\dot{x} = B(t)x\,, \tag{6.12}$$

whose transition operator is denoted by $\Phi : \mathbb{R} \times \mathbb{R} \to \mathbb{R}^{d\times d}$.

For an n-tuple of the same vector $x \in \mathbb{R}^d$, the abbreviation $x^n := (x, \dots, x) \in \mathbb{R}^{dn}$ will be used. Given $n \in \mathbb{N}$ and a linear subspace $V \subset \mathbb{R}^d$, the space $\mathcal{L}_n(\mathbb{R}^d, V)$ is the space of symmetric n-linear mappings from $(\mathbb{R}^d)^n$ to V, and write $\mathcal{L}_n(\mathbb{R}^d) := \mathcal{L}_n(\mathbb{R}^d, \mathbb{R}^d)$ and $\mathcal{L}(\mathbb{R}^d) := \mathcal{L}_1(\mathbb{R}^d)$; all these spaces are equipped with their canonical norm. For a mapping $X \in \mathcal{L}_n(\mathbb{R}^d)$, abbreviate $Xx_1 \cdots x_n := X(x_1, \dots, x_n)$, and with a matrix $T \in \mathbb{R}^{d\times d}$, define $X_T \in \mathcal{L}_n(\mathbb{R}^d)$ by

$$X_T x_1 \cdots x_n := X(Tx_1, \dots, Tx_n) \quad \text{for all } x_1, \dots, x_n \in \mathbb{R}^d,$$

and obtain the norm estimate (cf. ABRAHAM, MARSDEN & RATIU [**1**, p. 62])

$$\|X_T\| \le \|T\|^n \|X\| \quad \text{for all } n \in \mathbb{N}\,. \tag{6.13}$$

The mapping $F : \mathbb{R} \times \mathbb{R}^d \to \mathbb{R}^d$ is said to be *uniformly bounded* if it maps bounded subsets of $\mathbb{R}^d$ into bounded sets uniformly in $\mathbb{R}$, i.e., if for any bounded $V \subset \mathbb{R}^d$, there exists an $M \ge 0$ such that $\|F(t,x)\| \le M$ for all $t \in \mathbb{R}$ and $x \in V$. For integers $m \ge 0$, the function F is said to be *uniformly C^m-bounded*, if $D_2^m F$ is uniformly bounded and the functions $D_2^n F(\cdot, 0) : \mathbb{R} \to \mathcal{L}_n(\mathbb{R}^d)$ are bounded for $n \in \{1, \dots, m-1\}$. The mean value theorem then implies the uniform boundedness of $D_2^n F$ for $n \in \{1, \dots, m\}$.

HYPOTHESIS 6.14. Suppose that for some fixed $m \in \mathbb{N}$, the following statements holds.

(H1) *Hypothesis on linear part.* Suppose that the function $A : \mathbb{R} \to \mathbb{R}^{d\times d}$ is bounded and that the linear system (6.2) admits an exponential dichotomy on $\mathbb{R}$, i.e., there exist complementary invariant projectors $P_\pm : \mathbb{R} \to \mathbb{R}^{d\times d}$ such that the inequalities

$$\|\Phi(t,t_0)P_+(t_0)\| \le Ke^{\alpha_+(t-t_0)} \quad \text{for all } t \ge t_0\,,$$
$$\|\Phi(t,t_0)P_-(t_0)\| \le Ke^{\alpha_-(t-t_0)} \quad \text{for all } t \le t_0$$

hold with real constants $K \ge 1$ and $\alpha_- < \alpha_+$.

(H2) *Hypothesis on nonlinearity.* F is m-times continuously differentiable in the second argument, $F(t,0) \equiv 0$ on $\mathbb{R}$, the limit relation

$$\lim_{x\to 0} \sup_{t\in\mathbb{R}} \|D_2 F(t,x)\| = 0 \tag{6.14}$$

holds and F is uniformly C^m-bounded.

Due to Theorem 6.10, the hypotheses (H1)–(H2) imply the existence of local nonautonomous invariant manifolds for (6.11), i.e., there exists a neighborhood U of 0 and continuous functions $w^\pm : \mathbb{R} \times U \to \mathbb{R}^d$ such that the graphs

$$\mathcal{W}^\pm := \big\{(t, x + w^\pm(t,x)) \in \mathbb{R} \times \mathbb{R}^d : x \in U \cap \mathfrak{R}(P_\pm(t))\big\}$$

are local invariant manifolds. It is easy to see that local integral manifolds satisfy the following nonlinear first order partial differential equation, named as *invariance equation*

$$\begin{aligned} B(t)w^{\pm}(t,x)+&P_{\mp}(t)F(t,x+w^{\pm}(t,x)) \\ &=D_2w^{\pm}(t,x)(B(t)x+P_{\pm}(t)F(t,x+w^{\pm}(t,x)))+D_1w^{\pm}(t,x) \end{aligned} \tag{6.15}$$

for all $t \in \mathbb{R}$ and $x \in \mathfrak{R}(P_{\pm}(t)) \cap U$. Note that, although the nonlinearity is C^m-smooth with respect to the second variable, it depends on the gap conditions if the function $w^{\pm}$ describing the local invariant manifold is smooth (see Remark 6.6).

Since the nonautonomous invariant manifold $\mathcal{W}^{\pm}$ is the graph of a function $w^{\pm}$ (which can be assumed to be sufficiently smooth in its state space variable, provided the gap conditions are fulfilled), it is natural to approximate $w^{\pm}$ by its Taylor expansion. Note that the Taylor expansion is unique (and does not depend on the cut-off function used in the proof of Theorem 6.10), see PÖTZSCHE & RASMUSSEN [**192**]. In this section, necessary equations will be derived that the corresponding Taylor coefficients need to satisfy, and it will be proved that they are uniquely solvable. Firstly, some results (and notation) on higher order derivatives are needed.

4.1. Higher order derivatives. A version of the Leibniz (product) rule and the chain rule for higher order derivatives will be quoted in this subsection. To achieve a compact and convenient notation, (ordered) partitions of finite sets will be used. These partitions consist of tuples of subsets of a given finite set. More precisely, with $j, l \in \mathbb{N}$, write

$$P_j(l) := \left\{ (N_1,\dots,N_j) \,\middle|\, \begin{array}{l} N_i \subset \{1,\dots,l\} \text{ for } i \in \{1,\dots,j\}, \\ N_1 \cup \ldots \cup N_j = \{1,\dots,l\}, \\ N_i \cap N_k = \emptyset \text{ for } i \neq k,\ i,k \in \{1,\dots,j\} \end{array} \right\}$$

for the *partitions* of $\{1,\dots,l\}$ with *length* j. Moreover, the *ordered partitions* of $\{1,\dots,l\}$ with length j are given by

$$P_j^<(l) := \left\{ (N_1,\dots,N_j) \in P_j(l) \,\middle|\, \begin{array}{l} N_i \neq \emptyset \text{ for } i \in \{1,\dots,j\}, \\ \max N_i < \max N_{i+1} \text{ for } i \in \{1,\dots,j-1\} \end{array} \right\}.$$

Let $\bar{F} : \mathbb{R}^d \to \mathbb{R}^d$ be a smooth mapping. In case of a set $N = \{n_1,\dots,n_k\} \subset \{1,\dots,l\}$ for $k \in \mathbb{N}$, where $k \leq l$, abbreviate $D^k\bar{F}(x)x_N := D^k\bar{F}(x)x_{n_1}\cdots x_{n_k}$ for $x, x_1,\dots,x_l \in \mathbb{R}^d$. Here $\bar{F}$ is assumed to be k-times differentiable.

A proof of the Leibniz rule below can be found in ABRAHAM, MARSDEN & RATIU [**1**, pp. 95–96 and p. 112, Ex. 2.4C]. Here $\#N$ denotes the cardinality of a finite set $N \subset \mathbb{N}$.

LEMMA 6.15 (Leibniz rule). *Let $m, n \in \mathbb{N}$, and consider mappings $f_i : \mathbb{R}^d \to \mathbb{R}^d$, $i \in \{1,\dots,n\}$, which m-times differentiable. Then for a multilinear mapping $M : (\mathbb{R}^d)^n \to \mathbb{R}^d$, also the mapping $g : \mathbb{R}^d \to \mathbb{R}^d$, $g(x) := M(f_1(x),\dots,f_n(x))$ is m-times differentiable, and for $l \in \{1,\dots,m\}$, the derivatives have the representation*

$$D^l g(x)x_1\cdots x_l = \sum_{(N_1,\dots,N_n)\in P_n(l)} M\big(D^{\#N_1}f_1(x)x_{N_1},\dots,D^{\#N_n}f_n(x)x_{N_n}\big)$$

for any $x_1,\dots,x_l \in \mathbb{R}^d$.

The following chain rule can be found in RYBAKOWSKI [**207**, Theorem 2].

LEMMA 6.16 (Chain rule). *Given $m \in \mathbb{N}$ and mappings $f : \mathbb{R}^d \to \mathbb{R}^d$ and $g : \mathbb{R}^d \to \mathbb{R}^d$, which are m-times differentiable. Then the composition $f \circ g : \mathbb{R}^d \to \mathbb{R}^d$ is m-times differentiable, and for $l \in \{1, \dots, m\}$, the derivatives have the representation*

$$D^l(f \circ g)(x)x_1 \cdots x_l = \sum_{j=1}^{l} \sum_{(N_1,\dots,N_j)\in P_j^<(l)} D^j f(g(x)) D^{\#N_1} g(x) x_{N_1} \cdots D^{\#N_j} g(x) x_{N_j}$$

for any $x_1, \dots, x_l \in \mathbb{R}^d$.

4.2. Multilinear theory. A crucial tool in the analysis of Taylor approximations of nonautonomous invariant manifolds will be a particular class of linear ordinary differential equations in spaces of n-linear mappings. This subsection features some preliminaries on this aspect.

As a consequence of the Taylor expansion, linear differential equations in $\mathcal{L}_n(\mathbb{R}^d)$ of the form

$$\dot{X}_{P_\pm(t)} = L_{B(t)} X_{P_\pm(t)} \tag{6.16}$$

need to be analyzed, where the coefficient function $L_T \in \mathcal{L}(\mathcal{L}_n(\mathbb{R}^d))$ is given by

$$(L_T X)x_1 \cdots x_n := TXx_1 \cdots x_n - \sum_{j=1}^{n} Xx_1 \cdots x_{j-1} T x_j x_{j+1} \cdots x_n$$

for $T \in \mathcal{L}(\mathbb{R}^d)$ and vectors $x_1, \dots, x_n \in \mathbb{R}^d$. It is worth mentioning that these equations are not ordinary differential equations of the form (6.12), since the projectors $P_\pm(t)$ are noninvertible in general. It is easy to see that, given $t_0 \in \mathbb{R}$ and initial state $\Xi \in \mathcal{L}_n\big(\mathbb{R}^d, \mathfrak{R}(P_\mp(t_0))\big)$ with $\Xi_{P_\pm(t_0)} = \Xi$,

$$\Lambda_\pm(t, t_0)\Xi := \Phi(t, t_0)\Xi_{\Phi(t_0,t)P_\pm(t)} \quad \text{for all } t \in \mathbb{R} \tag{6.17}$$

defines the uniquely determined solution $\Lambda_\pm(\cdot, t_0)\Xi$ of equation (6.16) satisfying the relation $(\Lambda_\pm(t, t_0)\Xi)_{P_\pm(t)} = \Lambda_\pm(t, t_0)\Xi$ for $t \in \mathbb{R}$.

The following result deals with perturbations of linear systems (6.16) in $\mathcal{L}_n(\mathbb{R}^d)$. With reals γ and a fixed $t_0 \in \mathbb{R}$, a function $\phi : \mathbb{R} \to \mathbb{R}^d$ is said to be *γ-exponentially bounded* if $\|\phi\|_{t_0,\gamma} := \sup_{t\in\mathbb{R}} \|\phi(t)\| e^{\gamma(t_0 - t)} < \infty$ holds; this extends previously used concepts to the entire time axis.

LEMMA 6.17. *Suppose that (H1) is satisfied, let $n \in \mathbb{N}$, $t_0 \in \mathbb{R}$, $\gamma \in \mathbb{R}$, and assume $H^\pm : \mathbb{R} \to \mathcal{L}_n(\mathbb{R}^d)$ is γ-exponentially bounded with $H^\pm(t) \in \mathcal{L}_n\big(\mathbb{R}^d, \mathfrak{R}(P_\mp(t))\big)$ for $t \in \mathbb{R}$. Then for the ordinary differential equation*

$$\dot{X}_{P_\pm(t)} = L_{B(t)} X_{P_\pm(t)} + H^\pm(t)_{P_\pm(t)} \tag{6.18}$$

in $\mathcal{L}_n(\mathbb{R}^d)$, the following holds:

(i) *If $\gamma < \alpha_- - n\alpha_+$, then there exists a unique γ-exponentially bounded solution $\Gamma_+ : \mathbb{R} \to \mathcal{L}_n(\mathbb{R}^d)$ of (6.18) with*

$$\Gamma_+(t) = \Gamma_+(t)_{P_+(t)} \in \mathcal{L}_n\big(\mathbb{R}^d, \mathfrak{R}(P_-(t))\big) \quad \textit{for all } t \in \mathbb{R}, \tag{6.19}$$

given by

$$\Gamma_+(t) := -\int_t^\infty \Phi(t, s) H^+(s)_{\Phi(s,t)P_+(t)} \, \mathrm{d}s \tag{6.20}$$

and satisfying the estimate $\|\Gamma_+\|_{t_0,\gamma} \le \frac{K^{n+1}}{\alpha_- - \gamma + n\alpha_+}\|H^+\|_{t_0,\gamma}$.

(ii) *If* $\gamma > \alpha_+ - n\alpha_-$, *then there exists a unique* γ*-exponentially bounded solution* $\Gamma_- : \mathbb{R} \to \mathcal{L}_n(\mathbb{R}^d)$ *of* (6.18) *with* $\Gamma_-(t) = \Gamma_-(t)_{P_-(t)} \in \mathcal{L}_n(\mathbb{R}^d, \mathfrak{R}(P_+(t)))$ *for all* $t \in \mathbb{R}$, *given by*

$$\Gamma_-(t) := \int_{-\infty}^{t} \Phi(t,s)H^-(s)_{\Phi(s,t)P_-(t)}\,\mathrm{d}s$$

and satisfying the estimate $\|\Gamma_-\|_{t_0,\gamma} \le \frac{K^{n+1}}{\gamma + \alpha_+ - n\alpha_-}\|H^-\|_{t_0,\gamma}$.

PROOF. Only assertion (ii) will be proved, since the proof of (i) is similar. Let $t_0 \in \mathbb{R}$. The proof will be divided in two parts.
Part 1. Firstly, consider the special case $H^+(t) \equiv 0$ on $\mathbb{R}$. Then equation (6.18) coincides with (6.16). Let $\Gamma_+ : \mathbb{R} \to \mathcal{L}_n(\mathbb{R}^d)$ be a γ-exponentially bounded solution of (6.18) satisfying (6.19). Taking the limit $t \to \infty$ in the inequality

$$\begin{aligned}
&\|\Gamma_+(t_0)\| \\
&\overset{(6.17)}{=} \|\Phi(t_0,t)\Gamma_+(t)_{\Phi(t,t_0)P_+(t_0)}\| \overset{(6.19)}{\le} \|\Phi(t_0,t)P_-(t)\|\|\Gamma_+(t)_{\Phi(t,t_0)P_+(t_0)}\| \\
&\overset{(6.13)}{\le} \|\Phi(t_0,t)P_-(t)\|\|\Gamma_+(t)\|\|\Phi(t,t_0)P_+(t_0)\|^n \le K^{n+1}e^{(n\alpha_+ + \gamma - \alpha_-)(t-t_0)}\|\Gamma_+\|_{t_0,\gamma}
\end{aligned}$$

for all $t \ge t_0$ yields $\Gamma_+(t_0) = 0$. Since $t_0 \in \mathbb{R}$ was arbitrary, the zero solution of (6.18) is the only γ-exponentially bounded solution satisfying (6.19).
Part 2. Omit now the restriction on H^+ and note that the function Γ_+ from (6.20) is well-defined, since the estimate

$$\begin{aligned}
\|\Gamma_+(t)\| &\overset{(6.20)}{\le} \int_t^\infty \|\Phi(t,s)P_-(s)H^+(s)_{\Phi(s,t)P_+(t)}\|\,ds \\
&\overset{(6.13)}{\le} \int_t^\infty \|\Phi(t,s)P_-(s)\|\|H^+(s)\|\|\Phi(s,t)P_+(t)\|^n\,ds \\
&\le K^{n+1}e^{\gamma(t-t_0)}\|H^+\|_{t_0,\gamma}\int_t^\infty e^{(s-t)(\gamma+n\alpha_+-\alpha_-)}\,ds \\
&= \frac{K^{n+1}}{\alpha_- - n\alpha_+ - \gamma}\|H^+\|_{t_0,\gamma}e^{\gamma(t-t_0)} \quad \text{for all } t \in \mathbb{R}
\end{aligned}$$

holds, which in turn yields the claimed estimate for $\|\Gamma_+\|_{t_0,\gamma}$. Moreover, it is easy to see that Γ_+ satisfies (6.19). Γ_+ is a solution of (6.18), since the Leibniz rule (Lemma 6.15) yields

$$\begin{aligned}
&\dot\Gamma_+(t)_{P_+(t)} \equiv \Phi(t,t)H^+(t)_{\Phi(t,t)P_+(t)} - B(t)\int_t^\infty \Phi(t,s)H^+(s)_{\Phi(s,t)P_+(t)}\,\mathrm{d}s \\
&-\sum_{j=1}^n \int_t^\infty \Phi(t,s)H^+(s)\big(\Phi(s,t)P_+(t),\dots,\underbrace{\frac{\mathrm{d}}{\mathrm{d}t}(\Phi(s,t)P_+(t))}_{j\text{th position}},\dots,\Phi(s,t)P_+(t)\big)\,\mathrm{d}s \\
&\equiv H^+(t)_{P_+(t)} + B(t)\Gamma_+(t) - \sum_{j=1}^n \Gamma_+(t)(\mathbb{1},\dots,\underbrace{B(t)}_{j\text{th position}},\dots,\mathbb{1}) \\
&\equiv L_{B(t)}\Gamma_+(t)_{P_+(t)} + H^+(t)_{P_+(t)} \quad \text{on } \mathbb{R}\,.
\end{aligned}$$

Finally, the uniqueness statement results from Part 1, because the difference of any two γ-exponentially bounded solutions of (6.18) is a γ-exponentially bounded solution of (6.16) and therefore identically vanishing. □

4.3. Taylor expansion. In addition to (H1)–(H2), suppose that the following assumption is satisfied, which in particular holds if $B : \mathbb{R} \to \mathbb{R}^{d\times d}$ is a C^m-mapping and $F : \mathbb{R} \times \mathbb{R}^d \to \mathbb{R}^d$ is of class C^{m+1}.

HYPOTHESIS 6.18. Suppose that in addition to (H1)–(H2), the following statement holds.

(H3) $w^\pm$ is m-times continuously differentiable with respect to the second variable, and the partial derivatives $D_2^n w^\pm(\cdot, 0)$ are differentiable for $n \in \{2, \dots, m\}$.

Note that the differentiability of $w^\pm$ can be guaranteed under the gap conditions as described in Remark 6.6. The mapping $w^\pm$ will be approximated via a Taylor expansion in the following, yielding the representation

$$w^\pm(t,x) = \sum_{n=2}^{m} \frac{1}{n!} w_n^\pm(t)x^n + R_m^\pm(t,x) \tag{6.21}$$

with coefficient functions $w_n^\pm : \mathbb{R} \to \mathcal{L}_n(\mathbb{R}^d)$ given by $w_n^\pm(t) := D_2^n w^\pm(t,0)$ and a remainder $R_m^\pm$ satisfying $\lim_{x\to 0} R_m^\pm(x,t)/\|x\|^m = 0$. It can be shown (see [**192**]) that the functions $w_n^\pm$ are bounded, i.e., one has $\|w_n^\pm(t)\| \le \gamma_n$ for $t \in \mathbb{R}$ and $n \in \{2, \dots, m\}$. Before proceeding, some handy notational preparations are needed:

(i) The function $W^\pm : \mathbb{R} \times U \to \mathbb{R}^d$, $W^\pm(t,x) := P_+(t)x + w^\pm(t,x)$ satisfies

$$D_2 W^\pm(t,0) = P_\pm(t) \quad \text{and} \quad D_2^n W^\pm(t,0) = D_2^n w^\pm(t,0) \quad \text{for all } t \in \mathbb{R}$$

and $n \in \{2, \dots, m\}$. Hence, for $W_n^\pm(t) := D_2^n W^\pm(t,0)$, the estimates

$$\|W_1^\pm(t)\| \le K_\pm \quad \text{and} \quad \|W_n^\pm(t)\| \le \gamma_n \quad \text{for all } n \in \{2, \dots, m\} \tag{6.22}$$

hold.

(ii) Abbreviate $g^\pm(x,t) := B(t)P_\pm(t)x + P_\pm(t)F(t, W^\pm(x,t))$, and the chain rule from Lemma 6.16 yields that the partial derivatives $g_n^\pm(t) := D_2^n g^\pm(t,0)$ are given by (cf. (6.14))

$$g_1^\pm(t)x_1 = B(t)P_\pm(t)x_1,$$

$$g_n^\pm(t)x_1 \cdots x_n = \sum_{l=2}^{n} \sum_{(N_1,\dots,N_l)\in P_l^<(n)} P_\pm(t) D_2^l F(t,0) W_{\#N_1}^\pm(t)_{P_\pm(t)} x_{N_1} \cdots W_{\#N_l}^\pm(t)_{P_\pm(t)} x_{N_l}$$

for all $x_1, \dots, x_n \in \mathbb{R}^d$ and $n \in \{2, \dots, m\}$. Moreover, (H2)–(H3) and the estimate (6.22) imply that $g_n^\pm : \mathbb{R} \to \mathcal{L}_n(\mathbb{R}^d)$ is a bounded function.

Directly from the invariance equation (6.15), one obtains

$$\begin{aligned} &B(t)w^\pm(t,x) + P_\mp(t)F(t, P_\pm(t)x + w^\pm(t,x)) \\ &\quad = D_2 w^\pm(t,x)(B(t)P_\pm(t)x + P_\pm(t)F(t, P_\pm(t)x + w^\pm(t,x))) + D_1 w^\pm(t,x)\,. \end{aligned}$$

In the notation introduced above, this reads as

$$D_1 w^\pm(t,x) = B(t)w^\pm(t,x) + P_\mp(t)F(t, W^\pm(t,x)) - D_2 w^\pm(t,x) g(t,x)$$

for all $t \in \mathbb{R}$ and $x \in U$. By differentiating this identity, using Lemma 6.15 and Lemma 6.16, and setting $x = 0$, one gets the equation

$$\begin{aligned}\dot{w}_n^{\pm}(t)_{P_\pm(t)}x_1 \cdots x_n &= B(t)w_n^{\pm}(t)_{P_\pm(t)}x_1 \cdots x_n \\ &+ P_\mp(t)\sum_{j=2}^{n} \sum_{(N_1,\dots,N_j)\in P_j^<(n)} D_2^j F(t,0)W_{\#N_1}^{\pm}(t)_{P_\pm(t)}x_{N_1} \cdots W_{\#N_j}^{\pm}(t)x_{N_j} \\ &- \sum_{\substack{(N_1,N_2)\in P_2(n)\\ N_1,N_2\neq\emptyset}} w_{\#N_1+1}^{\pm}(t)_{P_\pm(t)}x_{N_1} \cdot g_{\#N_2}^{\pm}(t)_{P_\pm(t)}x_{N_2}\end{aligned}$$

for $n \in \{2,\dots,m\}$ and $x_1,\dots,x_n \in \mathbb{R}^d$. Therefore, the function $w_n^{\pm} : \mathbb{R} \to \mathcal{L}_n(\mathbb{R}^d)$ is a solution of the linear ordinary differential equation

$$\dot{X}_{P_\pm(t)} = L_{B(t)}X_{P_\pm(t)} + H_n^{\pm}(t)_{P_\pm(t)} \tag{6.23}$$

in $\mathcal{L}_n(\mathbb{R}^d)$, which is called *homological equation* for $\mathcal{W}^{\pm}$, with inhomogeneities $H_n^{\pm} : \mathbb{R} \to \mathcal{L}_n(\mathbb{R}^d)$, given by

$$\begin{aligned}H_n^{\pm}(t)x_1 \cdots x_n &:= P_\pm(t)D_2^n F(t,0)x_1 \cdots x_n \\ &+ P_\mp(t)\sum_{j=2}^{n-1} \sum_{(N_1,\dots,N_j)\in P_j^<(n)} D_2^j F(t,0)W_{\#N_1}^{\pm}(t)x_{N_1} \cdots W_{\#N_j}^{\pm}(t)x_{N_j} \\ &- \sum_{\substack{(N_1,N_2)\in P_2(n)\\ 0<\#N_1<n-1\\ N_2\neq\emptyset}} w_{\#N_1+1}^{\pm}(t)x_{N_1} \cdot g_{\#N_2}^{\pm}(t)x_{N_2}.\end{aligned} \tag{6.24}$$

Obviously, one has $H_2^{\pm}(t) = P_\mp(t)D_2^2F(t,0)$, and for $n \in \{3,\dots,m\}$, the values $H_n^{\pm}(t)$ only depend on $w_2^{\pm},\dots,w_{n-1}^{\pm}$. Therefore, (6.23) represents a hierarchy of linear differential equations for the coefficients $w_n^{\pm}$. These equations have to be solved step by step, starting with $n = 2$, and increasing n by 1 at each step. The solutions are given by the following theorem.

THEOREM 6.19. *Suppose (H1)–(H3) hold, and consider the mapping $w^{\pm} : \mathbb{R}\times U \to \mathbb{R}^d$. Then the following holds:*

(i) *The coefficients $w_n^+ : \mathbb{R} \to \mathcal{L}_n(\mathbb{R}^d)$, $n \in \{2,\dots,m\}$, in the Taylor expansion* (6.21) *of the mapping $w^+ : \mathbb{R}\times U \to \mathbb{R}^d$ can be determined recursively from the* Lyapunov–Perron integrals

$$w_n^+(t) = -\int_t^{\infty} \Phi(t,s)H_n^+(s)_{\Phi(s,t)P_+(t)}\,\mathrm{d}s \quad \textit{for all } n \in \{2,\dots,m\}. \tag{6.25}$$

(ii) *The coefficients $w_n^- : \mathbb{R} \to \mathcal{L}_n(\mathbb{R}^d)$, $n \in \{2,\dots,m\}$, in the Taylor expansion* (6.21) *of the mapping $w^- : \mathbb{R}\times U \to \mathbb{R}^d$ can be determined recursively from the* Lyapunov–Perron integrals

$$w_n^-(t) = \int_{-\infty}^{t} \Phi(t,s)H_n^-(s)_{\Phi(s,t)P_-(t)}\,\mathrm{d}s \quad \textit{for all } n \in \{2,\dots,m\}. \tag{6.26}$$

PROOF. In the explanations preceding Theorem 6.19, it became clear that $w_n^{\pm} : \mathbb{R} \to \mathcal{L}_n(\mathbb{R}^d)$ is a bounded solution of the homological equation (6.23). Moreover, it follows from (6.24), (H2) and (6.22) that each inhomogeneity $H_n^{\pm}$ is bounded, i.e., 0-exponentially bounded. Consequently, it follows from Lemma 6.17 that $w_n^{\pm}$ has the claimed appearance. □

REMARK 6.20. For an autonomous ordinary differential equation, the functions (6.25), (6.26) are constant and given as stationary solutions of the homological equation (6.23). Then (6.23) reduces to an algebraic problem as discussed, e.g., in BEYN & KLESS [**22**] and EIROLA & VON PFALER [**73**].

5. Reduction principle

Center manifolds are important in particular for stability and bifurcation theory. In this section, it will be shown that a stability analysis of a particular solution can be reduced in dimension provided an unstable manifold exists. In such a case, the stability is then determined by the equation reduced to the center manifold. This so-called *reduction principle* goes back to KELLEY [**107**] and PLISS [**180**].

For simplicity of notation, assume now that the linear part of a system of the form (6.5) is decoupled in a stable and center-unstable part. This can be reached via a so-called *kinematic similarity transformation* (see, e.g., COPPEL [**57**]). Consider a nonlinear ordinary differential equation

$$\begin{aligned} \dot{x}_+ &= B_+(t)x_+ + F_+(t, x_+, x_-) \\ \dot{x}_- &= B_-(t)x_- + F_-(t, x_+, x_-) \end{aligned} \tag{6.27}$$

with continuous functions $B_+ : \mathbb{R} \to \mathbb{R}^{d_+ \times d_+}$ and $B_- : \mathbb{R} \to \mathbb{R}^{d_- \times d_-}$, and C^1-functions $F_+ : \mathbb{R} \times \mathbb{R}^{d_+} \times \mathbb{R}^{d_-} \to \mathbb{R}^{d_+}$ and $F_- : \mathbb{R} \times \mathbb{R}^{d_+} \times \mathbb{R}^{d_-} \to \mathbb{R}^{d_-}$ such that $F_+(t,0,0) = 0$ and $F_-(t,0,0) = 0$ for all $t \in \mathbb{R}$. Assume in addition that the function B_- is bounded. The process induced by (6.27) will be denoted by $\phi = (\phi_+, \phi_-)$. In addition to (6.27), consider the corresponding linear ordinary differential equations, given by

$$\dot{x}_+ = B_+(t)x_+ \quad \text{and} \quad \dot{x}_- = B_-(t)x_- , \tag{6.28}$$

the transition operators of which are denoted by $\Phi_+ : \mathbb{R} \times \mathbb{R} \to \mathbb{R}^{d_+ \times d_+}$ and $\Phi_- : \mathbb{R} \times \mathbb{R} \to \mathbb{R}^{d_- \times d_-}$, respectively.

It will be assumed that the there exists a stable and a center-unstable manifold, which is guaranteed by the following hypothesis.

HYPOTHESIS 6.21. Suppose that the following two statements hold.

(H1) *Hypothesis on linear part.* The linear part of system (6.27) admits an exponential dichotomy on $\mathbb{R}$, i.e., the inequalities

$$\|\Phi_+(t, t_0)\| \le K e^{\alpha_+(t - t_0)} \quad \text{for all } t \ge t_0 ,$$
$$\|\Phi_-(t, t_0)\| \le K e^{\alpha_-(t - t_0)} \quad \text{for all } t \le t_0$$

hold with real constants $K \ge 1$, $\alpha_+ < 0$ and $\alpha_- > \alpha_+$.

(H2) *Hypothesis on nonlinearity.* The limit relation

$$\lim_{(x_+, x_-) \to 0} \sup_{t \in \mathbb{R}} \left(\|D_{(2,3)} F_+(t, x_+, x_-)\| + \|D_{(2,3)} F_-(t, x_+, x_-)\| \right) = 0$$

holds.

Due to this hypothesis, there exist a $\rho_0 > 0$ and a local center-unstable manifold $\mathcal{W}^-$, given as graph

$$\mathcal{W}^- = \left\{\left(t, c(t,x_-), x_-\right) \in \mathbb{R} \times \mathbb{R}^{d_+} \times \mathbb{R}^{d_-} : t \in \mathbb{R} \text{ and } x_- \in B_{\rho_0}(0)\right\}$$

of a differentiable function $c : \mathbb{R} \times B_{\rho_0}(0) \to \mathbb{R}^{d_+}$.

Firstly, observe that the mappings $\omega_+, \omega_-, \omega_c : [0, \rho_0) \to \mathbb{R}$, given by

$$\begin{aligned} \omega_+(\rho) &:= \sup_{(t,x_+,x_-)\in\mathbb{R}\times B_\rho(0)\times B_\rho(0)} \|D_{(2,3)}F_+(t,x_+,x_-)\| \,, \\ \omega_-(\rho) &:= \sup_{(t,x_+,x_-)\in\mathbb{R}\times B_\rho(0)\times B_\rho(0)} \|D_{(2,3)}F_-(t,x_+,x_-)\| \\ \text{and} \qquad \omega_c(\rho) &:= \sup_{(t,x_-)\in\mathbb{R}\times B_\rho(0)} \|D_2 c(t,x_-)\| \end{aligned}$$

are increasing (note that the center manifold is differentiable as outlined in Remark 6.6). It is easy to see that one has the limit relations

$$\lim_{\rho\to 0}\omega_+(\rho) = \lim_{\rho\to 0}\omega_+(\rho) = \lim_{\rho\to 0}\omega_c(\rho) = 0\,,$$

and the mean value theorem implies that for any $\rho \in [0, \rho_0)$, one has

$$\|c(t,x)\| \le \omega_c(\rho)\|x\| \quad \text{for all } t \in \mathbb{R} \text{ and } x \in B_\rho(0)\,. \tag{6.29}$$

The following lemma is a continuous-time version of PÖTZSCHE [**184**], which itself is based on a lemma by PALMER [**172**].

LEMMA 6.22. *Let $\gamma \in (0, \min\{0, \alpha_-\} - \alpha_+)$ and $t_0 < t_1$ be two reals, and choose $\rho \in (0, \rho_0/2)$ such that the estimates*

$$\begin{gathered} \max\{2\omega_-(\rho), \omega_c(2\rho)\} < 1\,, \\ K(\omega_+(\rho) + \omega_c(\rho)\omega_-(\rho)) \le \gamma\,, \\ K\big(\omega_-(2\rho)(1+\omega_c(2\rho)) + 2K\omega_+(\rho)\big) < \alpha_- - \alpha_+ - \gamma \end{gathered} \tag{6.30}$$

hold. Suppose that $\tilde{x} = (\tilde{x}_+, \tilde{x}_-)$ is a solution of (6.27) defined on $[t_0, t_1]$ such that $\|\tilde{x}_+(t)\| + \|\tilde{x}_-(t)\| < \rho$ for all $t \in [t_0, t_1]$, and let $\tilde{y}$ denote the maximal solution of the differential equation

$$\dot{y} = B_-(t)y + F_-(t, c(t,y), y) \tag{6.31}$$

fulfilling the initial condition $\tilde{y}(t_1) = \tilde{x}_-(t_1)$. Then the following statements hold:

(i) *$\tilde{y}$ is defined on the interval $[t_0, t_1]$,*
(ii) *$\|\tilde{y}(t)\| < 2\rho$ for all $t \in [t_0, t_1]$,*
(iii) *one has the estimate*

$$\begin{aligned} &\max\left\{\|\tilde{x}_+(t) - c(t,\tilde{y}(t))\|, \|\tilde{x}_-(t) - \tilde{y}(t)\|\right\} \\ &\le \left(K + \tfrac{1}{2}\right)\left\|\tilde{x}_-(t_0) - c(t_0, \tilde{x}_+(t_0))\right\| e^{(\alpha_+ + \gamma)(t-t_0)} \quad \textit{for all } t \in [t_0, t_1]\,. \end{aligned}$$

PROOF. Consider $z(t) := \tilde{x}_+(t) - c(t, \tilde{x}_-(t))$ for all $t \in [t_0, t_1]$. Applying the invariance equation from (6.15), one obtains

$$\begin{aligned} \dot{z}(t) = B_+(t)z(t) &+ F_+\big(t, \tilde{x}_+(t), \tilde{x}_-(t)\big) - F_+\big(t, c(t,\tilde{x}_-(t)), \tilde{x}_-(t)\big) \\ &- D_2 c(t, \tilde{x}_-(t))\big(F_-\big(t, \tilde{x}_+(t), \tilde{x}_-(t)\big) - F_-\big(t, c(t,\tilde{x}_-(t)), \tilde{x}_-(t)\big)\big)\,, \end{aligned}$$

so z is a solution to an inhomogeneous linear differential equation. The variation of constants formula thus yields

$$z(t) = \Phi_+(t,t_0)z(t_0) + \int_{t_0}^{t} \Phi_+(t,s)\Big(F_+\big(s,\tilde{x}_+(s),\tilde{x}_-(s)\big) - F_+\big(s,c(s,\tilde{x}_-(s)),\tilde{x}_-(s)\big) - D_2c(s,\tilde{x}_-(s))\big(F_-\big(s,\tilde{x}_+(s),\tilde{x}_-(s)\big) - F_-\big(s,c(s,\tilde{x}_-(s)),\tilde{x}_-(s)\big)\big)\Big)\,\mathrm{d}s\,.$$

The mean value inequality implies that $\|c(t,\tilde{x}_-(t))\| \le \omega_c(\rho)\|\tilde{x}_-(t)\| < \rho$. Apply this to obtain the estimate

$$\|z(t)\| \le Ke^{\alpha_+(t-t_0)}\|z(t_0)\| + \int_{t_0}^{t} Ke^{\alpha_+(t-t_0)}(\omega_+(\rho)+\omega_-(\rho)\omega_c(\rho))\|z(s)\|\,\mathrm{d}s$$

for all $t \in [t_0,t_1]$. Define $w(t) := e^{-\alpha_+(t-t_0)}\|z(t)\|$ and apply Gronwall's Lemma to the resulting inequality to obtain

$$\|z(t)\| \le Ke^{\alpha_+ + K(\omega_+(\rho)+\omega_-(\rho)\omega_c(\rho))}\|z(t_0)\|\,.$$

Hence, for $t \in [t_0,t_1]$, it follows that

$$\|\tilde{x}_+(t) - c(t,\tilde{x}_-(t))\| \le Ke^{(\alpha_+ + \gamma)(t-t_0)}\|\tilde{x}_+(t_0) - c(t_0,\tilde{x}_-(t_0))\|\,. \tag{6.32}$$

Let $\tau \in [t_0,t_1]$ be the least number such that $\tilde{x}$ is defined on $[\tau,t_1]$ fulfilling $\|\tilde{x}_+(t)\| + \|\tilde{x}_-(t)\| \le 2\rho$ on $[\tau,t_1]$. Then $u(t) := \tilde{x}_-(t) - \tilde{y}(t)$ solves the linear inhomogeneous differential equation

$$\dot{y} = B_-(t)y + F_-(t,\tilde{x}(t)) - F_-(t,c(t,\tilde{y}(t)),\tilde{y}(t))$$

on $[\tau,t_1]$ and satisfies the initial condition $u(t_1) = 0$. By the variation of constants formula, the solution u satisfies

$$u(t) = -\int_{t}^{t_1} \Phi_-(t,s)\Big(F_-(s,\tilde{x}_+(s),\tilde{x}_-(s)) - F_-(s,c(s,\tilde{x}_-(s)),\tilde{x}_-(s)) + F_-(s,c(s,\tilde{x}_-(s)),\tilde{x}_-(s)) - F_-(s,c(s,\tilde{y}(s)),\tilde{x}_-(s)) + F_-(s,c(s,\tilde{y}(s)),\tilde{x}_-(s)) - F_-(s,c(s,\tilde{y}(s)),\tilde{y}(s))\Big)\,\mathrm{d}s\,,$$

so that for $\tau \le t \le t_1$,

$$\begin{aligned}\|u(t)\| &\le K\int_{t}^{t_1} e^{\alpha_-(t-s)}(\omega_-(\rho)\|\tilde{x}_+(s) - c(s,\tilde{x}_-(s))\| + \omega_-(2\rho)(1+\omega_c(2\rho))\|u(s)\|)\,\mathrm{d}s \\ &\le \frac{K^2\omega_-(\rho)}{\alpha_- - \alpha_+ - \gamma}\|\tilde{x}_+(t_0) - c(t_0,\tilde{x}_-(t_0))\|e^{(\alpha_+ + \gamma)t} \\ &\quad + K\omega_-(2\rho)(1+\omega_c(2\rho))\int_{t}^{t_1} e^{\alpha_-(t-s)}\|u(s)\|\,\mathrm{d}s\,.\end{aligned}$$

Note that (6.32) was used here. Putting $w(t) := e^{-\alpha_- t}\|u(t)\|$ and apply Gronwall's Lemma to the resulting inequality to obtain

$$\begin{aligned}&\|u(t)\| \\ &\le \frac{K^2\omega_-(\rho)}{\alpha_- - \alpha_+ - \gamma - K\omega_-(2\rho)(1+\omega_c(2\rho))}e^{(\alpha_+ + \gamma)(t-\tau)}\|\tilde{x}_+(t_0) - c(t_0,\tilde{x}_-(t_0))\| \\ &\overset{(6.30)}{\le} \tfrac{1}{2}e^{(\alpha_+ + \gamma)(t-\tau)}\|\tilde{x}_+(t_0) - c(t_0,\tilde{x}_-(t_0))\|\,.\end{aligned}$$

This leads to

$$\begin{aligned}\|\tilde{y}(t)\| &\leq \|\tilde{x}_-(t)\| + \|\tilde{y}(t) - \tilde{x}_-(t)\| \\ &\leq \|\tilde{x}_-(t)\| + e^{(\alpha_+ + \gamma)(t-\tau)}\|\tilde{x}_+(t_0) - c(t_0, \tilde{x}_-(t_0))\| \\ &\leq \rho + \tfrac{1}{2}(\rho + \omega_c(\rho)\rho) < 2\rho \quad \text{for all } \tau \leq t \leq t_1 .\end{aligned}$$

In particular, $\|\tilde{y}(\tau)\| < 2\rho$, and so, $\tau = t_0$. Hence, $\tilde{y}$ is defined on $[t_0, t_1]$, which yields (i) and (ii). Finally,

$$\begin{aligned}&\max\big\{\|\tilde{x}_+(t) - c(t, \tilde{y}(t))\|, \|\tilde{x}_-(t) - \tilde{y}(t)\|\big\} \\ &\leq \max\big\{\|\tilde{x}_+(t) - c(t, \tilde{x}_-(t))\| + \|c(t, \tilde{x}_-(t)) - c(t, \tilde{y}(t))\|, \|\tilde{x}_-(t) - \tilde{y}(t)\|\big\} \\ &\leq \max\big\{Ke^{(\alpha_+ + \gamma)(t-t_0)}\|\tilde{x}_+(t_0) - c(t_0, \tilde{x}_-(t_0))\| + \omega_c(2\rho)\|\tilde{x}_-(t) - \tilde{y}(t)\|, \\ &\qquad \|\tilde{x}_-(t) - \tilde{y}(t)\|\big\} \\ &\leq \max\big\{K + \tfrac{1}{2}\omega_c(2\rho), \tfrac{1}{2}\big\}e^{(\alpha_+ + \gamma)(t-t_0)}\|\tilde{x}_+(t_0) - c(t_0, \tilde{x}_-(t_0))\| \\ &\leq \big(K + \tfrac{1}{2}\big)e^{(\alpha_+ + \gamma)(t-t_0)}\|\tilde{x}_+(t_0) - c(t_0, \tilde{x}_-(t_0))\| \quad \text{for all } t \in [t_0, t_1] ,\end{aligned}$$

which concludes the proof. □

Fix now for the rest of this section $\gamma \in (0, \min\{0, \alpha_-\} - \alpha_+)$, and choose $\rho \in (0, \rho_0/2)$ such that the three estimates in (6.30) hold. An immediate consequence of Lemma 6.22 is that small bounded solutions of (6.27) must lie on the center-unstable manifold.

COROLLARY 6.23. *The following statements hold:*

(i) *If* (6.27) *has a solution* $\tilde{x} = (\tilde{x}_+, \tilde{x}_-) : \mathbb{R} \to \mathbb{R}^{d_+} \times \mathbb{R}^{d_-}$ *satisfying* $\|\tilde{x}_+(t)\| + \|\tilde{x}_-(t)\| < \rho$ *for all* $t \in \mathbb{R}$, *then* $\tilde{x}_+(t) = c(t, \tilde{x}_-(t))$ *holds for all* $t \in \mathbb{R}$.
(ii) *If* $\mathcal{S} \subset \mathbb{R} \times B_\rho(0) \times B_\rho(0) \subset \mathbb{R} \times \mathbb{R}^{d_+} \times \mathbb{R}^{d_-}$ *is a nonautonomous set such that for any* $(t_0, x_+^0, x_-^0) \in \mathcal{S}$, *the solution* $\phi(\cdot, t_0, x_+^0, x_-^0)$ *exists on* $\mathbb{R}$ *and satisfies* $\|\phi(t, t_0, x_+^0, x_-^0)\| < \rho$ *for all* $t \in \mathbb{R}$, *then* $\mathcal{S} \subset \mathcal{W}^-$.

PROOF. (i) Putting $t = t_1$ in Lemma 6.22 (iii) yields

$$\begin{aligned}\|\tilde{x}_+(t_1) - c(t_1, \tilde{x}_-(t_1))\| &\leq \big(K + \tfrac{1}{2}\big)\big\|\tilde{x}_-(t_0) - c(t_0, \tilde{x}_+(t_0))\big\|e^{(\alpha_+ + \gamma)(t_1 - t_0)} \\ &\leq 2\rho\big(K + \tfrac{1}{2}\big)e^{(\alpha_+ + \gamma)(t_1 - t_0)} ,\end{aligned}$$

where the last inequality follows from (6.29). Since $\alpha_+ + \gamma < 0$, the limit $t_0 \to -\infty$ yields $\tilde{x}_+(t_1) = c(t_1, \tilde{x}_-(t_1))$, and since $t_1 \in \mathbb{R}$ can be chosen arbitrarily, the assertion follows.
(ii) This is an immediate consequence of (i), which is applied to individual solutions $\phi(\cdot, t_0, x_+^0, x_-^0)$ of (6.27) with $(t_0, x_+^0, x_-^0) \in \mathcal{S}$. □

The asymptotic behavior of small solutions of (6.27) will now be related to solutions of the reduced equation (6.31), which guarantees that the center-unstable manifold is exponentially attractive.

PROPOSITION 6.24 (Asymptotic phase). *Let* $t_0 \in \mathbb{R}$ *and* $\tilde{x} = (\tilde{x}_+, \tilde{x}_-) : [t_0, \infty) \to \mathbb{R}^{d_+} \times \mathbb{R}^{d_-}$ *be a solution of* (6.27) *satisfying* $\|\tilde{x}_+(t)\| + \|\tilde{x}_-(t)\| < \rho$ *for all* $t \geq t_0$.

Then there exists a solution $\tilde{y} : [t_0, \infty) \to \mathbb{R}^{d_-}$ *of the reduced equation* (6.31) *such that*

$$\max\left\{\|\tilde{x}_+(t) - c(t, \tilde{y}(t))\|, \|\tilde{x}_-(t) - \tilde{y}(t)\|\right\}$$
$$\leq \left(K + \tfrac{1}{2}\right)\left\|\tilde{x}_-(t_0) - c(t_0, \tilde{x}_+(t_0))\right\|e^{(\alpha_+ + \gamma)(t-t_0)} \quad \textit{for all } t \geq t_0 .$$

PROOF. For $m \in \mathbb{Z}$ with $m > t_0$, let $\tilde{y}_m$ denote the solution of the reduced equation (6.31) fulfilling the initial condition $\tilde{y}_m(m) = \tilde{x}_-(m)$. Then Lemma 6.22 implies that $\tilde{y}_m$ is defined on $[t_0, m]$ with $\|\tilde{y}_m(t)\| < 2\rho$ for all $t \in [t_0, m]$, and define in addition, $\tilde{y}_m(t) := 0$ for all $t > m$. It also follows from this lemma that the inequality

$$\max\left\{\|\tilde{x}_+(t) - c(t, \tilde{y}_m(t))\|, \|\tilde{x}_-(t) - \tilde{y}_m(t)\|\right\}$$
$$\leq \left(K + \tfrac{1}{2}\right)\left\|\tilde{x}_-(t_0) - c(t_0, \tilde{x}_+(t_0))\right\|e^{(\alpha_+ + \gamma)(t-t_0)} \quad \text{for all } t \in [t_0, m]$$

holds. In addition, one has for $t \in [t_0, m]$,

$$\|\dot{\tilde{y}}_m(t)\| = \|B_-(t)\tilde{y}_m(t) + F_-(t, c(t, \tilde{y}_m(t)), \tilde{y}_m(t))\|$$
$$\leq 2\rho\Big(\sup_{t\in\mathbb{R}} \|B_-(t)\| + \omega_-(2\rho)(1 + \omega_c(2\rho))\Big),$$

and hence, the Theorem of Arzelà–Ascoli implies the existence of a subsequence $(\tilde{y}_{m_k})_{k\in\mathbb{N}}$ converging to a function $\tilde{y}$ uniformly on compact subintervals of $[t_0, \infty)$. The function $\tilde{y}$ is a solution of (6.31), and the claimed estimate follows from the above estimate in the limit $m_k \to \infty$. □

Having this proposition available, it is not difficult to maintain stability properties of (6.27) from the corresponding properties of the reduced equation (6.31).

THEOREM 6.25 (Reduction principle). *The trivial solution of* (6.31) *is locally pullback (forward, uniformly, respectively) attractive if and only if the trivial solution of* (6.27) *is locally pullback (forward, uniformly, respectively) attractive.*

EXERCISE 6.26. Prove Theorem 6.25, and formulate and prove the theorem with the notions of stability instead of attractivity, which will be introduced in the next chapter (see Definition 7.2).

Endnotes. The construction of invariant manifolds goes back to both Poincaré [**182**] and Hadamard [**88**]. There is a huge quantity of publications on invariant manifolds for autonomous systems, see, e.g., Bates & Jones [**18**], Bates, Lu & Zeng [**19**], Hirsch, Pugh & Shub [**94**], Kelley [**108**], Kirchgraber & Palmer [**109**], Shub [**222**] and Wiggins [**236**]. For center manifold theory, see Carr [**34**], Chow, Li & Wang [**47**], Vanderbauwhede [**232**]. The construction of the invariant manifolds in Section 1 follows closely Pötzsche [**188**], and see Aulbach & Wanner [**13**], Sell [**219**] and Yi [**238**] for further literature on nonautonomous invariant manifolds (invariant manifolds and foliations for quasi-periodic systems have been constructed by Chow & Lu [**48**]). Infinite-dimensional nonautonomous center manifolds have been constructed by means of evolution operators in Chicone & Latushkin [**44**]. The Taylor approximation from Section 4 is based on Pötzsche & Rasmussen [**191, 192**]. The proof of the reduction principle in Section 5 is based on discrete results obtained in Pötzsche [**184, 188**] which itself is based on autonomous results from Palmer [**172**] (see also Aulbach [**10**]). A nonautonomous version of the Theorem of Hartman–Grobman is not discussed in this book; results can be found in Aulbach & Wanner [**14, 15**] and Pötzsche [**187**].

CHAPTER 7

Lyapunov functions

A Lyapunov function of a dynamical system on a metric space X is a real-valued function $V : X \to \mathbb{R}$, which decreases along solutions. For this reason, sublevel sets of Lyapunov functions describe forward invariant areas of the phase space, which leads to an effective practical as well as theoretical tool for the investigation of stability properties of dynamical systems. Converse results ensuring the existence of a Lyapunov function that characterizes a particular type of stability property, such as the uniform asymptotic stability of a global attractor of an autonomous dynamical system, have been particularly useful in numerical dynamics and perturbation analysis.

The first section of this chapter deals with basic theorems of Lyapunov's second method in the nonautonomous case, which is also referred to as the *direct method of Lyapunov*, because it is possible to detect asymptotic properties of a given equation directly, i.e., without knowing explicit expressions of the solutions. It is one of the building blocks of the so-called *qualitative theory of dynamical systems*. The construction of a Lyapunov function for a global attractor of an autonomous difference equation is then presented in Section 2. Amongst other things, this result is useful for investigating nonautonomous perturbations of an autonomous system. In addition, in Section 3, a construction of a Lyapunov function is given, which characterizes the pullback attractor of a skew product flow generated by a nonautonomous differential equation that is driven by an autonomous dynamical system on a metric space. Finally, the last section of this chapter contains the treatment of Lyapunov functions for nonautonomous Morse decompositions.

1. Lyapunov functions for solutions

Consider a nonautonomous differential equation

$$\dot{x} = f(t, x)\,, \tag{7.1}$$

where $f : \mathbb{R} \times \mathbb{R}^d \to \mathbb{R}^d$ is assumed to fulfill conditions for the global existence and uniqueness of solutions. The differential equation (7.1) gives rise to a process $\phi : \mathbb{R} \times \mathbb{R} \times \mathbb{R}^d \to \mathbb{R}^d$.

A Lyapunov function for a differential equation is nonincreasing along solutions. In many cases, such a function $V : \mathbb{R} \times \mathbb{R}^d \to \mathbb{R}$ is continuously differentiable. In this case, the decreasing property is characterized by the non-positivity of the so-called *orbital derivative* of V with respect to (7.1), which is defined by

$$\dot{V}(t, x) := \frac{\partial V}{\partial t}(t, x) + \left\langle \frac{\partial V}{\partial x}(t, x), f(t, x) \right\rangle .$$

Note that for a solution $x : \mathbb{R} \to \mathbb{R}^d$, one has $\frac{\mathrm{d}}{\mathrm{d}t} V(t, x(t)) = \dot{V}(t, x(t))$, so the function $\dot{V}$ describes the derivative of V along solutions of (7.1).

DEFINITION 7.1. A continuously differentiable function $V : \mathbb{R} \times \mathbb{R}^d \to \mathbb{R}$ is called *Lyapunov function* of the differential equation (7.1) if the following two conditions are fulfilled:

(i) V is bounded below on sets of the form $\mathbb{R} \times K$, where $K \subset \mathbb{R}^d$ is a compact set,
(ii) the orbital derivative $\dot{V}$ fulfills $\dot{V}(t, x) \leq 0$ for all $(t, x) \in \mathbb{R} \times \mathbb{R}^d$.

Note that in case of an autonomous Lyapunov function of an autonomous differential equation, the first condition in Definition 7.1 is always fulfilled.

In addition to the notions of local attractivity introduced in Section 8 of Chapter 3, the following concepts of Lyapunov stability will be studied.

DEFINITION 7.2. Consider the differential equation (7.1) with the induced process ϕ. Then a solution $x : \mathbb{R} \to \mathbb{R}^d$ of (7.1) is called

(i) *forward stable* if for all $\varepsilon > 0$, there exists a $\delta > 0$ such that for all $t_0 \geq 0$ and $x_0 \in B_\delta(x(t_0))$, one has
$$\|\phi(t, t_0, x_0) - x(t)\| < \varepsilon \quad \text{for all } t \geq t_0 \,.$$
(ii) *pullback stable* if for all $\varepsilon > 0$, there exists a $\delta > 0$ such that for all $t_0 \leq 0$ and $x_0 \in B_\delta(x(t_0))$, one has
$$\|\phi(t, t_0, x_0) - x(t)\| < \varepsilon \quad \text{for all } t \in [t_0, 0] \,.$$
(iii) *uniformly stable* if for all $\varepsilon > 0$, there exists a $\delta > 0$ such that for all $t_0 \in \mathbb{R}$ and $x_0 \in B_\delta(x(t_0))$, one has
$$\|\phi(t, t_0, x_0) - x(t)\| < \varepsilon \quad \text{for all } t \geq t_0 \,.$$

If a solution is uniformly stable, then it obviously both forward and pullback stable. The converse, however, does not hold.

EXERCISE 7.3. Show via a counterexample that a solution, which is both forward and pullback stable, does not need to be uniformly stable.

Some results of the so-called *Lyapunov's second method* are developed in the following. Assume from now on that the right-hand side f of (7.1) fulfills $f(t, 0) = 0$ for all $t \in \mathbb{R}$, which implies that (7.1) has the trivial solution (the stability and attractivity of which will be examined). This situation can be reached by considering the equation of perturbed motion, see Subsection 1.3 of Chapter 2.

A criterion for stability is given as follows.

THEOREM 7.4. *Let $V : \mathbb{R} \times \mathbb{R}^d \to \mathbb{R}^d$ be a Lyapunov function for (7.1) fulfilling the following conditions for some unbounded interval $\mathbb{I}$:*

(i) *$V(t, 0) = 0$ for all $t \in \mathbb{I}$,*
(ii) *there exist a $\rho > 0$ and continuous functions u_1, $u_2 : B_\rho(0) \to \mathbb{R}_0^+$ with $u_1(0) = u_2(0) = 0$ and*
$$0 < u_1(x) \leq V(t, x) < u_2(x) \quad \textit{for all } t \in \mathbb{I} \textit{ and } x \in B_\rho(0) \setminus \{0\} \,.$$

Then the trivial solution of (7.1) *is*

(a) *forward stable if* $\mathbb{I}$ *is unbounded above,*
(b) *pullback stable if* $\mathbb{I}$ *is unbounded below,*
(c) *uniformly stable if* $\mathbb{I} = \mathbb{R}$.

PROOF. The proof will only be given for (a), since the other cases can be treated similarly. Let $\varepsilon > 0$, define $\nu := \min\{u_1(x) : \|x\| = \varepsilon\}$ and set $M := \{x \in B_\varepsilon(0) : u_2(x) < \nu\}$. Then $\nu > 0$, because u_1 is continuous, and the continuity of u_2 implies that M is a neighborhood of 0. Hence, $\delta := \min\{\|x\| : x \in \partial M\}$ is positive. Assume that there exist $t \geq t_0 \geq 0$ and $x_0 \in B_\delta(x(t_0))$ such that $\|\phi(t, t_0, x_0)\| \geq \varepsilon$. We may assume that $\|\phi(t, t_0, x_0)\| = \varepsilon$, since $\delta < \varepsilon$. Then one has $V(t, \phi(t, t_0, x_0)) \geq \nu$ and $V(t_0, x_0) < \nu$. This contradicts the fact that V decreases along solutions and finishes the proof of this theorem. □

Having more conditions on the Lyapunov function available, attractivity can be proved.

THEOREM 7.5. *Let* $V : \mathbb{R} \times \mathbb{R}^d \to \mathbb{R}^d$ *be a Lyapunov function for* (7.1) *fulfilling the conditions of Theorem 7.4 for some unbounded interval* $\mathbb{I}$, *and assume that there exists a continuous function* $u : B_\rho(0) \to \mathbb{R}_0^+$ *with* $u(0) = 0$ *and*

$$\dot{V}(t, x) \leq -u(x) < 0 \quad \textit{for all } (t, x) \in \mathbb{I} \times (B_\rho(0) \setminus \{0\}) \,.$$

Then the trivial solution of (7.1) *is*

(a) *locally forward attractive if* $\mathbb{I}$ *is unbounded above,*
(b) *locally pullback attractive if* $\mathbb{I}$ *is unbounded below,*
(c) *locally uniformly attractive if* $\mathbb{I} = \mathbb{R}$.

PROOF. The proof will only be given for (a), since the other cases can be treated similarly. Due to Theorem 7.4, there exists a $\eta > 0$ such that $\|\phi(t, t_0, x_0)\| \leq \rho$ for any $t \geq t_0$ in $\mathbb{I}$ and $x_0 \in B_\eta(0)$. Similarly, for all $\varepsilon \in (0, \eta]$, there exists a $\delta = \delta(\varepsilon) \in \left(0, \frac{1}{2}\varepsilon\right)$ such that $\|\phi(t, t_0, x_0)\| \leq \varepsilon$ for any $t \geq t_0$ in $\mathbb{I}$ and $x_0 \in B_\delta(0)$. It will be shown now that for all $\varepsilon \in (0, \eta]$, there exists a $T = T(\varepsilon) > 0$ with

$$\phi\big(t, t_0, B_\varepsilon(0)\big) \subset B_{\delta(\varepsilon)}(0) \quad \text{for all } t_0 \in \mathbb{I} \text{ and } t \geq t_0 + T \,. \tag{7.2}$$

Repeated application of this relation then yields the assertion (note that $\delta(\varepsilon) < \frac{1}{2}\varepsilon$). Choose $\varepsilon > 0$ arbitrarily. Due to Theorem 7.4, there exists a $\tilde{\delta} > 0$ such that

$$\phi\big(t, t_0, B_{\tilde{\delta}}(0)\big) \subset B_{\delta(\varepsilon)}(0) \quad \text{for all } t \geq t_0 \text{ in } \mathbb{I} \,. \tag{7.3}$$

Define

$$c_1 := \sup\{u_2(x) : \tilde{\delta} \leq \|x\| \leq \rho\} \quad \text{and} \quad c_2 := \inf\{u(x) : \tilde{\delta} \leq \|x\| < \rho\}$$

and $T := T(\varepsilon) := \frac{c_1}{c_2}$. Suppose that there exist $t_0 \in \mathbb{I}$ and $x_0 \in B_\varepsilon(0)$ such that $\|\phi(t, t_0, x_0)\| \geq \tilde{\delta}$ for all $t \in [t_0, t_0 + T]$. Note that one has by (i) that

$$\dot{V}(t, \phi(t, t_0, x_0)) \leq -u(\phi(t, t_0, x_0)) < 0 \quad \text{for all } t \geq t_0 \,.$$

This means that the function $v(t) := V(t, \phi(t, t_0, x_0))$, $t \geq t_0$, satisfies

$$v(t) \leq v(t_0) - \int_{t_0}^{t} u(\phi(s, t_0, x_0))\, \mathrm{d}s \quad \text{for all } t \geq t_0 \,. \tag{7.4}$$

Since $\|\phi(t,t_0,x_0)\| \geq \tilde{\delta}$ for all $t \in [t_0, t_0+T]$, one has $v(t_0+T) \leq v(t_0) - \frac{c_1}{c_2}c_2 < 0$. This is a contradiction, and hence, there exists $\tilde{T} \in [t_0, t_0+T]$ such that $\|\phi(t,t_0,x_0)\| \geq \tilde{\delta}$. Together with (7.3), this implies (7.2) and finishes the proof of this theorem. □

EXERCISE 7.6. Let $V : \mathbb{R} \times \mathbb{R}^d \to \mathbb{R}^d$ be a Lyapunov function for (7.1) fulfilling the following conditions for some $a, b, c, m, \rho > 0$:

(i) $a\|x\|^m \leq V(t,x) \leq b\|x\|^m$ for all $(t,x) \in \mathbb{R} \times B_\rho(0)$,
(ii) $\dot{V}(t,x) \leq -c\|x\|^m$ for all $(t,x) \in \mathbb{R} \times B_\rho(0)$.

Show that the trivial solution is exponentially attractive, i.e., there exist $\eta, K > 0$ and $\gamma < 0$ with

$$\|\phi(t,t_0,x_0)\| \leq Ke^{\gamma(t-t_0)} \quad \text{for all } t \geq t_0 \text{ and } x_0 \in B_\gamma(0)\,.$$

2. Lyapunov functions for autonomous attractors

Consider an autonomous dynamical system on a compact metric space (X,d), which is generated by an autonomous difference equation

$$x_{n+1} = f(x_n)\,, \tag{7.5}$$

where $f : X \to X$ is globally Lipschitz continuous with Lipschitz constant $L > 0$, i.e.,

$$d(f(x), f(y)) \leq Ld(x,y) \quad \text{for all } x, y \in X\,.$$

A nonempty and compact proper subset A of X is called *globally uniformly asymptotically stable* if it is both

(i) *Lyapunov stable*, i.e., for all $\varepsilon > 0$, there exists a $\delta > 0$ such that
$$\operatorname{dist}(x,A) < \delta \Rightarrow \operatorname{dist}(f^n(x),A) < \varepsilon \quad \text{for all } n \in \mathbb{N}_0\,, \tag{7.6}$$
(ii) *globally uniformly attracting*, i.e., for all $\varepsilon > 0$, there exists an integer $N > 1$ such that
$$\operatorname{dist}(f^n(x),A) < \varepsilon \quad \text{for all } x \in X \text{ and } n \geq N\,. \tag{7.7}$$

Note that such a set A is the global attractor for the system generated by the difference equation (7.5) if A is invariant.

Global uniform asymptotical stability is characterized in terms of a Lyapunov function by the following necessary and sufficient conditions. The following theorem is taken from DIAMOND & KLOEDEN [**70**].

THEOREM 7.7. *Let $f : X \to X$ be globally Lipschitz continuous, and let A be a nonempty compact subset of X. Then A is globally uniformly asymptotically stable with respect to the dynamical system generated by* (7.5) *if and only if there exist*

(i) *a Lyapunov function $V : X \to \mathbb{R}_0^+$,*
(ii) *monotone increasing continuous functions $\alpha, \beta : \mathbb{R}_0^+ \to \mathbb{R}_0^+$ with $\alpha(0) = \beta(0) = 0$ and $0 < \alpha(r) < \beta(r)$ for all $r > 0$, and*
(iii) *constants $K > 0$ and $0 \leq q < 1$ such that for all $x, y \in X$, one has*
 (1) *$|V(x) - V(y)| \leq Kd(x,y)$,*
 (2) *$\alpha(\operatorname{dist}(x,A)) \leq V(x) \leq \beta(\operatorname{dist}(x,A))$ and*

(3) $V(f(x)) \le qV(x)$.

PROOF. *Sufficiency.* Let V be a Lyapunov function as described in the theorem. Choose $\varepsilon > 0$ arbitrarily and define $\delta := \beta^{-1}(\alpha(\varepsilon)/q)$, which means that $\alpha(\varepsilon) = q\beta(\delta)$. This implies that

$$\alpha(\operatorname{dist}(f^n(x), A)) \le V(f^n(x)) \le q^n V(x) \le qV(x) \le q\beta(\operatorname{dist}(x, A)),$$

so that

$$\operatorname{dist}(f^n(x), A) \le \alpha^{-1}\left(q\beta(\operatorname{dist}(x, A))\right) \le \alpha^{-1}(\alpha(\varepsilon)) \le \varepsilon \quad \text{for all } n \in \mathbb{N}$$

when $\operatorname{dist}(x, A) < \delta$. Thus, A is Lyapunov stable. Now define

$$N := \max\left\{1, 1 + \left\lfloor \frac{\ln(\alpha(\varepsilon)/V_0)}{\ln q} \right\rfloor\right\},$$

where $V_0 := \max_{x\in X} V(x)$ is finite by continuity of V_0 and compactness of X. For $n \ge N$, one has $q^n \le q^N$, since $0 \le q < 1$. Since, from above,

$$\alpha(\operatorname{dist}(f^n(x), A)) \le q^n V(x) \le q^n V_0 \le q^N V_0 \le \alpha(\varepsilon) \quad \text{for all } n \ge N,$$

one obtains $\operatorname{dist}(f^n(x), A) < \varepsilon$ for $n \ge N$ and $x \in X$. This means that A is globally uniformly attracting and hence globally uniformly asymptotically stable.
Necessity. Let A be globally *uniformly asymptotically stable*, i.e., for given $\varepsilon > 0$, there exists $\delta = \delta(\varepsilon)$ such that (7.6) holds, and for given $\varepsilon > 0$, there exists $N = N(\varepsilon)$ such that (7.7) holds. Define $G_k : \mathbb{R}_0^+ \to \mathbb{R}_0^+$ for $k \in \mathbb{N}$ by

$$G_k(r) := \begin{cases} r - \frac{1}{k} & : \quad r \ge \frac{1}{k}, \\ 0 & : \quad 0 \le r < \frac{1}{k} \end{cases} \qquad \text{for all } r \in \mathbb{R}_0^+.$$

Then

$$|G_k(r) - G_k(s)| \le |r - s| \quad \text{for all } r, s \ge 0.$$

Now choose q so that $0 < q < \min\{1, L\}$, where L is the Lipschitz constant of the mapping f, and define

$$g_k := \left(\frac{q}{L}\right)^{N(1/k)} \quad \text{for all } k \in \mathbb{N}$$

and

$$V_k(x) = g_k \sup_{n\in\mathbb{N}_0} q^{-n} G_k(\operatorname{dist}(f^n(x), A)) \quad \text{for all } k \in \mathbb{N}.$$

Then the following statements hold:

(i) $V_k(x) = 0$ if and only if $\operatorname{dist}(x, A) < \delta(1/k)$, because of Lyapunov stability.
(ii) Since $|\operatorname{dist}(x, A) - \operatorname{dist}(y, A)| \le d(x, y)$ (see Exercise 7.8 below) and

$$d(f^n(x), f^n(y)) \le Ld(f^{n-1}(x), f^{n-1}(y)) \le \cdots \le L^n d(x, y),$$

it follows that

$$\begin{aligned}
&|V_k(x) - V_k(y)| \\
&\leq g_k \sup_{n\geq 0} q^{-n} \left| G_k(\operatorname{dist}(f^n(x), A)) - G_k(\operatorname{dist}(f^n(y), A))\right| \\
&\leq g_k \sup_{0\leq n\leq N(1/k)} q^{-n} \left| G_k(\operatorname{dist}(f^n(x), A)) - G_k(\operatorname{dist}(f^n(y), A))\right| \\
&\leq g_k \sup_{0\leq n\leq N(1/k)} q^{-n} d\left(f^n(x), f^n(y)\right) \\
&\leq g_k \sup_{0\leq n\leq N(1/k)} q^{-n} L^n d(x,y) = d(x,y)\,.
\end{aligned}$$

(iii) From above, it follows that $V_k(x) \leq V_k(y) + d(x,y)$. For all $y \in A$, one obtains that $V_k(y) = 0$ and $V_k(x) \leq d(x,y)$, and since A is compact, the minimum over all $y \in A$ is attained and $V_k(x) \leq \operatorname{dist}(x, A)$.

(iv) $V_k(f(x)) \leq qV_k(x)$, since

$$\begin{aligned}
V_k(f(x)) &\leq g_k \sup_{n\geq 0} q^{-n} G_k(\operatorname{dist}(f^n(f(x)), A)) \\
&= q g_k \sup_{n\geq 0} q^{-n-1} G_k(\operatorname{dist}(f^{n+1}(x), A)) \\
&= q g_k \sup_{n\geq 1} q^{-n} G_k(\operatorname{dist}(f^{n}(x), A)) \\
&\leq q g_k \sup_{n\geq 0} q^{-n} G_k(\operatorname{dist}(f^{n}(x), A)) = qV_k(x)\,.
\end{aligned}$$

Finally, define $V(x) := \sum_{k=1}^{\infty} 2^{-k} V_k(x)$. Then the following statements hold:

(a) $V(x) = 0$ if and only if $\operatorname{dist}(x, A) = 0$, i.e., if and only if $x \in A$.

(b) One has

$$\begin{aligned}
&|V(x) - V(y)| \\
&= \left| \sum_{k=1}^{\infty} 2^{-k} \left(V_k(x) - V_k(y)\right)\right| \\
&\leq \sum_{k=1}^{\infty} 2^{-k} \left|V_k(x) - V_k(y)\right| \leq \sum_{k=1}^{\infty} 2^{-k} d(x,y) = d(x,y)\,.
\end{aligned}$$

(c) $V(x) \leq \sum_{k=1}^{\infty} 2^{-k} V_k(y) + d(x,y) \leq \sum_{k=1}^{\infty} 2^{-k} \operatorname{dist}(x, A) = \operatorname{dist}(x, A)$.

(d) $V(f(x)) = \sum_{k=1}^{\infty} 2^{-k} V_k(f(x)) \leq q \sum_{k=1}^{\infty} 2^{-k} V_k(x) = qV(x)$.

(e) For $\operatorname{dist}(x, A) \geq 1$, one gets

$$V(x) \geq \frac{1}{2} V_1(x) \geq \frac{1}{2} g_1 G_1(\operatorname{dist}(x, A)) \geq \frac{1}{2} g_1 (\operatorname{dist}(x, A) - 1)$$

whereas for $1/(k-1) \geq \operatorname{dist}(x, A) \geq 1/k$, one has

$$\begin{aligned}
V(x) &\geq \frac{1}{2^{k+1}} V_{k+1}(x) \geq \frac{1}{2^{k+1}} g_{k+1} G_{k+1}(\operatorname{dist}(x, A)) \\
&\geq \frac{1}{2^{k+1}} g_{k+1} \left(\operatorname{dist}(x, A) - \frac{1}{k+1}\right) \\
&\geq \frac{1}{2^{k+1}} g_{k+1} \frac{1}{k(k+1)}\,.
\end{aligned}$$

It can then be shown as in YOSHIZAWA [**239**, p. 98] that there exists a continuous monotone increasing function $a : \mathbb{R}_0^+ \to \mathbb{R}_0^+$ with $a(0) = 0$ such that $a(\text{dist}(x, A)) \leq V(x)$. Setting $\alpha(r) = a(r)$ and $\beta(r) \equiv 1$, one sees that V is a Lyapunov function with the desired properties listed in the theorem. □

EXERCISE 7.8. Show for a compact set $A \subset X$ that

$$|\text{dist}(x, A) - \text{dist}(y, A)| \leq d(x, y) \quad \text{for all } x, y \in X .$$

YOSHIZAWA [**239**] gives various necessary and sufficient conditions involving Lyapunov functions for a compact set to be uniformly asymptotically stable for an autonomous differential equation

$$\dot{x} = F(x) \tag{7.8}$$

with a right hand side $F : \mathbb{R}^d \to \mathbb{R}^d$. The following theorem is a reformulation of [**239**, Theorem 22.5] in the global context here. Its proof is similar to that above for Theorem 7.7.

THEOREM 7.9. *Suppose that $F : \mathbb{R}^d \to \mathbb{R}^d$ is globally Lipschitz continuous. Then a nonempty and compact set $A \subset \mathbb{R}^d$ is globally uniformly asymptotically stable for the differential equation* (7.8) *if and only if there exists a function $V : \mathbb{R}^d \to \mathbb{R}_0^+$, which satisfies the following conditions:*

(i) *V is globally Lipschitz continuous, i.e., there exists an $L > 0$ such that*

$$|V(x) - V(y)| \leq L\|x - y\| \quad \text{for all } x, y \in \mathbb{R}^d ,$$

(ii) *there exist continuous strictly increasing functions $\alpha, \beta : \mathbb{R}_0^+ \to \mathbb{R}_0^+$ with $\alpha(0) = \beta(0) = 0$ and $0 < \alpha(r) < \beta(r)$ for all $r > 0$ such that*

$$\alpha(\text{dist}(x, A)) \leq V(x) \leq \beta(\text{dist}(x, A)) \quad \text{for all } x \in \mathbb{R}^d ,$$

(iii) *there exists a constant $c > 0$ such that*

$$D^+_{(7.8)} V(x) \leq -cV(x) \quad \text{for all } x \in \mathbb{R}^d .$$

Here $D^+_{(7.8)} V$ is the upper left Dini derivative *of V with respect to the differential equation* (7.8)*, which is defined by*

$$D^+_{(7.8)} V(x) = \limsup_{h \to 0+} \frac{V(x + hF(x)) - V(x)}{h} . \tag{7.9}$$

Theorems 7.7 and 7.9 can be used to show the existence of a pullback attractor of nonautonomous perturbations of an autonomous system with a global attractor (see Chapter 11).

3. Lyapunov functions for pullback attractors

Consider a parameterized differential equation

$$\dot{x} = f(p, x)$$

on $\mathbb{R}^d$, where the parameter p is allowed to vary with time in a certain way. In particular, let P be a topological space and consider a group $\theta = (\theta_t)_{t \in \mathbb{R}}$ of mappings $\theta_t : P \to P$ for each $t \in \mathbb{R}$ such that $(t, p) \mapsto \theta_t(p)$ is continuous. The autonomous dynamical system θ on P acts as a driving mechanism that generates the time variation in the parameter p in the parameterized differential equation above to form a nonautonomous differential equation

$$\dot{x} = f(\theta_t(p), x) \tag{7.10}$$

on $\mathbb{R}^d$ for each $p \in P$. It will be assumed amongst other things (see later) that $f : P \times \mathbb{R}^d \to \mathbb{R}^d$ is continuous, that $f(p, \cdot)$ is globally Lipschitz continuous on $\mathbb{R}^d$ for each $p \in P$ and that the global forwards existence and uniqueness of solutions of (7.10) holds (e.g., due to an additional dissipativity structural assumption). The solution mapping $\varphi : \mathbb{R}_0^+ \times P \times \mathbb{R}^d \to \mathbb{R}^d$ of (7.10), for which

$$\frac{\mathrm{d}}{\mathrm{d}t}\varphi(t, p, x_0) = f\big(\theta_t(p), \varphi(t, p, x_0)\big) \quad \text{for all } t \geq 0\,, p \in P \text{ and } x_0 \in \mathbb{R}^d$$

and the *initial condition property* $\varphi(0, p, x_0) = x_0$ for all $x_0 \in \mathbb{R}^d$ and $p \in P$ hold, yields a skew product flow (θ, φ) on $P \times \mathbb{R}^d$.

The main result in this section provides the existence of a Lyapunov function that characterizes pullback attraction and pullback attractors. This will be done in terms of attracting universes $\mathfrak{D}$, i.e., collection of nonautonomous sets $\mathcal{D}$ with fibers D_p in $\mathbb{R}^d$, $p \in P$, with a pullback attractor $\mathcal{A}$ defined as in Definition 3.24 by

$$\lim_{t \to \infty} \operatorname{dist}\big(\varphi(t, \theta_{-t}(p), D_{\theta_{-t}(p)}), A_p\big) = 0\,.$$

THEOREM 7.10. *Let $\mathcal{A}$ be a pullback attractor with respect to an attracting universe $\mathfrak{D}$ for the skew product flow (θ, φ) generated by the differential equation* (7.10), *where*

(i) *$(p, x) \mapsto f(p, x)$ is continuous in $(p, x) \in P \times \mathbb{R}^d$,*
(ii) *$x \mapsto f(p, x)$ is globally Lipschitz continuous with Lipschitz constant $L(p)$ for each $p \in P$,*
(iii) *$p \mapsto L(p)$ is continuous,*
(iv) *$(t, p) \mapsto \theta_t(p)$ is continuous.*

Define $\mathfrak{D}(p) := \bigcup\{D_p : \mathcal{D} \in \mathfrak{D}\}$ for each $p \in P$. Then there exists a Lyapunov function $V : \bigcup_{p \in P}\{p\} \times \mathfrak{D}(p) \to \mathbb{R}_0^+$ having the following properties:

(P1) Upper bound. *For all $p \in P$ and $x_0 \in \mathfrak{D}(p)$, one has*

$$V(p, x_0) \leq \operatorname{dist}(x_0, A_p)\,. \tag{7.11}$$

(P2) Lower bound. *For all $p \in P$, there exists a function $a(p, \cdot) : \mathbb{R}_0^+ \to \mathbb{R}_0^+$ with $a(p, 0) = 0$ and $a(p, r) > 0$ for $r > 0$, which is increasing in r such that*

$$a\big(p, \operatorname{dist}(x_0, A_p)\big) \leq V(p, x_0) \quad \textit{for all } x_0 \in \mathfrak{D}(p)\,. \tag{7.12}$$

(P3) Lipschitz condition. *For all $p \in P$ and $x_0, y_0 \in \mathfrak{D}(p)$, one has*

$$|V(p, x_0) - V(p, y_0)| \leq \|x_0 - y_0\|\,.$$

(P4) Pullback convergence. *For all $p \in P$ and $\mathcal{D} \in \mathfrak{D}$, one has*

$$\limsup_{t \to \infty} \sup_{z \in D_{\theta_{-t}(p)}} V\big(p, \varphi(t, \theta_{-t}(p), z)\big) = 0\,.$$

(P5) Forward convergence. *There exists a compact nonautonomous set* $\mathcal{D} \in \mathfrak{D}$, *which is* φ*-positively invariant in the sense that*

$$\varphi(t, p, D_p) \subseteq D_{\theta_t(p)} \quad \textit{for all } t \geq 0 \textit{ and } p \in P$$

and satisfies $A_p \subset \operatorname{int} D_p$ *for each* $p \in P$ *as well as*

$$V(\theta_t(p), \varphi(t, p, x_0)) \leq e^{-t} V(p, x_0) \tag{7.13}$$

for all $x_0 \in D_p$ *and* $t \geq 0$.

The proof of Theorem 7.10 will be given at the end of this section, because some preparations for the proof are in order.

Firstly, however, note that the forward convergence inequality (7.13) of the pullback Lyapunov function in Theorem 7.10 does not imply the usual forwards Lyapunov stability or asymptotic stability. Although the inequality

$$a\big(\theta_t(p), \operatorname{dist}\left(\varphi(t, p, x_0), A_{\theta_t(p)}\right)\big) \leq e^{-t} V(p, x_0)$$

then holds, $\operatorname{dist}(\varphi(t, p, x_0), A_{\theta_t p})$ need not become small as $t \to \infty$. The reason for this is that, without additional assumptions on the dynamical behavior, it is possible that $\inf_{t \geq 0} a(\theta_t(p), r) = 0$ for some $r > 0$ and $p \in P$.

EXAMPLE 7.11. Consider the differential equation

$$\dot{x} = 2tx$$

with the general solution $x(t, t_0, x_0) = x_0 e^{t^2 - t_0^2}$ yielding the cocycle mapping

$$\varphi(t, t_0, x_0) = x_0 e^{(t+t_0)^2 - t_0^2} \quad \text{for all } t, t_0, x_0 \in \mathbb{R}.$$

Here the parameter $p = t_0 \in P = \mathbb{R}$ and $\theta_t(t_0) = t + t_0$. The pullback attractor has the fibers $A_{t_0} = \{0\}$ for each $t_0 \in \mathbb{R}$, and the pullback attraction is global, i.e., there is no restriction on the bounded subsets that are considered in the attracting universe. A Lyapunov function satisfying the properties of Theorem 7.10 is given by

$$V(t, x) = |x| e^{-t - t^2 - \frac{1}{4}} \quad \text{for all } t, x \in \mathbb{R}.$$

The properties (P1) with $a(t_0, r) = re^{-|t_0| - t_0^2 - \frac{1}{4}}$ and (P2) are immediate, while (P3) follows from

$$\begin{aligned} V(t_0, \varphi(t; t_0 - t, x_0)) &= \left| x_0 e^{(t+t_0-t)^2 - (t_0-t)^2} \right| e^{-t_0 - t_0^2 - \frac{1}{4}} \\ &= e^{-(t_0-t)^2 - t_0 - \frac{1}{4}} |x_0| \to 0 \quad \text{as } t \to \infty. \end{aligned}$$

In addition, V satisfies inequality (7.13), since

$$\begin{aligned} V(t_0 + t, \varphi(t; t_0, x_0)) &= \left| x_0 e^{(t+t_0)^2 - t_0^2} \right| e^{-(t_0+t) - (t_0+t)^2 - \frac{1}{4}} \\ &= e^{-t} V(t_0, x_0) \to 0 \quad \text{as } t \to \infty. \end{aligned}$$

However, the zero solution is obviously not forwards Lyapunov stable.

On the other hand, if the pullback attractor is uniformly pullback attracting, then it is also uniformly forward attracting and is thus a uniform attractor. In this case the property (P2) of the Lyapunov function in Theorem 7.10 has a uniform lower bound. See Chapter 11 for a modification to uniform nonautonomous attractors.

COROLLARY 7.12. *Suppose in addition to the assumptions of Theorem 7.10 that the parameter space P is compact and that the pullback attractor $\mathcal{A}$ is even a uniform attractor. Then the Lyapunov function V satisfies the properties* (P1)–(P5) *as well as the following uniform version of* (P2)*:*

(P2') Uniform lower bound. *There exists an increasing continuous function $a : \mathbb{R}_0^+ \to \mathbb{R}_0^+$ with $a(0) = 0$ and $a(r) > 0$ for $r > 0$ such that for all $x_0 \in \mathbb{R}^d$ and $p \in P$, one has*

$$a(\operatorname{dist}(x_0, A_p)) \leq V(p, x_0)\,.$$

The proof of this corollary will be given after the proof of Theorem 7.10 at the end of this section.

Next it is shown that a pullback attractor has a so-called pullback absorbing neighborhood system, which is defined as follows.

DEFINITION 7.13. Let $\mathcal{A}$ be a pullback attractor of φ with respect to an attraction universe $\mathfrak{D}$. A nonautonomous set $\mathcal{B} \in \mathfrak{D}$ whose fibers are nonempty compact subsets of $\mathbb{R}^N$ with nonempty interior is called a *pullback absorbing neighborhood system* if it is φ-positively invariant and if it *pullback absorbs* all $\mathcal{D} \in \mathfrak{D}$, i.e., for each $\mathcal{D} \in \mathfrak{D}$ and $p \in P$, there exists a $T > 0$ such that

$$\varphi\big(t, \theta_{-t}(p), D_{\theta_{-t}(p)}\big) \subset \operatorname{int} B_p \quad \text{for all } t \geq T\,.$$

A pullback absorbing neighborhood system $\mathcal{B}$ obviously satisfies $\mathcal{A} \subset \mathcal{B} \in \mathfrak{D}$. Moreover, by positive invariance and the cocycle property,

$$\varphi\big(s+t, \theta_{-s-t}(p), B_{\theta_{-s-t}(p)}\big) \subset \varphi\big(t, \theta_{-t}(p), B_{\theta_{-t}(p)}\big)$$

for all $s, t \geq 0$ and $p \in P$, from which it follows that

$$A_p = \bigcap_{t \geq 0} \varphi\big(t, \theta_{-t}(p), B_{\theta_{-t}(p)}\big) \quad \text{for all } p \in P\,. \tag{7.14}$$

The following lemma shows that there always exists such a pullback absorbing neighborhood system for any given pullback attractor. This will be required for the construction of the Lyapunov function for the proof of Theorem 7.10.

LEMMA 7.14. *Let $\mathcal{A}$ be a pullback attractor with respect to an attraction universe $\mathfrak{D}$ for a skew product flow (θ, φ) for which the mapping $(t, p, x) \mapsto \varphi(t, \theta_{-t}(p), x)$ is continuous. Then there exists a pullback absorbing neighborhood system $\mathcal{B} \subset \mathfrak{D}$ of $\mathcal{A}$.*

PROOF. Since $A_p \subset \operatorname{int} \mathfrak{D}(p)$, there exists $\delta_p \in (0, 1]$ such that for the $2\delta_p$-neighborhood of A_p, one has $B[A_p, 2\delta_p] \subset \operatorname{int} \mathfrak{D}(p)$ for each $p \in P$. Define

$$B_p := \overline{\bigcup_{t \geq 0} \varphi\big(t, \theta_{-t}(p), B\big[A_{\theta_{-t}(p)}, \delta_{\theta_{-t}(p)}\big]\big)} \quad \text{for all } p \in P\,.$$

Obviously, $A_p \subset B[A_p, \delta_p] \subset B_p \subset \mathfrak{D}(p)$ for each $p \in P$. By the cocycle property,

$$\begin{aligned}
\varphi(t,p,B_p) &\subseteq \overline{\bigcup_{s\geq 0} \varphi\big(t,p,\varphi\big(s,\theta_{-s}(p),B\big[A_{\theta_{-s}(p)},\delta_{\theta_{-s}(p)}\big]\big)\big)} \\
&= \overline{\bigcup_{s\geq 0} \varphi\big(s+t,\theta_{-s}(p),B\big[A_{\theta_{-s}(p)},\delta_{\theta_{-s}(p)}\big]\big)} \\
&= \overline{\bigcup_{r\geq t} \varphi\big(r,\theta_{-r+t}(p),B\big[A_{\theta_{-r+t}(p)},\delta_{\theta_{-r+t}(p)}\big]\big)} \\
&\subseteq \overline{\bigcup_{r\geq 0} \varphi\big(r,\theta_{-r}\theta_t(p),B\big[A_{\theta_{-s}\theta_t(p)},\delta_{\theta_{-r}\theta_t(p)}\big]\big)} = B_{\theta_t(p)}
\end{aligned}$$

for all $t \geq 0$, which implies $\varphi(t,p,B_p) \subseteq B_{\theta_t(p)}$ and means that the so-defined nonautonomous set $\mathcal{B}$ is φ-positively invariant. Now by pullback convergence, there exists a $T > 0$ such that

$$\varphi\big(t,\theta_{-t}(p),B\big[A_{\theta_{-t}(p)},\delta_{\theta_{-t}(p)}\big]\big) \subset B\big[A_p,\delta_p\big] \subset B_p \quad \text{for all } t \geq T\,.$$

Hence, one obtains

$$\begin{aligned}
B_p &= \overline{\bigcup_{t\geq 0} \varphi\big(t,\theta_{-t}(p),B\big[A_{\theta_{-t}(p)},\delta_{\theta_{-t}(p)}\big]\big)} \\
&\subseteq B[A_p,\delta_p] \cup \overline{\bigcup_{t\in[0,T]} \varphi\big(t,\theta_{-t}(p),B\big[A_{\theta_{-t}(p)},\delta_{\theta_{-t}(p)}\big]\big)} \\
&= \overline{\bigcup_{t\in[0,T]} \varphi\big(t,\theta_{-t}(p),B\big[A_{\theta_{-t}(p)},\delta_{\theta_{-t}(p)}\big]\big)} \\
&\subseteq \overline{\bigcup_{t\in[0,T]} \varphi\big(t,\theta_{-t}(p),B\big[A_{\theta_{-t}(p)},1\big]\big)} \\
&\subseteq \overline{\bigcup_{t\in[0,T]} \varphi\big(t,\theta_{-t}(p),B'\big)} =: U_{p,T}\,.
\end{aligned}$$

Here the set

$$B' := \overline{\bigcup_{t\in[0,T]} B\big[A_{\theta_{-t}(p)},1\big]}$$

is compact by the continuity of the mapping $t \mapsto A_{\theta_{-t}(p)}$ and the compactness of the sets $B\big[A_{\theta_{-t}(p)},1\big]$. The compactness of the set $U_{p,T}$ then follows by the continuity of the set-valued mapping $t \mapsto \varphi\big(t,\theta_{-t}(p),B'\big)$. Hence B_p is compact for each $p \in P$. To see that $\mathcal{B}$ is pullback absorbing from $\mathfrak{D}$, let $\mathcal{D} \in \mathfrak{D}$ and fix $p \in P$. Since $\mathcal{A}$ is pullback attracting, there exists a $T > 0$ such that

$$\operatorname{dist}\big(\varphi\big(t,\theta_{-t}(p),D_{\theta_{-t}(p)}\big),A_p\big) < \delta_p \quad \text{for all } t \geq T\,,$$

that is, $\varphi\big(t,\theta_{-t}(p),D_{\theta_{-t}(p)}\big) \subset B[A_p,\delta_p]$ for all $t \geq T$. But $B[A_p,\delta_p] \subset \operatorname{int} B_p$, so $\varphi\big(t,\theta_{-t}(p),D_{\theta_{-t}(p)}\big) \subset \operatorname{int} B_p$ for all $t \geq T$. Hence, $\mathcal{B}$ is pullback absorbing as required. □

PROOF OF THEOREM 7.10. Define $V(p,x_0)$ for all $p \in P$ and $x_0 \in \mathfrak{D}(p)$ by

$$V(p,x_0) := \sup_{t\geq 0} e^{-T_{p,t}} \operatorname{dist}\big(x_0,\varphi\big(t,\theta_{-t}(p),B_{\theta_{-t}(p)}\big)\big)\,,$$

where

$$T_{p,t} = t + \int_0^t L(\theta_{-s}(p))\,\mathrm{d}s \quad \text{with } T_{p,0} = 0\,.$$

The integral here exists due to the continuity assumptions. Note that both $T_{p,t} \geq t$ and

$$T_{\theta_t(p),s+t} := T_{p,s} + t + \int_0^t L(\theta_r(p))\,\mathrm{d}r \quad \text{for all } s,t \geq 0 \text{ and } p \in P$$

hold, the latter follows from

$$\begin{aligned} T_{\theta_t(p),s+t} &= s + t + \int_0^{s+t} L(\theta_{-r}\theta_t(p))\,\mathrm{d}r \\ &= s + \int_t^{s+t} L(\theta_{-r+t}(p))\,\mathrm{d}r + t + \int_0^t L(\theta_{-r+t}(p))\,\mathrm{d}r \\ &= s + \underbrace{\int_0^s L(\theta_{-u}(p))\,\mathrm{d}u}_{u=r-t} + t - \underbrace{\int_t^0 L(\theta_v(p))\,\mathrm{d}v}_{v=t-r} \\ &= T_{p,s} + t + \int_0^t L(\theta_v(p))\,\mathrm{d}v\,. \end{aligned}$$

The properties (P1)–(P5) will be proved in the following.
(P1): Since $e^{-T_{p,t}} \leq 1$ for all $t \geq 0$ and since $\operatorname{dist}\big(x_0, \varphi(t, \theta_{-t}(p), B_{\theta_{-t}(p)})\big)$ is monotonically increasing from $0 \leq \operatorname{dist}(x_0, \varphi(0,p,B_p)) = \operatorname{dist}(x_0, B_p)$ at $t = 0$ to $\operatorname{dist}(x_0, A_p)$ in the limit $t \to \infty$, it follows that

$$V(p, x_0) = \sup_{t \geq 0} e^{-T_{p,t}} \operatorname{dist}\big(x_0, \varphi(t, \theta_{-t}(p), B_{\theta_{-t}(p)})\big) \leq 1 \cdot \operatorname{dist}(x_0, A_p)\,.$$

(P2): By property (P1), one has $V(p, x_0) = 0$ for $x_0 \in A_p$. Assume instead that $x_0 \in \mathfrak{D}(p) \setminus A_p$. Now the supremum in

$$V(p, x_0) = \sup_{t \geq 0} e^{-T_{p,t}} \operatorname{dist}\big(x_0, \varphi(t, \theta_{-t}(p), B_{\theta_{-t}(p)})\big)$$

involves the product of an exponentially decreasing quantity bounded below by zero and a bounded increasing function, since the $\varphi\big(t, \theta_{-t}(p), B_{\theta_{-t}(p)}\big)$ are a nested family of compact sets decreasing to A_p with increasing t. Hence, there exists a $T' > 0$ such that

$$\frac{1}{2} \operatorname{dist}(x_0, A_p) \leq \operatorname{dist}\big(x_0, \varphi(t, \theta_{-t}(p), B_{\theta_{-t}(p)})\big)$$

for all $t \geq T'$, but not for $t < T'$. Thus, from above, the relation

$$V(p, x_0) \geq e^{-T_{p,T'}} \operatorname{dist}\big(x_0, \varphi(T', \theta_{-T'}(p), B_{\theta_{-T'}(p)})\big) \geq \frac{1}{2} e^{-T_{p,T'}} \operatorname{dist}(x_0, A_p)$$

follows. Define

$$\hat{T}(p, r) := \sup\big\{T'(p, x_0) : x_0 \in \mathfrak{D}(p) \text{ and } \operatorname{dist}(x_0, A_p) = r\big\}\,.$$

Then $\hat{T}(p,r) < \infty$. To see this note that by the triangle inequality

$$\begin{aligned} &\operatorname{dist}(x_0, A_p) \\ &\leq \operatorname{dist}\big(x_0, \varphi(t, \theta_{-t}(p), B_{\theta_{-t}(p)})\big) + \operatorname{dist}\big(\varphi(t, \theta_{-t}(p), B_{\theta_{-t}(p)}), A_p\big)\,. \end{aligned}$$

In addition, by pullback convergence, there exists a finite $T(p, r/2)$ such that

$$\operatorname{dist}\left(\varphi\big(t, \theta_{-t}(p), B_{\theta_{-t}(p)}\big), A_p\right) < \frac{1}{2}r \quad \text{for all } t \geq T(p, r/2)\,.$$

Hence,

$$r \leq \operatorname{dist}\left(x_0, \varphi\big(t, \theta_{-t}(p), B_{\theta_{-t}(p)}\big)\right) + \frac{1}{2}r$$

for $\operatorname{dist}(x_0, A_p) = r$ and $t \geq T(p, r/2)$, that is

$$\frac{1}{2}r \leq \operatorname{dist}\left(x_0, \varphi\big(t, \theta_{-t}(p), B_{\theta_{-t}(p)}\big)\right).$$

Thus, $\hat{T}(p,r) \leq T(p,r/2) < \infty$. In addition, $\hat{T}(p,r)$ is obviously nondecreasing in r as $r \to 0$. Finally, define

$$a(p,r) := \frac{1}{2}r\, e^{-T_{p,\hat{T}(p,r)}}\,, \tag{7.15}$$

which satisfies the stated properties.
(P3): It follows from the definition that

$$\begin{aligned}
|V(p,x_0) - V(p,y_0)| &= \Big|\sup_{t\geq 0} e^{-T_{p,t}} \operatorname{dist}\left(x_0, \varphi\big(t, \theta_{-t}(p), B_{\theta_{-t}(p)}\big)\right) \\
&\qquad - \sup_{t\geq 0} e^{-T_{p,t}} \operatorname{dist}\left(y_0, \varphi\big(t, \theta_{-t}(p), B_{\theta_{-t}(p)}\big)\right)\Big| \\
&\leq \sup_{t\geq 0} e^{-T_{p,t}} \Big| \operatorname{dist}\left(x_0, \varphi\big(t, \theta_{-t}(p), B_{\theta_{-t}(p)}\big)\right) \\
&\qquad\qquad - \operatorname{dist}\left(y_0, \varphi\big(t, \theta_{-t}(p), B_{\theta_{-t}(p)}\big)\right)\Big| \\
&\leq \sup_{t\geq 0} e^{-T_{p,t}} \|x_0 - y_0\| \leq \|x_0 - y_0\|\,.
\end{aligned}$$

(P4): Assume that the opposite of (P4) holds. Then there exist an $\varepsilon_0 > 0$, a sequence $t_j \to \infty$ in $\mathbb{R}_0^+$ and points $x_j \in \varphi\big(t_j, \theta_{-t_j}(p), D_{\theta_{-t_j}(p)}\big)$ such that $V(p, x_j) \geq \varepsilon_0$ for all $j \in \mathbb{N}$. Since $\mathcal{D} \in \mathfrak{D}$ and $\mathcal{B}$ is pullback absorbing, there exists a $T \geq 0$ such that

$$\varphi\big(t_j, \theta_{-t_j}(p), D_{\theta_{-t_j}(p)}\big) \subset B_p \quad \text{for all } t_j \geq T\,.$$

Hence, $x_j \in B_p$ for all j with $t_j \geq T$. Since B_p is a compact set, there exists a convergent subsequence $x_{j'} \to x^* \in B_p$, but one has both

$$x_{j'} \in \overline{\bigcup_{t\geq t_{j'}} \varphi\big(t, \theta_{-t}(p), D_{\theta_{-t}(p)}\big)}$$

and

$$\bigcap_{t_{j'}} \overline{\bigcup_{t\geq t_{j'}} \varphi\big(t, \theta_{-t}(p), D_{\theta_{-t}(p)}\big)} \subseteq A_p$$

by (7.14) and the definition (and existence) of a pullback absorbing system. Hence, both $x^* \in A_p$ and $V(p, x^*) = 0$ hold, but V is Lipschitz continuous in its second variable by (P3), so

$$\varepsilon_0 \leq V(p, x_{j'}) = |V(p, x_{j'}) - V(p, x^*)| \leq \|x_{j'} - x^*\|\,,$$

which contradicts the convergence $x_{j'} \to x^*$. This finishes the proof of (P4).
(P5): Define $D_p := B_p$ for each $p \in P$. The so-defined nonautonomous set $\mathcal{D}$ is positively invariant. It remains to establish the exponential decay inequality

(7.13). Note that the cocycle mapping φ, considered as the solution mapping of the nonautonomous differential equation (7.10), satisfies the Lipschitz condition

$$\|\varphi(t,p,x_0)-\varphi(t,p,y_0)\| \le e^{\int_0^t L(\theta_s(p))\,\mathrm{d}s}\|x_0-y_0\| \quad \text{for all } x_0,y_0\in\mathbb{R}^d,$$

from which it follows that

$$\operatorname{dist}\big(\varphi(t,p,x_0),\varphi(t,p,C_p)\big) \le e^{\int_0^t L(\theta_s(p))\,\mathrm{d}s}\operatorname{dist}(x_0,C_p)$$

for all nonempty compact subset C_p of $\mathbb{R}^d$. Now $\varphi(t,p,x_0)\in D_{\theta_t(p)}$ when $x_0\in D_p$. Re-indexing, then using the cocycle property and the above Lipschitz condition, thus gives

$$\begin{aligned}
&V(\theta_t(p),\varphi(t,p,x_0))\\
&= \sup_{s\ge 0} e^{-T_{\theta_t(p),s+t}}\operatorname{dist}\big(\varphi(t,p,x_0),\varphi(s+t,\theta_{-s}(p),B_{\theta_{-s}(p)})\big)\\
&= \sup_{s\ge 0} e^{-T_{\theta_t(p),s+t}}\operatorname{dist}\big(\varphi(t,p,x_0),\varphi(t,p,\varphi(s,\theta_{-s}(p),B_{\theta_{-s}(p)}))\big)\\
&\le \sup_{s\ge 0} e^{-T_{\theta_t(p),s+t}}e^{\int_0^t L(\theta_r(p))\,\mathrm{d}r}\operatorname{dist}\big(x_0,\varphi(s,\theta_{-s}(p),B_{\theta_{-s}(p)})\big)
\end{aligned}$$

However, $T_{\theta_t(p),s+t}=T_{p,s}+t+\int_0^t L(\theta_r(p))\,\mathrm{d}r$, so

$$\begin{aligned}
&V(\theta_t(p),\varphi(t,p,x_0))\\
&\le \sup_{s\ge 0} e^{-T_{p,s}-t-\int_0^t L(\theta_r(p))\,\mathrm{d}r+\int_0^t L(\theta_r(p))\,\mathrm{d}r}\operatorname{dist}\big(x_0,\varphi(s,\theta_{-s}(p),B_{\theta_{-s}(p)})\big)\\
&= \sup_{s\ge 0} e^{-T_{p,s}-t}\operatorname{dist}\big(x_0,\varphi(s,\theta_{-s}(p),B_{\theta_{-s}(p)})\big)\\
&= e^{-t}\sup_{s\ge 0} e^{-T_{p,s}}\operatorname{dist}\big(x_0,\varphi(s,\theta_{-s}(p),B_{\theta_{-s}(p)})\big) = e^{-t}V(p,x_0),
\end{aligned}$$

which is the desired inequality. This completes the proof of Theorem 7.10. □

PROOF OF COROLLARY 7.12. The above construction of the Lyapunov function for a pullback attractor will be modified here for uniform pullback attractors and a compact parameter space P. The essential change is to show that there exists a lower bound function $a(r)$ in (P2) that is independent of the parameter $p\in P$. As above, define $V(p,x_0)$ for all $p\in P$ and $x_0\in\mathbb{R}^d$ by

$$V(p,x_0) := \sup_{t\ge 0} e^{-T_{p,t}}\operatorname{dist}\big(x_0,\varphi(t,\theta_{-t}(p),B_{\theta_{-t}(p)})\big),$$

where $\mathcal{B}$ is a pullback absorbing neighborhood system for the pullback attractor $\mathcal{A}$ and

$$T_{p,t} = t+\int_0^t L(\theta_{-s}(p))\,\mathrm{d}s \quad \text{with} \quad T_{p,0}=0.$$

Note that $T_{p,t}$ satisfies $t\le T_{p,t}\le (1+L_{\max})t$ for all $t\ge 0$ and $p\in P$, where $L_{\max} := \max_{p\in P} L(p)$ is finite by the continuity assumptions and the compactness of P. By (P1), one gets $V(p,x_0)=0$ for all $x_0\in A_p$. Assume instead that $x_0\in\mathbb{R}^d\setminus A_p$. Now the supremum in the definition of $V(p,x_0)$ involves the product of an exponentially decreasing quantity bounded below by zero and a bounded increasing function, since the $\varphi(t,\theta_{-t}(p),B_{\theta_{-t}(p)})$ are a nested family of compact sets decreasing to A_p in the limit $t\to\infty$. Hence there exists a $T'\ge 0$ such that

$$\frac{1}{2}\operatorname{dist}(x_0,A_p) \le \operatorname{dist}\big(x_0,\varphi(t,\theta_{-t}(p),B_{\theta_{-t}(p)})\big)$$

for all $t \geq T'$, but not for $t < T'$. Thus, from above, one gets

$$V(p, x_0) \geq e^{-T_{p,T'}} \operatorname{dist}\left(x_0, \varphi(T', \theta_{-T'}(p), B_{\theta_{-T'}(p)})\right) \geq \frac{1}{2} e^{-T_{p,T'}} \operatorname{dist}(x_0, A_p).$$

The lower bound $a(p, r)$ in (P2) above was defined by

$$a(p, r) := \frac{1}{2} r\, e^{-T_{p,\widehat{T}(p,r)}},$$

where

$$\widehat{T}(p, r) := \sup\left\{T'(p, x_0) : x_0 \in \mathbb{R}^d \text{ and } \operatorname{dist}(x_0, A_p) = r\right\},$$

which is finite. Now by uniform pullback convergence, there exists a finite $T\left(\frac{r}{2}\right)$, which is independent of $p \in P$ and can be chosen to be nonincreasing in r with

$$h\left(\varphi(t, \theta_{-t}(p), B_{\theta_{-t}(p)}), A_p\right) < \frac{r}{2}$$

for all $t \geq T\left(\frac{r}{2}\right)$ and all $p \in P$. Hence

$$r \leq \operatorname{dist}\left(x_0, \varphi(t, \theta_{-t}(p), B_{\theta_{-t}(p)})\right) + \frac{r}{2}$$

for $\operatorname{dist}(x_0, A_p) = r$ and $t \geq T\left(\frac{r}{2}\right)$, i.e.,

$$\frac{r}{2} \leq \operatorname{dist}\left(x_0, \varphi(t, \theta_{-t}(p), B_{\theta_{-t}(p)})\right).$$

Thus, $\widehat{T}(p, r) \leq T\left(\frac{r}{2}\right) < \infty$ and so $T_{p,\widehat{T}(p,r)} \leq T_{p,T(r/2)} \leq (1 + L_{\max})T\left(\frac{r}{2}\right) < \infty$ for all $p \in P$. Finally, define

$$a(r) := \frac{r}{2}\, e^{-(1+L_{\max})T(r/2)} \leq a(p, r),$$

which satisfies the stated properties. □

4. Lyapunov functions for Morse decompositions

In this section, Lyapunov functions for Morse decompositions, which are constant on the Morse sets and which strictly decrease outside them are obtained explicitly.

Consider the situation from Chapter 4, more precisely, let $(t, t_0, x_0) \mapsto \phi(t, t_0, x_0)$ be an invertible process on a compact and invariant phase space $\mathcal{X}$, and let $\mathbb{T} = \mathbb{Z}$ (discrete time) or $\mathbb{T} = \mathbb{R}$ (continuous time).

Firstly, the case that the past Morse decomposition is given by a past attractor-repeller pair is treated.

LEMMA 7.15. *Let $(\mathcal{R}^*, \mathcal{R})$ be a past attractor-repeller pair. Then there exists a continuous function $V : \mathcal{X} \to [0, 1]$ such that $V|_{\mathcal{R}^*} \equiv 0$, $V|_{\mathcal{R}} \equiv 1$ and*

$$V\left(t, \phi(t, t_0, x_0)\right) < V(t_0, x_0) \quad \text{for all } t > 0 \text{ and } (t_0, x_0) \in \mathcal{X} \setminus (\mathcal{R}^* \cup \mathcal{R})$$

is satisfied.

PROOF. Define the function $W : \mathcal{X} \to [0, 1]$ by

$$W(t, x) := \frac{\operatorname{dist}(x, R_t^*)}{\operatorname{dist}(x, R_t^*) + \operatorname{dist}(x, R_t)} \quad \text{for all } (t, x) \in \mathcal{X}.$$

This function is continuous and fulfills $W|_{R^*} \equiv 0$, $W|_R \equiv 1$, but is not necessarily decreasing along solutions. To reach this, define by

$$\bar{W}(t,x) := \inf_{s \le t} W\big(s, \phi(s,t,x)\big) \quad \text{for all } (t,x) \in \mathcal{X}$$

a function $\bar{W} : \mathcal{X} \to [0,1]$, which obviously satisfies $\bar{W}|_{\mathcal{R}^*} \equiv 0$, $\bar{W}|_{\mathcal{R}} \equiv 1$ and $\bar{W}(t, \phi(t,t_0,x_0)) \le \bar{W}(t_0,x_0)$ for all $(t_0,x_0) \in \mathcal{X}$. The proof that $\bar{W}$ is continuous is left as exercise to the reader. Note that $\bar{W}$ is not strictly decreasing along solutions in $\mathcal{X} \setminus (\mathcal{R}^* \cup \mathcal{R})$. Therefore, define V to be a weighted average of $\bar{W}$ over the backward solution:

$$V(t,x) := \int_0^\infty e^{-s} \bar{W}\big(t-s, \phi(t-s,t,x)\big)\, \mathrm{d}s \quad \text{for all } (t,x) \in \mathcal{X}\,.$$

This function is obviously continuous, and for all $t \ge t_0$,

$$\begin{aligned} V(t, \phi(t,t_0,x_0)) &= \int_0^\infty e^{-s} \bar{W}\big(t-s, \phi(t-s,t,\phi(t,t_0,x_0))\big)\, \mathrm{d}s \\ &\le \int_0^\infty e^{-s} \bar{W}\big(t_0-s, \phi(t_0-s,t_0,x_0)\big)\, \mathrm{d}s = V(t_0,x_0)\,. \end{aligned}$$

To prove that V is also strictly decreasing along solutions in $\mathcal{X} \setminus (\mathcal{R}^* \cup \mathcal{R})$, suppose that one has $V\big(t, \phi(t,t_0,x_0)\big) = V(t_0,x_0)$ for some $x_0 \in \mathcal{X} \setminus (\mathcal{R}^* \cup \mathcal{R})$ and $t > t_0$. Then $\bar{W}\big(t_0-s, \phi(t_0-s,t_0,x_0)\big) = \bar{W}\big(t-s, \phi(t-s,t_0,x_0)\big)$ for all $s \ge 0$. This is impossible, since $\bar{W}(t_0,x_0) \in (0,1)$ and $\lim_{s\to\infty} \bar{W}\big(t_0-s, \phi(t_0-s,t_0,x_0)\big) = 0$ due to Theorem 4.4. □

In the next theorem, the Lyapunov function is extended from past attractor-repeller pairs to past Morse decompositions.

THEOREM 7.16. *Consider a past Morse decomposition* $\{\mathcal{M}^1, \dots, \mathcal{M}^n\}$. *Then there exists a function* $V : \mathcal{X} \to [0,1]$, *which is continuous with respect to* $x \in X$ *such that* $V|_{\mathcal{M}^i} \equiv \frac{i-1}{n-1}$ *for* $i \in \{1, \dots, n\}$ *and*

$$V\big(t, \phi(t,t_0,x_0)\big) < V(t_0,x_0) \quad \textit{for all } t > t_0 \textit{ and } (t_0,x_0) \notin \mathcal{M}^1 \cup \dots \cup \mathcal{M}^n$$

is satisfied.

PROOF. Let $\mathcal{X} = \mathcal{R}^0 \supsetneq \mathcal{R}^1 \supsetneq \dots \supsetneq \mathcal{R}^n = \emptyset$ be the sequence of local forward repellers leading to the given past Morse decomposition, i.e.,

$$\mathcal{M}^i = \mathcal{R}^{i,*} \cap \mathcal{R}^{i-1} \quad \text{for all } i \in \{1, \dots, n\}\,.$$

Furthermore, let V_i, $i \in \{1, \dots, n-1\}$, be the Lyapunov function corresponding to the past attractor-repeller pair $(\mathcal{R}^{i,*}, \mathcal{R}^i)$ as introduced in Lemma 7.15. Define

$$V(t,x) := \frac{1}{n-1} \sum_{i=1}^{n-1} V_i(t,x) \quad \text{for all } (t,x) \in \mathcal{X}\,.$$

Choose $(t,x) \in \mathcal{M}^i$ arbitrarily, and let $j \in \{1, \dots, n-1\}$. Then $(t,x) \in \mathcal{R}^j$ if and only if $j \in \{1, \dots, i-1\}$, and $(t,x) \in \mathcal{R}^{j,*}$ if and only if $j \in \{i, \dots, n-1\}$. This implies $V|_{\mathcal{M}^i} \equiv \frac{i-1}{n-1}$ for all $i \in \{1, \dots, n\}$. Now choose an element $(t,x) \in \mathcal{X} \setminus (\mathcal{M}^1 \cup \dots \cup \mathcal{M}^n)$. Then there exists a $j \in \{1, \dots, n\}$ with $(t,x) \notin \mathcal{R}^{j,*} \cup \mathcal{R}^j$. This means that $V_j\big(t, \phi(t,t_0,x_0)\big) < V_j(t_0,x_0)$ for all $t > t_0$ and finishes the proof of this theorem. □

Endnotes. Lyapunov functions have been studied extensively in the literature (see, e.g., Bhatia & Szegö [**23**], Sibirsky [**223**] and Stuart & Humphries [**226**], and see Giesl [**82**] for approximations of Lyapunov functions). For the nonautonomous case in particular, the references Grüne, Kloeden, Siegmund & Wirth [**85**], Kato, Martynyuk & Shestakov [**104**], Sell [**216**] and Yoshizawa [**239**] are valuable sources of necessary and sufficient conditions for Lyapunov functions for different kinds of stability and boundedness properties. The sufficient conditions for stability and attractivity of solutions from Section 1 can be found in Hahn [**89**] and Knobloch & Kappel [**147**]. The construction of the Lyapunov function for the global attractor of an autonomous difference equation (Section 2) is from Diamond & Kloeden [**70**] and based on a more complicated construction for ordinary differential equations by Yoshizawa [**239**]. The construction of the Lyapunov function for pullback attractors (Section 3) is from Kloeden [**114, 115**]. Further developments can be found in Grüne, Kloeden, Siegmund & Wirth [**85**]. The construction of a Lyapunov function for nonautonomous Morse decompositions (Section 4) is due to Rasmussen [**195**] and is similar to the autonomous case discussed in Conley [**56**]. Finally, note that Lyapunov's second method in the context of random dynamical systems is treated in Arnold & Schmalfuß [**8**].

CHAPTER 8

Bifurcations

Just as simple examples played a critical role in the development of bifurcation theory for autonomous systems by encapsulating the salient features and mechanisms of various typical bifurcation scenarios, they also provide much useful insight into what might happen when nonautonomous dynamical systems undergo a bifurcation.

Bifurcation theory for nonautonomous systems has been mainly developed for low-dimensional systems so far, and there are only few general results in higher dimensions.

1. Nonautonomous Bernoulli equations

The autonomous scalar differential equation

$$\dot{x} = ax - bx^3 \tag{8.1}$$

with $b > 0$ has the global attractor $A_a = \{0\}$ for $a < 0$. For $a = 0$, the equilibrium point 0 loses attractivity for the linearized equation (however, it remains stable there), but the set $A_0 = \{0\}$ is still a global attractor for the nonlinear system. To see this, use the Lyapunov function $V(x) = x^2$ to obtain

$$\frac{\mathrm{d}}{\mathrm{d}t} V(x_0(t)) = -2bx_0(t)^4 = -2bV(x_0(t))^2 \,,$$

and hence,

$$V(x_0(t)) = \frac{V(x_0(0))}{1 + 2bV(x_0(0))t} \to 0 \quad \text{as } t \to \infty \,,$$

where $x_0(t)$ is any solution of the differential equation (8.1) for $a = 0$. Note that this Lyapunov function can also be used for $a < 0$. For $a > 0$, however, there are the three equilibrium points 0 and $\pm\sqrt{a/b}$, and the global attractor is now given by

$$A_a = \left[-\sqrt{\tfrac{a}{b}}, \sqrt{\tfrac{a}{b}}\right] \quad \text{for all } a \geq 0 \,,$$

i.e., a supercritical pitchfork bifurcation has occurred at $a = 0$.

Now consider a nonautonomous analogue of the scalar differential equation (8.1), namely

$$\dot{x} = ax - b(t)x^3, \tag{8.2}$$

with the function $b : \mathbb{R} \to \mathbb{R}$ is continuous satisfying $b(t) \in [b_0, b_1]$ for all $t \in \mathbb{R}$ with $0 < b_0 < b_1 < \infty$. Assuming $b(t)$ is nonconstant, the nonautonomous differential equation has exactly one equilibrium point, namely 0, and this exists for all values of a. Supposing that $a \le 0$, using the Lyapunov function $V(x) = x^2$ shows that

$$\begin{aligned}\frac{\mathrm{d}}{\mathrm{d}t}V(x_a(t)) &= 2ax_a(t)^2 - 2b(t)x_a(t)^4 \le 2aV(x_a(t)) - 2b_0V(x_a(t))^2 \\ &\le -2b_0V(x_a(t))^2\,,\end{aligned}$$

which means that

$$V(x_a(t)) \le \frac{V(x_a(t_0))}{1 + 2b_0V(x_a(t_0))(t-t_0)} \to 0 \quad \text{as } t \to \infty\,,$$

where $x_a(t)$ is any solution of the differential equation (8.2) for $a \le 0$. Moreover, the equilibrium point 0 loses stability in the linearized equation at $a = 0$ and is unstable for $a > 0$.

Unlike the autonomous case above, no new equilibrium points come into existence when $a > 0$. Instead, it will be shown that there is a family of time-dependent sets

$$A_a(t) = \left[-\xi_a(t), \xi_a(t)\right] \quad \text{for all } t \in \mathbb{R}\,, \tag{8.3}$$

which is pullback attracting and also uniformly Lyapunov asymptotically stable. Here ξ_a is given by

$$\xi_a(t) = \frac{1}{\sqrt{2\int_{-\infty}^{t} b(s)e^{-2a(t-s)}\,\mathrm{d}s}}\,, \tag{8.4}$$

and both ξ_a and $-\xi_a$ are entire solutions of the nonautonomous differential equation (8.2) with $\xi_a(t) \in \left[\sqrt{a/b_1}, \sqrt{a/b_0}\right]$ for all $t \in \mathbb{R}$. In particular, this means that

$$\operatorname{dist}\left(A_a(t), \{0\}\right) \le \sqrt{a/b_0} \to 0 \quad \text{as } a \to 0 \quad \text{for all } t \in \mathbb{R}\,.$$

Note that ξ_a is almost periodic when the coefficient b is almost periodic by Theorem 3.46 and Theorem 3.47.

The proofs of the above assertions are instructive. Firstly, note that (8.2) is a Bernoulli equation, so it can be transformed into a linear differential equation

$$\dot{v} + 2av = 2b(t)$$

with the substitution $v = x^{-2}$ (recall that the singular point $x = 0$ is an equilibrium of the Bernoulli equation (8.2)), which can be integrated to give

$$\frac{1}{\phi_a(t,t_0,x_0)^2} = \frac{1}{x_0^2}e^{-2a(t-t_0)} + 2\int_{t_0}^{t} b(s)e^{-2a(t-s)}\,\mathrm{d}s\,. \tag{8.5}$$

Now hold t and x_0 fixed in (8.5) and take the *pullback* limit as $t_0 \to -\infty$, to obtain the pullback limit solution

$$\xi_a(t) = \lim_{t_0 \to -\infty} \phi_a(t,t_0,x_0)$$

given by (8.4). This is itself a solution of the differential equation (8.2) and thus satisfies (8.5) with $x_0 = \xi_a(t_0)$, specifically,

$$\frac{1}{\xi_a(t)^2} = \frac{1}{\xi_a(t_0)^2}e^{-2a(t-t_0)} + 2\int_{t_0}^{t} b(s)e^{-2a(t-s)}\,\mathrm{d}s. \tag{8.6}$$

To show that $\xi_a(t)$ is asymptotically stable for all $x_0 > 0$, subtract (8.5) from (8.6) to obtain

$$\frac{1}{\xi_a(t)^2} - \frac{1}{x_a(t,t_0,x_0)^2} = \left(\frac{1}{\xi_a(t_0)^2} - \frac{1}{x_0^2}\right) e^{-2a(t-t_0)} \to 0 \quad \text{as } t \to \infty\,,$$

from which the result follows. It is in fact uniformly asymptotically stable because, using the fact that $\xi_a(t) \in \left[\sqrt{a/b_1}, \sqrt{a/b_0}\right]$, for every $\varepsilon > 0$ and $x_0 > \xi_a(t_0)$, there is a $T(x_0,\varepsilon) \geq 0$ such that

$$\left(\frac{1}{\xi_a(t_0)^2} - \frac{1}{x_0^2}\right) e^{-2a(t-t_0)} \leq \left(\frac{b_1}{a} - \frac{1}{x_0^2}\right) e^{-2a(t-t_0)} < \varepsilon \tag{8.7}$$

for $t \geq T(x_0,\varepsilon) + t_0$, and similarly if $0 < x_0 < \xi_a(t_0)$. The corresponding result holds for the solution $-\xi_a(t)$ for all $x_0 < 0$.

Note here that the nonzero solutions $\phi_a(t,t_0,x_0)$ of the differential equation (8.2) converge to $\operatorname{sgn}(x_0)\xi_a(t)$ in both the usual forward sense (i.e., $t \to \infty$ with t_0 fixed) as well as in the pullback sense (i.e., $t_0 \to -\infty$ with t fixed).

2. One-dimensional bifurcation patterns

This section is devoted to the study of nonautonomous analogues of both the transcritical and pitchfork bifurcation. The nonautonomous bifurcations occur, in particular, when under variation of the parameter, either a radius of attraction/repulsion shrinks down to zero, or a nontrivial attractor/repeller becomes trivial.

2.1. Nonautonomous transcritical bifurcation. In case of the transcritical bifurcation, only the bifurcation of a radius of attraction/repulsion can be observed.

THEOREM 8.1 (Nonautonomous transcritical bifurcation). *Let $x_- < 0 < x_+$ and $\nu_- < \nu_+$ be in $\overline{\mathbb{R}}$ and $\mathbb{I}$ be an unbounded interval of the form $\mathbb{R}_0^+$, $\mathbb{R}_0^-$ or $\mathbb{R}$, respectively, and consider the nonautonomous differential equation*

$$\dot{x} = a(t,\nu)x + b(t,\nu)x^2 + r(t,x,\nu) \tag{8.8}$$

with continuous functions $a : \mathbb{I} \times (\nu_-,\nu_+) \to \mathbb{R}$, $b : \mathbb{I} \times (\nu_-,\nu_+) \to \mathbb{R}$ and $r : \mathbb{I} \times (x_-,x_+) \times (\nu_-,\nu_+) \to \mathbb{R}$ fulfilling $r(\cdot,0,\cdot) \equiv 0$. Let $\Phi_\nu : \mathbb{I} \times \mathbb{I} \to \mathbb{R}$ denote the transition operator of the linearized equation $\dot{x} = a(t,\nu)x$, and assume that there exists an $\nu_0 \in (\nu_-,\nu_+)$ such that the following hypotheses hold:

(i) Hypothesis on linear part. *There exist two functions $\beta_1, \beta_2 : (\nu_-,\nu_+) \to \mathbb{R}$ which are either both monotone increasing or both monotone decreasing and $K \geq 1$ such that $\lim_{\nu\to\nu_0} \beta_1(\nu) = \lim_{\nu\to\nu_0} \beta_2(\nu) = 0$ and*

$$\Phi_\nu(t,s) \leq Ke^{\beta_1(\nu)(t-s)} \quad \text{for all } \nu \in (\nu_-,\nu_+) \text{ and } t,s \in \mathbb{I} \text{ with } t \geq s\,,$$
$$\Phi_\nu(t,s) \leq Ke^{\beta_2(\nu)(t-s)} \quad \text{for all } \nu \in (\nu_-,\nu_+) \text{ and } t,s \in \mathbb{I} \text{ with } t \leq s\,.$$

(ii) Hypothesis on nonlinearity. *The quadratic term either fulfills*

$$0 < \liminf_{\nu\to\nu_0} \inf_{t\in\mathbb{I}} b(t,\nu) \leq \limsup_{\nu\to\nu_0} \sup_{t\in\mathbb{I}} b(t,\nu) < \infty \tag{8.9}$$

or

$$-\infty < \liminf_{\nu\to\nu_0} \inf_{t\in\mathbb{I}} b(t,\nu) \leq \limsup_{\nu\to\nu_0} \sup_{t\in\mathbb{I}} b(t,\nu) < 0\,, \tag{8.10}$$

and the remainder satisfies

$$\lim_{x\to 0}\sup_{\nu\in(\nu_0-|x|,\nu_0+|x|)}\sup_{t\in\mathbb{I}}\frac{|r(t,x,\nu)|}{|x|^2}=0 \tag{8.11}$$

and

$$\limsup_{\nu\to\nu_0}\limsup_{x\to 0}\sup_{t\in\mathbb{I}}\frac{2K|r(t,x,\nu)|}{|x|\max\{\beta_1(\nu),\beta_2(\nu)\}}<1\,. \tag{8.12}$$

Then there exist $\hat{\nu}_- < 0 < \hat{\nu}_+$ such that the following statements are fulfilled:

(i) *In case the functions β_1 and β_2 are monotone increasing, the trivial solution is locally forward (pullback, uniformly, respectively) attractive for $\nu \in (\hat{\nu}_-, \nu_0)$ and locally forward (pullback, uniformly, respectively) repulsive for $\nu \in (\nu_0, \hat{\nu}_+)$. The radii of attraction $\mathfrak{A}^\nu$ and the radii of repulsion $\mathfrak{R}^\nu$ satisfy*

$$\lim_{\nu\nearrow\nu_0}\mathfrak{A}_0^\nu=0 \quad \text{and} \quad \lim_{\nu\searrow\nu_0}\mathfrak{R}_0^\nu=0\,.$$

(ii) *In case the functions β_1 and β_2 are monotone decreasing, the trivial solution is locally forward (pullback, uniformly, respectively) repulsive for $\nu \in (\hat{\nu}_-, \nu_0)$ and locally forward (pullback, uniformly, respectively) attractive for $\nu \in (\nu_0, \hat{\nu}_+)$. The radii of repulsion $\mathfrak{R}^\nu$ and the radii of attraction $\mathfrak{A}^\nu$ satisfy*

$$\lim_{\nu\nearrow\nu_0}\mathfrak{R}_0^\nu=0 \quad \text{and} \quad \lim_{\nu\searrow\nu_0}\mathfrak{A}_0^\nu=0\,.$$

The two cases of the preceding theorem are summarized in the following table.

trivial sol.	$\nu \in (\hat{\nu}_-, \nu_0)$	$\nu \in (\nu_0, \hat{\nu}_+)$
β_1, β_2 incr.	attractive, $\lim_{\nu\nearrow\nu_0}\mathfrak{A}_0^\nu = 0$	repulsive, $\lim_{\nu\searrow\nu_0}\mathfrak{R}_0^\nu = 0$
β_1, β_2 decr.	repulsive, $\lim_{\nu\nearrow\nu_0}\mathfrak{R}_0^\nu = 0$	attractive, $\lim_{\nu\searrow\nu_0}\mathfrak{A}_0^\nu = 0$

PROOF. First of all, one may assume that $K > 1$. Let ϕ_ν be the process induced by (8.8). Only the proof of assertion (i) will be given, since the proof of (ii) is similar. Hence, the functions β_1 and β_2 are monotone increasing, and only the case (8.9) is treated. Choose $\hat{\nu}_- < \nu_0 < \hat{\nu}_+$ such that

$$0<\inf_{\nu\in(\hat{\nu}_-,\hat{\nu}_+),\,t\in\mathbb{I}} b(t,\nu)\le \sup_{\nu\in(\hat{\nu}_-,\hat{\nu}_+),\,t\in\mathbb{I}} b(t,\nu)<\infty \tag{8.13}$$

(cf. (8.9)) and

$$\limsup_{x\to 0}\sup_{t\in\mathbb{I}}\frac{|r(t,x,\nu)|}{|x|}<\frac{-\min\{\beta_1(\nu),-\beta_2(\nu)\}}{2K} \quad \text{for all } \nu\in(\hat{\nu}_-,\hat{\nu}_+)$$

(cf. (8.12)). Because of these two relations, Theorem 3.54 can be applied, and the attractivity and repulsivity of the trivial solutions as stated in the theorem follows. Assume to the contrary that

$$\eta := \limsup_{\nu\nearrow\nu_0}\mathfrak{A}_0^\nu > 0$$

holds. Due to (8.13) and (8.11), there exist $\tilde{\nu}_- \in (\hat{\nu}_-, \nu_0)$, $x_0 \in (0, \eta)$ and $L \in (0, \frac{x_0}{4K})$ with

$$b(t,\nu)x^2+r(t,x,\nu)>L \quad \text{for all } t\in\mathbb{I},\ \nu\in(\tilde{\nu}_-,\nu_0) \text{ and } x\in\left[\frac{x_0}{2K^2},x_0\right]. \tag{8.14}$$

Fix $\hat{\nu} \in (\tilde{\nu}_-, \nu_0)$ such that $\mathfrak{A}_0^{\hat{\nu}} > x_0$ and $\beta_2(\hat{\nu}) \ge \beta := -\frac{2KL}{x_0} > -\frac{1}{2}$. For arbitrary $t_0 \in \mathbb{I}$, the solution $\mu_{t_0}(\cdot) := \phi_{\hat{\nu}}(\cdot, t_0, x_0)$ of (8.8) is also a solution of the inhomogeneous linear differential equation

$$\dot{x} = a(t,\hat{\nu})x + b(t,\hat{\nu})(\mu_{t_0}(t))^2 + r(t, \mu_{t_0}(t), \hat{\nu}) \,. \tag{8.15}$$

Since $\mathfrak{A}_0^{\hat{\nu}} > x_0 = \mu_t(t)$ for all $t \in \mathbb{I}$, there exist $t_0, t_2 \in \mathbb{I}$, $t_0 < t_2$, with $\mu_{t_0}(t_2) \le \frac{x_0}{2K^2}$. Choose t_2 minimal with this property, i.e., $\mu_{t_0}(t) \ge \frac{x_0}{2K^2}$ for all $t \in [t_0, t_2]$. Furthermore, choose $t_1 \in [t_0, t_2]$ such that

$$\mu_{t_0}(t_1) = \frac{x_0}{2K} \quad \text{and} \quad \mu_{t_0}(t) \in \left[\frac{x_0}{2K^2}, x_0\right] \quad \text{for all } t \in [t_1, t_2] \,.$$

Therefore, and due to (8.14) and the variation of constants formula, applied to (8.15), the relation

$$\begin{aligned}
&\mu_{t_0}(t_2) \\
&= \Phi_{\hat{\nu}}(t_2, t_1)\mu_{t_0}(t_1) + \int_{t_1}^{t_2} \Phi_{\hat{\nu}}(t_2, t)\big(b(t,\hat{\nu})(\mu_{t_0}(t))^2 + r(t, \mu_{t_0}(t), \hat{\nu})\big)\, dt \\
&> \frac{x_0}{2K^2} e^{\beta(t_2 - t_1)} + \frac{L}{K}\int_{t_1}^{t_2} e^{\beta(t_2 - t)}\, dt \\
&= e^{\beta(t_2 - t_1)} \underbrace{\left(\frac{x_0}{2K^2} + \frac{L}{K\beta}\right)}_{=0} - \frac{L}{K\beta} = \frac{x_0}{2K^2}
\end{aligned}$$

holds ($K > 1$ implies $t_1 < t_2$). This is a contradiction and proves the relation $\lim_{\nu \nearrow \nu_0} \mathfrak{A}_0^{\nu} = 0$. Analogously, one can show $\lim_{\nu \searrow \nu_0} \mathfrak{R}_0^{\nu} = 0$ and treat the case (8.10). □

In the limit $\nu \to \nu_0$, the attractivity or repulsivity of the trivial solution is only lost in one direction. For instance, in case the functions β_1, β_2 are monotone increasing and (8.9) is satisfied, there exists a $\gamma < 0$ such that $(\gamma, 0]$ is attracted by the trivial solution of (8.11) for $\nu \in (\hat{\nu}_-, \nu_0)$ in the sense of forward, pullback or uniform attractivity, respectively. Since the loss of stability is only one-sided, there are no attractor and repeller bifurcations connected with the transcritical bifurcation, contrary to the pitchfork bifurcation in the next section.

REMARK 8.2. The hypothesis on the linear part implies that the future (past, all-time, respectively) dichotomy spectrum of the linearization $\dot{x} = a(t,\nu)x$ converges to $\{0\}$ in Hausdorff distance in the limit $\nu \to \nu_0$. In addition, condition (8.12) is only used to obtain the attractivity or repulsivity of the trivial solution by applying Theorem 3.54. Alternatively, one can directly postulate that the trivial solution changes their stability at the parameter value ν_0 from, say, attractivity to repulsivity.

Note that the above bifurcation result is essentially the combination of two scenarios which are independent of each other. This means that it is possible to consider (8.8) only for $\nu > \nu_0$ or $\nu < \nu_0$, respectively, in order to obtain the results which apply for these parameter values.

The following example shows that Theorem 8.1 is indeed a nonautonomous generalization of the well-known autonomous result.

EXAMPLE 8.3. Let $x_- < 0 < x_+$ and $\nu_- < 0 < \nu_+$ be in $\overline{\mathbb{R}}$, and consider the autonomous differential equation

$$\dot{x} = f(x, \nu), \tag{8.16}$$

where the C^4-function $f : (x_-, x_+) \times (\nu_-, \nu_+) \to \mathbb{R}$ satisfies the following assumptions:

(i) $f(0, \nu) = 0 \quad \text{for all } \nu \in (\nu_-, \nu_+)$,
(ii) $D_1 f(0, 0) = 0$,
(iii) $D_1 D_2 f(0, 0) \neq 0$,
(iv) $D_1^2 f(0, 0) \neq 0$.

Note that (i) implies $D_2^n f(0, \nu) = 0$ for all $\nu \in (\nu_-, \nu_+)$ and $n \in \mathbb{N}$. Then (8.16) admits an autonomous transcritical bifurcation (see, e.g., WIGGINS [**235**, p. 265 f.]), i.e., there exist a neighborhood $U \times V$ of $(0, 0)$ in $\mathbb{R}^2$ and a C^1-function $h : U \to V$ with $h(0) = 0$ and

$$f(x, h(x)) = 0 \quad \text{for all } x \in U.$$

Except the trivial equilibria and the equilibria described by h, there are no other equilibria in $U \times V$. It will be shown that this example fulfills the hypotheses of Theorem 8.1. Thereto, write the second order Taylor expansion of f:

$$f(x, \nu) = \underbrace{D_1 D_2 f(0, 0)\nu}_{=:\bar{a}(\nu)} x + \underbrace{\frac{1}{2} D_1^2 f(0, 0)}_{=:\bar{b}(\nu)} x^2 + r(x, \nu),$$

where

$$r(x, \nu) = \int_0^1 \frac{(1-t)^2}{2} \big(D_1^3 f(tx, t\nu)x^3 + 3D_1^2 D_2 f(tx, t\nu)x^2\nu + 3D_1 D_2^2 f(tx, t\nu)x\nu^2 + D_2^3 f(tx, t\nu)\nu^3\big)\, dt.$$

Obviously, the hypothesis on the linear part are fulfilled (with $\beta_1(\nu) := \beta_2(\nu) := \bar{a}(\nu)$ and $K := 1$), and (8.9) or (8.10) holds, since the above defined function $\bar{b}$ is constant. Furthermore, the representation for the remainder implies that

$$\lim_{x \to 0} \sup_{\nu \in (-|x|, |x|)} \frac{|r(x, \nu)|}{|x|^2} = 0$$

and

$$\limsup_{x \to 0} \frac{|r(x, \nu)|}{|x|} \leq \nu^2 \int_0^1 \frac{(1-t)^2}{2} \big(|3D_1 D_2^2 f(0, t\nu)| + t|D_1 D_2^3 f(0, t\nu)\nu|\big)\, dt.$$

This means that (8.12) holds, since $\max\{\beta_1(\nu), \beta_2(\nu)\}$ depends linearly in ν. Hence, all hypotheses of Theorem 8.1 are fulfilled, and thus, this example shows that Theorem 8.1 is a proper generalization of the well-known autonomous transcritical bifurcation pattern.

2.2. Nonautonomous pitchfork bifurcation. This section is devoted to a nonautonomous generalization of the classical pitchfork bifurcation. In addition to the bifurcations of radii of attraction and repulsion, also attractor and repeller bifurcations can be observed.

THEOREM 8.4 (Nonautonomous pitchfork bifurcation). *Let $x_- < 0 < x_+$ and $\nu_- < \nu_+$ be in $\overline{\mathbb{R}}$ and $\mathbb{I}$ be an unbounded interval of the form $\mathbb{R}_0^+$, $\mathbb{R}_0^-$ or $\mathbb{R}$, respectively, and consider the nonautonomous differential equation*

$$\dot{x} = a(t,\nu)x + b(t,\nu)x^3 + r(t,x,\nu) \tag{8.17}$$

with continuous functions $a : \mathbb{I} \times (\nu_-, \nu_+) \to \mathbb{R}$, $b : \mathbb{I} \times (\nu_-, \nu_+) \to \mathbb{R}$ and $r : \mathbb{I} \times (x_-, x_+) \times (\nu_-, \nu_+) \to \mathbb{R}$ fulfilling $r(\cdot, 0, \cdot) \equiv 0$. Let $\Phi_\nu : \mathbb{I} \times \mathbb{I} \to \mathbb{R}$ be the transition operator of the linearized equation $\dot{x} = a(t,\nu)x$, and assume that there exists an $\nu_0 \in (\nu_-, \nu_+)$ such that the following hypotheses hold:

(i) Hypothesis on linear part. *There exist two functions $\beta_1, \beta_2 : (\nu_-, \nu_+) \to \mathbb{R}$ which are either both monotone increasing or both monotone decreasing and $K \geq 1$ such that $\lim_{\nu \to \nu_0} \beta_1(\nu) = \lim_{\nu \to \nu_0} \beta_2(\nu) = 0$ and*

$$\Phi_\nu(t,s) \leq Ke^{\beta_1(\nu)(t-s)} \quad \text{for all } \nu \in (\nu_-, \nu_+) \text{ and } t, s \in \mathbb{I} \text{ with } t \geq s\,,$$
$$\Phi_\nu(t,s) \leq Ke^{\beta_2(\nu)(t-s)} \quad \text{for all } \nu \in (\nu_-, \nu_+) \text{ and } t, s \in \mathbb{I} \text{ with } t \leq s\,.$$

(ii) Hypothesis on nonlinearity. *The cubic term either fulfills*

$$0 < \liminf_{\nu \to \nu_0} \inf_{t \in \mathbb{I}} b(t,\nu) \leq \limsup_{\nu \to \nu_0} \sup_{t \in \mathbb{I}} b(t,\nu) < \infty \tag{8.18}$$

or

$$-\infty < \liminf_{\nu \to \nu_0} \inf_{t \in \mathbb{I}} b(t,\nu) \leq \limsup_{\nu \to \nu_0} \sup_{t \in \mathbb{I}} b(t,\nu) < 0\,, \tag{8.19}$$

and the remainder satisfies

$$\lim_{x \to 0} \sup_{\nu \in (\nu_0 - x^2, \nu_0 + x^2)} \sup_{t \in \mathbb{I}} \frac{|r(t,x,\nu)|}{|x|^3} = 0 \tag{8.20}$$

and

$$\limsup_{\nu \to \nu_0} \limsup_{x \to 0} \sup_{t \in \mathbb{I}} \frac{2K|r(t,x,\nu)|}{|x| \max\{\beta_1(\nu), \beta_2(\nu)\}} < 1\,.$$

Then there exist $\hat{\nu}_- < 0 < \hat{\nu}_+$ such that the following statements are fulfilled:

(i) *In case (8.18) and the functions β_1 and β_2 are monotone increasing, the trivial solution is locally forward (pullback, uniformly, respectively) attractive for $\nu \in (\hat{\nu}_-, \nu_0)$ and locally forward (pullback, uniformly, respectively) repulsive for $\nu \in (\nu_0, \hat{\nu}_+)$. The corresponding radii of attraction $\mathfrak{A}^\nu$ satisfy*

$$\lim_{\nu \nearrow \nu_0} \mathfrak{A}^\nu = 0\,.$$

If, in addition, $\mathbb{I} = \mathbb{R}_0^+$, then, for $\nu \in (\hat{\nu}_-, \nu_0)$, there exists a nontrivial local forward repeller $\mathcal{R}^\nu \subset \mathbb{I} \times \mathbb{R}$, which admits a bifurcation of the form

$$\lim_{\nu \nearrow \nu_0} h\big(R_t^\nu, \{0\}\big) = 0 \quad \text{for all } t \in \mathbb{I}\,.$$

(ii) *In case (8.19) and the functions β_1 and β_2 are monotone increasing, the trivial solution is locally forward (pullback, uniformly, respectively) attractive for $\nu \in (\hat{\nu}_-, \nu_0)$ and locally forward (pullback, uniformly, respectively) repulsive for $\nu \in (\nu_0, \hat{\nu}_+)$. The corresponding radii of repulsion $\mathfrak{R}^\nu$ satisfy*

$$\lim_{\nu \searrow \nu_0} \mathfrak{R}^\nu = 0\,.$$

If, in addition, $\mathbb{I} = \mathbb{R}_0^-$*, then, for* $\nu \in (\nu_0, \hat{\nu}_+)$*, there exists a nontrivial local pullback attractor* $\mathcal{A}^\nu \subset \mathbb{I} \times \mathbb{R}$*, which admits a bifurcation of the form*

$$\lim_{\nu \searrow \nu_0} h\big(A_t^\nu, \{0\}\big) = 0 \quad \text{for all } t \in \mathbb{I}\,.$$

(iii) *In case* (8.18) *and the functions* β_1 *and* β_2 *are monotone decreasing, the trivial solution is locally forward (pullback, uniformly, respectively) repulsive for* $\nu \in (\hat{\nu}_-, \nu_0)$ *and locally forward (pullback, uniformly, respectively) attractive for* $\nu \in (\nu_0, \hat{\nu}_+)$*. The corresponding radii of attraction* $\mathfrak{A}^\nu$ *satisfy*

$$\lim_{\nu \searrow \nu_0} \mathfrak{A}^\nu = 0\,.$$

If, in addition, $\mathbb{I} = \mathbb{R}_0^+$*, then, for* $\nu \in (\nu_0, \hat{\nu}_+)$*, there exists a nontrivial local forward repeller* $\mathcal{R}^\nu \subset \mathbb{I} \times \mathbb{R}$*, which admits a bifurcation of the form*

$$\lim_{\nu \searrow \nu_0} h\big(R_t^\nu, \{0\}\big) = 0 \quad \text{for all } t \in \mathbb{I}\,.$$

(iv) *In case* (8.19) *and the functions* β_1 *and* β_2 *are monotone decreasing, the trivial solution is locally forward (pullback, uniformly, respectively) repulsive for* $\nu \in (\hat{\nu}_-, \nu_0)$ *and locally forward (pullback, uniformly, respectively) attractive for* $\nu \in (\nu_0, \hat{\nu}_+)$*. The corresponding radii of repulsion* $\mathfrak{R}^\nu$ *satisfy*

$$\lim_{\nu \nearrow \nu_0} \mathfrak{R}^\nu = 0\,.$$

If, in addition, $\mathbb{I} = \mathbb{R}_0^-$*, then, for* $\nu \in (\hat{\nu}_-, \nu_0)$*, there exists a nontrivial local pullback attractor* $\mathcal{A}^\nu \subset \mathbb{I} \times \mathbb{R}$*, which admits a bifurcation of the form*

$$\lim_{\nu \nearrow \nu_0} h\big(A_t^\nu, \{0\}\big) = 0 \quad \text{for all } t \in \mathbb{I}\,.$$

The four cases of the preceding theorem are summarized in the following table.

trivial sol.	$\nu \in (\hat{\nu}_-, \nu_0)$	$\nu \in (\nu_0, \hat{\nu}_+)$
β_1, β_2 incr., (8.18)	attractive, $\lim_{\nu \nearrow \nu_0} \mathfrak{A}_0^\nu = 0$	repulsive
β_1, β_2 incr., (8.19)	attractive	repulsive, $\lim_{\nu \searrow \nu_0} \mathfrak{R}_0^\nu = 0$
β_1, β_2 decr., (8.18)	repulsive	attractive, $\lim_{\nu \searrow \nu_0} \mathfrak{A}_0^\nu = 0$
β_1, β_2 decr., (8.19)	repulsive, $\lim_{\nu \nearrow \nu_0} \mathfrak{R}_0^\nu = 0$	attractive

PROOF. The first part of this theorem concerning the bifurcation of the attraction or repulsion radii, respectively, can be proved using the same methods as in the proof of Theorem 8.1. Write $\tilde{\nu}_-$ and $\tilde{\nu}_+$ for the constants $\hat{\nu}_-$ and $\hat{\nu}_+$ used in this proof. For the proof of the attractor and repeller transitions, only the case (ii) is considered, i.e., $\mathbb{I} = \mathbb{R}_0^-$, condition (8.19) holds and the functions β_1 and β_2 are monotone increasing. Denote the process induced by (8.17) by ϕ_ν and define

$$b_+ := \frac{1}{2} \sup_{t \in \mathbb{I},\, \nu \in (\tilde{\nu}_-, \tilde{\nu}_+)} b(t, \nu) < 0\,.$$

Due to (8.20), there exists a $\rho > 0$ such that

$$|r(t, x, \nu)| \leq -b_+ |x|^3 \quad \text{for all } x \in [-\rho, \rho], \nu \in \big(\nu_0 - x^2, \nu_0 + x^2\big) \text{ and } t \in \mathbb{I}\,.$$

The remaining proof is divided into two parts.

Part 1. It will be shown that for given $x_1, x_2, x_3 \leq \rho$ such that $0 < x_1 \leq x_2 \leq \frac{x_3}{2K}$, there exists a uniquely determined constant

$$\nu^* = \nu^*(x_1, x_2, x_3) \in \left(\nu_0, \min\left\{\nu_0 + x_1^2, \tilde{\nu}_+\right\}\right]$$

with the following properties:

- $\phi_\nu(t, \tau, [-x_2, x_2]) \subset (-x_3, x_3)$ for all $\tau \leq t \leq 0$ and $\nu \in (\nu_0, \nu^*)$,
- there exists a constant $T^* > 0$ such that for all $\nu \in (\nu_0, \nu^*)$ and $\tau \leq -T^*$, there exist $t_+, t_- \in [0, T^*]$ with

$$\phi_\nu(\tau + t_+, \tau, x_2) = x_1 \quad \text{and} \quad \phi_\nu(\tau + t_-, \tau, -x_2) = -x_1\,,$$

- ν^* is chosen maximal, i.e., for all bigger ν^*, one of the two above properties is violated.

Only the existence of a constant ν^* will be proved such that

(a) $\phi_\nu(t, \tau, x_2) \leq x_3$ for all $\tau \leq t \leq 0$ and $\nu \in (\nu_0, \nu^*)$

(b) there exists a constant $T^* > 0$ such that for all $\nu \in (\nu_0, \nu^*)$ and $\tau \leq -T^*$, there exists a $t_+ \in [0, T^*]$ with $\phi_\nu(\tau + t_+, \tau, x_2) = x_1$,

since the extension to the above assertion follows similarly and by taking the supremum of all such ν^*. Firstly, note that for arbitrary $\tau \in \mathbb{I}$, the solution $\mu_\tau(\cdot) := \phi_\nu(\cdot, \tau, x_2)$ of (8.17) is also a solution of the inhomogeneous linear differential equation

$$\dot{x} = a(t, \nu)x + b(t, \nu)(\mu_\tau(t))^3 + r(t, \mu_\tau(t), \nu)\,. \tag{8.21}$$

Concerning the expression

$$s(\nu, T) := Ke^{\beta_1(\nu)T}x_2 + \frac{b_+ x_1^3}{K}T \quad \text{for all } \nu \in (\nu_0, \tilde{\nu}_+) \text{ and } T \geq 0\,,$$

there exist $\nu^* \in \left(\nu_0, \min\left\{\nu_0 + x_1^2, \tilde{\nu}_+\right\}\right]$ and $T^* > 0$ such that

$$s(\nu, T^*) < 0 \quad \text{and} \quad s(\nu, T) \leq 2Kx_2 \quad \text{for all } \nu \in (\nu_0, \nu^*] \text{ and } T \in [0, T^*]\,.$$

This follows by choosing $T^* > 0$ such that $\frac{b_+ x_1^3}{K}T^* \leq -2Kx_2$ and ν^* such that $\exp\left(\beta_1(\nu^*)T^*\right) \leq 2$. Choose $\nu \in (\nu_0, \nu^*]$ and $\tau, \tau^* \leq 0$ with $\tau \leq \tau^*$, and assume that $x_1 \leq \mu_\tau(t) \leq x_3$ for all $t \in [\tau, \tau^*]$. Then the variation of constants formula, applied to (8.21), yields the relation

$$\begin{aligned}
&\mu_\tau(\tau^*) \\
&= \Phi_\nu(\tau^*, \tau)x_2 + \int_\tau^{\tau^*} \underbrace{\Phi_\nu(\tau^*, s)}_{\geq \frac{1}{K}\exp(\beta_2(\nu)(\tau^* - s))} \underbrace{\left(b(s, \nu)(\mu_\tau(s))^3 + r(s, \mu_\tau(s), \nu)\right)}_{\leq b_+ x_1^3 < 0} \mathrm{d}s \\
&\leq Ke^{\beta_1(\nu)(\tau^* - \tau)}x_2 + \int_\tau^{\tau^*} \frac{1}{K}e^{\beta_2(\nu)(\tau^* - s)}b_+ x_1^3\,\mathrm{d}s \\
&= Ke^{\beta_1(\nu)(\tau^* - \tau)}x_2 + \frac{b_+ x_1^3}{K\beta_2(\nu)}\left(e^{\beta_2(\nu)(\tau^* - \tau)} - 1\right) \\
&\leq Ke^{\beta_1(\nu)(\tau^* - \tau)}x_2 + \frac{b_+ x_1^3}{K}(\tau^* - \tau) = s(\nu, \tau^* - \tau)\,.
\end{aligned}$$

Since $s(\nu,T) \le 2Kx_2 \le x_3$ for all $T \in [0,T^*]$, the assumption $\mu_\tau(t) \le x_3$ for all $t \in [\tau,\tau^*]$ is justified. This proves (a). Because of $s(\nu,T^*) < 0$, also (b) is fulfilled.
Part 2. There exists an $\hat\nu_+ \in (\nu_0,\tilde\nu_+)$ such that for all $\nu \in (\nu_0,\hat\nu_+)$, there exists a nontrivial past attractor $\mathcal{A}^\nu \subset \mathbb{I}\times\mathbb{R}$ of (8.17) which fulfills

$$\lim_{\nu\searrow\nu_0} h\big(A_t^\nu,\{0\}\big) = 0 \quad \text{for all } t \in \mathbb{I}\,.$$

For $x_3 := \frac{\rho}{K}$ and $x_2 := \frac{x_3}{2K}$, consider the function $\gamma : (0,x_2) \to (\nu_0,\nu_+)$, defined by

$$\gamma(x_1) := \nu^*(x_1,x_2,x_3) \quad \text{for all } x_1 \in (0,x_2)\,,$$

where ν^* stems from Part 1. Set $\bar\nu := \gamma\big(\frac{x_2}{2}\big)$ and define

$$\delta(\nu) := \inf\big\{x_1 \in (0,x_2) : \gamma(x_1) \ge \nu\big\} \quad \text{for all } \nu \in (\nu_0,\bar\nu]\,.$$

Due to $\nu_0 < \nu^*(x_1,x_2,x_3) \le \nu_0 + x_1^2$, $\lim_{x_1\to 0}\gamma(x_1) = \nu_0$, and since γ is monotone increasing, this implies that δ is monotone increasing, $\delta(\nu) > 0$ for all $\nu \in (\nu_0,\bar\nu]$ and

$$\lim_{\nu\searrow\nu_0} \delta(\nu) = 0\,. \tag{8.22}$$

Define

$$\bar x_3(\nu) := 3K\delta(\nu) \quad \text{and} \quad \bar x_2(\nu) := \bar x_1(\nu) := \frac{3}{2}\delta(\nu) \quad \text{for all } \nu \in (\nu_0,\bar\nu]\,,$$

and consider the function $\bar\gamma : (\nu_0,\bar\nu] \to (\nu_0,\nu_+)$, defined by

$$\bar\gamma(\nu) := \nu^*\big(\bar x_1(\nu),\bar x_2(\nu),\bar x_3(\nu)\big) \quad \text{for all } \nu \in (\nu_0,\bar\nu]\,,$$

where ν^* is taken from Part 1 again. Moreover, define

$$M := [-x_2,x_2] \quad \text{and} \quad B^\nu := [-\bar x_3(\nu),\bar x_3(\nu)] \quad \text{for all } \nu \in (\nu_0,\bar\nu]$$

and fix a $\beta \in (\nu_0,\bar\nu]$ and an $\nu \in \big(\nu_0,\min\{\bar\gamma(\beta),\beta\}\big)$. Since $\nu \le \beta$ and $x_2 \ge \frac{3}{2}\delta(\beta)$, and due to the definition of δ, there exists a $T^* > 0$ such that for all $\tau \le -T^*$, there exist $t_+,t_- \in [0,T^*]$ with

$$\phi_\nu(t^+,\tau,x_2) = \frac{3}{2}\delta(\beta) = \bar x_2(\beta) \quad \text{and} \quad \phi_\nu(t^-,\tau,-x_2) = -\frac{3}{2}\delta(\beta) = -\bar x_2(\beta)\,.$$

Moreover, since $\nu < \bar\gamma(\beta)$, for all $\tau \le t \le 0$, one has

$$\phi_\nu\big(t,\tau,[-\bar x_2(\beta),\bar x_2(\beta)]\big) \subset (-\bar x_3(\beta),\bar x_3(\beta))\,.$$

This means that, considering equation (8.17), $B^\beta \times \mathbb{I}$ is pullback absorbing with respect to a universe containing $\{M\times\mathbb{I}\}$. Then due to Theorem 3.27, there exists a pullback attractor $\mathcal{A}^\nu \subset B^\beta \times \mathbb{I}$ with respect to $\{M\times\mathbb{I}\}$, i.e.,

$$\lim_{t\to\infty} \operatorname{dist}\big(\phi(\tau,\tau-t,M),A_\tau^\nu\big) = 0 \quad \text{for all } \tau \in \mathbb{I}\,.$$

Since M is neighborhood of $B^\beta\times\mathbb{I} \supset \mathcal{A}^\nu$, this pullback attractor is also a local pullback attractor. The limit relation

$$\lim_{\nu\nearrow\nu_0} h\big(A_t^\nu,\{0\}\big) = 0 \quad \text{for all } t \in \mathbb{I}$$

follows from $\mathcal{A}^\nu \subset B^\beta\times\mathbb{I}$ for all $\nu < \min\{\bar\gamma(\beta),\beta\}$ and (8.22). By setting $\hat\nu_+ := \bar\gamma(\bar\nu)$, all assertions of this theorem are proved. □

This bifurcation result is essentially the combination of two scenarios which are independent of each other. This means that it is possible to consider (8.17) only for $\nu > \nu_0$ or $\nu < \nu_0$, respectively, in order to obtain the results which apply for these parameter values.

REMARK 8.5. The hypothesis on the linear part implies that the future (past, all-time, respectively) dichotomy spectrum of the linearization $\dot{x} = a(t,\nu)x$ converges to $\{0\}$ in Hausdorff distance in the limit $\nu \to \nu_0$.

As in Example 8.3, one can show that Theorem 8.4 is a proper generalization of the well-known autonomous pitchfork bifurcation (see, e.g., WIGGINS [**235**, p. 267 f.]). Case (i) of the above theorem is now compared with the equivalent autonomous bifurcation.

EXAMPLE 8.6. Let $x_- < 0 < x_+$ and $\nu_- < 0 < \nu_+$ be in $\overline{\mathbb{R}}$, and consider the autonomous differential equation

$$\dot{x} = f(x,\nu)\,, \tag{8.23}$$

where the C^4-function $f : (x_-, x_+) \times (\nu_-, \nu_+) \to \mathbb{R}$ satisfies the following assumptions:

(i) $f(0,\nu) = 0 \quad \text{for all } \nu \in (\nu_-, \nu_+)\,,$
(ii) $D_1 f(0,0) = 0\,,$
(iii) $D_1 D_2 f(0,0) > 0\,,$
(iv) $D_1^2 f(0,0) = 0\,,$
(v) $D_1^3 f(0,0) > 0\,.$

Then (8.23) admits an autonomous pitchfork bifurcation (see, e.g., WIGGINS [**235**, p. 268 f.]). There exist a neighborhood $U \times V$ of $(0,0)$ in $\mathbb{R}^2$ and a C^2-function $h : U \to V$ with $h(0) = 0$ and

$$f(x, h(x)) = 0 \quad \text{for all } x \in U\,.$$

Except the trivial equilibria and the equilibria described by h, there are no other equilibria in $U \times V$, and the function h is maximal at $x = 0$. It can be verified that this situation fits into case (i) of Theorem 8.4: The functions β_1 and β_2 can be chosen to be increasing, since $D_1 D_2 f(0,0) > 0$ by (iii), and (8.18) holds, since $D_1^3 f(0,0) > 0$ by (v). Due to (iii), the trivial equilibrium of (8.23) is attractive for $\nu < 0$ and repulsive for $\nu > 0$, and this carries over to nonautonomous notions of attractivity and repulsivity. The function h describes repulsive equilibria of (8.23), and these equilibria are the boundary of the domain of attraction of the trivial equilibria. Since $\lim_{x\to 0} h(x) = 0$, this gives a nonautonomous bifurcation in form of a shrinking radius of attraction.

3. Higher-dimensional Bernoulli-like equations

The local bifurcation behavior of the zero solution of a vector-valued generalization of the scalar Bernoulli equation (8.2) is considered here for equations of the form

$$\dot{x} = A(\varepsilon)x + b(t)H_m(x,\varepsilon)\,, \tag{8.24}$$

where $x \in \mathbb{R}^d$. $A(0)$ has a simple eigenvalue equal to zero and all other eigenvalues with real parts on the left side of the complex plane and bounded away from the imaginary axis, $b : \mathbb{R} \to \mathbb{R}$ is an almost periodic function which is strictly positive or strictly negative, and H_m is homogeneous in x of order $m \geq 2$. Analogous results to the scalar case above are obtained.

3.1. Formulation and assumptions. Consider the ordinary differential equation (8.24) with the following three assumptions.

Assumption 1. *$A(\varepsilon) = A_0 + \varepsilon A_1$ with $\varepsilon \in [-1, 1]$ is matrix in $\mathbb{R}^{d\times d}$ such that A_0 has a simple real-valued eigenvalue $\lambda_1 = 0$ with left and right eigenvectors v_1 and u_1 normalized such that $v_1^T u_1 = 1$, and all other eigenvalues λ_j for $j \in \{2, \ldots, d\}$ satisfy $\operatorname{Re}\lambda_j \leq -\eta_0$ for some $\eta_0 > 0$.*

Thus, for small enough $\varepsilon > 0$, the matrix $A(\varepsilon)$ has leading eigenvalue

$$\lambda_1(\varepsilon) = \lambda_{1,1}\varepsilon + \lambda_{1,2}\varepsilon^2 + \ldots \tag{8.25}$$

with left and right eigenvectors $v_1(\varepsilon)$ and $u_1(\varepsilon)$ which are continuous in ε (in particular, so $v_1(0) = v_1$ and $u_1(0) = u_1$) and are normalized so that $v_1(\varepsilon)^T u_1(\varepsilon) = 1$. For simplicity, assume that $\lambda_{1,1} \neq 0$.

Assumption 2. *$H_m : \mathbb{R}^d \times [-1, 1] \to \mathbb{R}^d$ is a continuous function which is homogeneous in x of order $m \geq 2$ and satisfies*

$$H_m(u_1(\varepsilon), \varepsilon) = \alpha(\varepsilon)u_1(\varepsilon)\,, \tag{8.26}$$

where $\alpha(\varepsilon) := v_1(\varepsilon)^T H_m(u_1(\varepsilon), \varepsilon)$.

By the Euler formula for homogeneous functions, one has

$$\nabla H_m(u_1(\varepsilon), \varepsilon)u_1(\varepsilon) = mH_m(u_1(\varepsilon), \varepsilon)\,,$$

so (8.26) is equivalent to $\nabla H_m(u_1(\varepsilon), \varepsilon)u_1(\varepsilon) = m\alpha(\varepsilon)u_1(\varepsilon)$.

In the final assumption, the definition of almost periodicity due to Bohr is used (see the Appendix).

Assumption 3. *$b \in C(\mathbb{R}, \mathbb{R})$ is almost periodic with $\beta_0 \leq b(t) \leq \beta_1$ for all $t \in \mathbb{R}$, where either $\beta_0 > 0$ or $\beta_1 < 0$.*

Finally, define $\alpha_1 := \alpha(0)$, so that $\alpha_1 := v_1^T H_m(u_1, 0)$ and $\alpha(\varepsilon) \to \alpha_1$ as $\varepsilon \to 0$.

Theorem 8.7. *Suppose that Assumptions 1–3 hold and that $\lambda_{1,1} \neq 0$ and $\alpha_1 \neq 0$. Then there exist positive constants ε_0 and R_0 such that the following statements hold:*

(i) *If m is even, then for $\varepsilon \in (0, \varepsilon_0)$, the nonautonomous differential equation (8.24) has a unique nontrivial almost periodic solution in $B_{R_0}(0)$, which is exponentially stable if $\lambda_{1,1} > 0$ and exponentially unstable if $\lambda_{1,1} < 0$.*
(ii) *If m is odd and $\alpha_1\beta_0\lambda_{1,1} < 0$, then for $\varepsilon \in (0, \varepsilon_0)$, the nonautonomous differential equation (8.24) has exactly two nontrivial almost periodic solutions in $B_{R_0}(0)$, which are exponentially stable if $\lambda_{1,1} > 0$ and exponentially unstable if $\lambda_{1,1} < 0$.*
(iii) *If m is odd and $\alpha_1\beta_0\lambda_{1,1} > 0$, then for $\varepsilon \in (0, \varepsilon_0)$, the nonautonomous differential equation (8.24) has no nontrivial almost periodic solutions in $B_{R_0}(0)$.*

Later, changed coordinates will be used such that the left and right eigenvectors $v_1(\varepsilon)$ and $u_1(\varepsilon)$ are the unit vector e_1. By homogeneity, the components $H_{m,j}$ of H_m consist of linear combinations of terms of the form

$$x_1^{j_1} x_2^{j_2} \cdots x_d^{j_d} \quad \text{with } j_1 + j_2 + \cdots + j_d = m$$

in these new coordinates. The equation (8.26) of Assumption 2 and the assumption that $\alpha_1 \neq 0$ mean that the first component of the homogeneous function, given by $H_{m,1}$, has a non-vanishing term of the form $\alpha(\varepsilon)x_1^m$, while the other components $H_{m,j}$ for $j \in \{2, \dots, d\}$ have no non-vanishing term of the form $\alpha_j(\varepsilon)x_1^m$, but consist instead of linear combinations of terms of the form $x_1^{j_1} x_2^{j_2} \cdots x_d^{j_d}$ with the condition $j_2 + \cdots + j_d \geq 1$.

The full proof of Theorem 8.7 can be found in KLOEDEN [**119**]. The key ideas of the proof will be sketched for a simplified case in the following subsections.

3.2. Existence of nontrivial stationary solutions. Consider the scalar Bernoulli differential equation

$$\dot{x} = \lambda_1(\varepsilon)x + \gamma(\varepsilon)^{m-1}\alpha(\varepsilon)b(t)x^m \,, \tag{8.27}$$

where $\alpha(\varepsilon) := v_1(\varepsilon)^T H_m(u_1(\varepsilon), \varepsilon)$ is nonzero with the same sign as α_1 for $\varepsilon > 0$ sufficiently small and $\gamma(\varepsilon)$ is a positive constant which depends on ε.

PROPOSITION 8.8. *Let $\chi_\varepsilon(t)$ be a nontrivial solution of the Bernoulli equation* (8.27). *Then*

$$x_\varepsilon(t) := \gamma(\varepsilon)\chi_\varepsilon(t)u_1(\varepsilon) \tag{8.28}$$

is a nontrivial solution of the differential equation (8.24).

PROOF. By (8.26) and the homogeneity of H_m in x,

$$\begin{aligned}
&\frac{\mathrm{d}}{\mathrm{d}t}x_\varepsilon(t) \\
&= \gamma(\varepsilon)u_1(\varepsilon)\frac{\mathrm{d}}{\mathrm{d}t}\chi_\varepsilon(t) \\
&= \gamma(\varepsilon)u_1(\varepsilon)\big(\lambda_1(\varepsilon)\chi_\varepsilon(t) + \gamma(\varepsilon)^{m-1}\alpha(\varepsilon)b(t)\chi_\varepsilon(t)^m\big) \\
&= \gamma(\varepsilon)\left[\lambda_1(\varepsilon)u_1(\varepsilon)\right]\chi_\varepsilon(t) + b(t)\gamma(\varepsilon)^m\chi_\varepsilon(t)^m\alpha(\varepsilon)u_1(\varepsilon) \\
&= \gamma(\varepsilon)\left[A(\varepsilon)u_1(\varepsilon)\right]\chi_\varepsilon(t) + b(t)\gamma(\varepsilon)^m\chi_\varepsilon(t)^m H_m(u_1(\varepsilon), \varepsilon) \\
&= A(\varepsilon)x_\varepsilon(t) + b(t)H_m(\gamma(\varepsilon)\chi_\varepsilon(t)u_1(\varepsilon), \varepsilon) = A(\varepsilon)x_\varepsilon(t) + b(t)H_m(x_\varepsilon(t), \varepsilon)\,,
\end{aligned}$$

since $u_1(\varepsilon)$ is a right eigenvector of the matrix $A(\varepsilon)$ corresponding to the eigenvalue $\lambda_1(\varepsilon)$. □

The scalar Bernoulli equation (8.27) reduces to

$$\dot{\eta}(t) + (m-1)\lambda_1(\varepsilon)\eta(t) = -(m-1)\gamma(\varepsilon)^{m-1}\alpha(\varepsilon)b(t)\,, \tag{8.29}$$

with the substitution $\eta = \chi_\varepsilon^{1-m}$. This equation has explicit solution

$$\eta(t) = \eta(t_0)e^{-(m-1)\lambda_1(\varepsilon)(t-t_0)} - (m-1)\gamma(\varepsilon)^{m-1}\alpha(\varepsilon)\int_{t_0}^{t} e^{-(m-1)\lambda_1(\varepsilon)(t-s)}b(s)\,\mathrm{d}s\,.$$

Take $\gamma(\varepsilon) = \varepsilon^{1/(m-1)}$. Then, since the function b is bounded, when $\lambda_1(\varepsilon) > 0$, the pullback limit

$$\bar{\eta}(t) = -(m-1)\varepsilon\alpha(\varepsilon)\operatorname{sgn}(\beta_0)\left|\int_{-\infty}^{t} e^{-(m-1)\lambda_1(\varepsilon)(t-s)}b(s)\,\mathrm{d}s\right| \tag{8.30}$$

exists and is a solution of the nonlinear differential equation (8.24). Similarly, when $\lambda_1(\varepsilon) < 0$, changing the roles of t_0 and t gives the solution

$$\bar{\eta}(t) = (m-1)\varepsilon\alpha(\varepsilon)\operatorname{sgn}(\beta_0)\left|\int_{t}^{\infty} e^{-(m-1)\lambda_1(\varepsilon)(t-s)}b(s)\,\mathrm{d}s\right|. \tag{8.31}$$

The absolute values of the integrals in (8.30) and (8.31) lie in the interval

$$\frac{1}{(m-1)|\lambda_1(\varepsilon)|}\left[|\beta_0|\wedge|\beta_1|, |\beta_0|\vee|\beta_1|\right]$$

where $|\lambda_1(\varepsilon)| \approx |\lambda_{1,1}\varepsilon|$, so the solutions are $O(1)$ with respect to ε. Moreover, they are almost periodic, since b is almost periodic.

The Bernoulli equation (8.27) thus has formal solutions

$$\bar{\chi}_\varepsilon(t) = \left(-(m-1)\varepsilon\alpha(\varepsilon)\operatorname{sgn}(\beta_0)\left|\int_{-\infty}^{t} e^{-(m-1)\lambda_1(\varepsilon)(t-s)}b(s)\,\mathrm{d}s\right|\right)^{-1/(m-1)} \tag{8.32}$$

when $\lambda_{1,1} > 0$, and

$$\bar{\chi}_\varepsilon(t) = \left((m-1)\varepsilon\alpha(\varepsilon)\operatorname{sgn}(\beta_0)\left|\int_{t}^{\infty} e^{-(m-1)\lambda_1(\varepsilon)(t-s)}b(s)\,\mathrm{d}s\right|\right)^{-1/(m-1)} \tag{8.33}$$

when $\lambda_{1,1} < 0$. Now $\alpha(\varepsilon)$ has the same sign as α_1 when $\varepsilon > 0$ is small. Thus, for m even, there is one solution of the form (8.32)–(8.33), whereas for m odd and $\alpha_1\beta_0\lambda_{1,1} > 0$, there is no solution of the form (8.32)–(8.33), but for m odd and $\alpha_1\beta_0\lambda_{1,1} < 0$, there are two distinct solutions of the form (8.32)–(8.33). Moreover, these solutions are almost periodic.

Finally, returning to the original vector-valued differential equation (8.24), it follows from Proposition 8.8 that it has a nontrivial almost periodic solution of the form

$$x_\varepsilon(t) = \varepsilon^{1/(m-1)}\bar{\chi}_\varepsilon(t)u_1(\varepsilon), \tag{8.34}$$

for each of almost periodic solutions $\bar{\chi}_\varepsilon(t)$ of the form (8.32)–(8.33) above. From the choice of $\gamma(\varepsilon) = \varepsilon^{1/(m-1)}$, it follows that the solutions $x_\varepsilon(t)$ are of order $O(\varepsilon^{1/(m-1)})$, since both $\bar{\chi}_\varepsilon(t)$ and $u_1(\varepsilon)$ are of order $O(1)$.

3.3. Stability analysis of the almost periodic solutions. The stability analysis of the nontrivial almost periodic solutions $x_\varepsilon(t)$ is divided into several parts.

3.3.1. *A property of the almost periodic solutions.* Firstly, consider any one of the nontrivial almost periodic solutions $x_\varepsilon(t) = \chi_\varepsilon(t)\gamma(\varepsilon)u_1(\varepsilon)$ of the nonlinear system (8.24), i.e., with a time-dependent function $\chi_\varepsilon(t)$ satisfying the Bernoulli equation (8.27). With the above choice of $\gamma(\varepsilon)$, this can be rewritten as

$$\frac{1}{\chi_\varepsilon(t)}\frac{\mathrm{d}}{\mathrm{d}t}\chi_\varepsilon(t) - \lambda_1(\varepsilon) = \varepsilon\alpha(\varepsilon))b(t)\chi_\varepsilon(t)^{m-1}, \tag{8.35}$$

recalling that in the new coordinate system $\alpha(\varepsilon) = H_{m,1}(e_1,\varepsilon) \neq 0$ with $\operatorname{sgn}(\alpha(\varepsilon)) = \operatorname{sgn}(\alpha(0)) = \operatorname{sgn}(\alpha_1)$ for small enough $\varepsilon > 0$. Recall also that $\lambda_1(\varepsilon)$

has the same sign as $\lambda_{1,1}$ for small enough $\varepsilon > 0$, and integrate both sides of (8.35) to obtain

$$\ln\left(\frac{|\chi_\varepsilon(t)|}{|\chi_\varepsilon(0)|}\right) - \lambda_1(\varepsilon)\, t = \varepsilon\alpha(\varepsilon)\int_0^t b(s)\chi_\varepsilon(s)^{m-1}\,\mathrm{d}s\,. \tag{8.36}$$

The almost periodic function $\chi_\varepsilon(t)$ is uniformly bounded and bounded away from zero, so the logarithmic term in (8.36) is bounded in time t. Moreover, the coefficient function $b(t)$ is either always positive or always negative with its sign determined by that of β_0. Distinguish now the following two cases.

Case m odd. One has $\chi_\varepsilon(t)^{m-1} > 0$ for all t, so

(i) for $\lambda_{1,1} > 0$, the left side tends to $-\infty$ as $t \to \infty$, which is only possible if $\alpha_1\beta_0 < 0$,
(ii) for $\lambda_{1,1} < 0$, the left side tends to ∞ as $t \to \infty$, which is only possible if $\alpha_1\beta_0 > 0$.

Case m even. In this case, $\chi_\varepsilon(t)^{m-1}$ has the same sign as $\chi_\varepsilon(t)$ for all t, this being the opposite of that of $\alpha_1\beta_0\lambda_{1,1}$. Thus,

(iii) for $\lambda_{1,1} > 0$, the left side tends to $-\infty$ as $t \to \infty$, which is only possible if $\alpha_1\beta_0 < 0$,
(iv) for $\lambda_{1,1} < 0$, the left side tends to ∞ as $t \to \infty$, which is only possible if $\alpha_1\beta_0 > 0$.

For later use, note that

$$\begin{aligned}\lambda_1(\varepsilon)\, t &+ m\varepsilon\alpha(\varepsilon)\int_0^t b(s)\chi_\varepsilon(s)^{m-1}\,\mathrm{d}s\\ &= \ln\left(\frac{|\chi_\varepsilon(t)|}{|\chi_\varepsilon(0)|}\right) + (m-1)\varepsilon\alpha(\varepsilon)\int_0^t b(s)\chi_\varepsilon(s)^{m-1}\,\mathrm{d}s\end{aligned}$$

goes to $\pm\infty$ as $t \to \infty$ according to whether $\mp\lambda_{1,1} > 0$.

3.3.2. *The linearized equation of an almost periodic solution.* Now consider the equation linearized about $x_\varepsilon(t) = \gamma(\varepsilon)\chi_\varepsilon(t)u_1(\varepsilon)$, namely

$$\dot z = A(\varepsilon)z + b(t)\nabla H_m(x_\varepsilon(t),\varepsilon)z\,,$$

which can be rewritten as

$$\dot z = A(\varepsilon)z + \varepsilon b(t)\chi_\varepsilon(t)^{m-1}\nabla H_m(u_1(\varepsilon),\varepsilon)z \tag{8.37}$$

from the definition of $\gamma(\varepsilon)$ and the fact that ∇H_m is homogeneous of degree $m-1$. Using the new coordinate system $z = (\eta, y)$, the linear equation (8.37) has the form

$$\dot\eta = \lambda_1(\varepsilon)\eta + \varepsilon b(t)\chi_\varepsilon(t)^{m-1}\frac{\partial H_{m,1}}{\partial x_1}(e_1,\varepsilon)\eta + \varepsilon b(t)\chi_\varepsilon(t)^{m-1}\sum_{j=2}^{d}\frac{\partial H_{m,1}}{\partial x_j}(e_1,\varepsilon)y_j\,,$$

$$\dot y = B_{d-1}(\varepsilon)y + \varepsilon b(t)\chi_\varepsilon(t)^{m-1}D(\varepsilon)y + \varepsilon b(t)\chi_\varepsilon(t)^{m-1}\frac{\partial \widehat{H}_m}{\partial x_1}(e_1,\varepsilon)\eta\,,$$

where $D(\varepsilon)$ has components $d_{i,j}(\varepsilon) = \partial H_{m,i}/\partial x_j(e_1,\varepsilon)$ for $i,j \in \{2,\dots,d\}$ (moreover, $d_{1,j}(\varepsilon)$ is also defined in this way below) and $\widehat{H}_m = (H_{m,2},\dots,H_{m,d})^T$. Now

$$\frac{\partial H_m}{\partial x_1}(e_1,\varepsilon) = mH_m(e_1,\varepsilon) = m\alpha(\varepsilon)e_1$$

by (8.26) of Assumption 2 and the Euler formula for homogeneous functions, so $\partial\widehat{H}_m/\partial x_1(e_1,\varepsilon) = 0$ (in $\mathbb{R}^{d-1}$). Hence, the above equation reduces to

$$\dot{\eta} = \lambda_1(\varepsilon)\eta + m\varepsilon\alpha(\varepsilon)b(t)\chi_\varepsilon(t)^{m-1}\eta + \varepsilon b(t)\chi_\varepsilon(t)^{m-1}\sum_{j=2}^{d} d_{1,j}(\varepsilon)y_j\,, \tag{8.38}$$

$$\dot{y} = B_{d-1}(\varepsilon)y + \varepsilon b(t)\chi_\varepsilon(t)^{m-1}D(\varepsilon)y\,. \tag{8.39}$$

By the stability property of the matrix $B_{d-1}(\varepsilon)$, there exists a symmetric positive definite matrix P such that

$$PB_{d-1}(\varepsilon) + B_{d-1}(\varepsilon)^T P \le -\frac{1}{2}\mathbb{1}_{d-1}$$

for all $\varepsilon > 0$ sufficiently small. Define $V(y) = y^T Py$. Then

$$\begin{aligned}
\frac{\mathrm{d}}{\mathrm{d}t}V(y(t)) &= \nabla V(y(t))^T\frac{\mathrm{d}y}{\mathrm{d}t} = \frac{\mathrm{d}y}{\mathrm{d}t}^T Py(t) + y(t)^T P\frac{\mathrm{d}y}{\mathrm{d}t}\\
&= y(t)^T\left[PB_{d-1}(\varepsilon) + B_{d-1}(\varepsilon)^T P\right]y(t)\\
&\quad + \varepsilon b(t)\chi_\varepsilon(t)^{m-1}y(t)^T\left[PD(\varepsilon) + D(\varepsilon)^T P\right]y(t)\\
&\le -\frac{1}{2}y(t)^T y(t) + \varepsilon b(t)\chi_\varepsilon(t)^{m-1}y(t)^T\left[PD(\varepsilon) + D(\varepsilon)^T P\right]y(t)\\
&\le -2\gamma_0 V(y(t)) + \varepsilon b(t)\chi_\varepsilon(t)^{m-1}y(t)^T\left[PD(\varepsilon) + D(\varepsilon)^T P\right]y(t)\,,
\end{aligned}$$

where $\gamma_0 := \frac{1}{4}\lambda_{\max}(P)$. Therefore, by the uniform boundedness of the term $b(t)\chi_\varepsilon(t)^{m-1}$, it follows that

$$\frac{\mathrm{d}}{\mathrm{d}t}V(y(t)) + 2\gamma_0 V(y(t)) \le \varepsilon KV(y(t))$$

for an appropriate constant $K > 0$. Choosing $\varepsilon > 0$ small enough so that $K\varepsilon \le \gamma_0$ gives

$$\frac{\mathrm{d}}{\mathrm{d}t}V(y(t)) + \gamma_0 V(y(t)) \le 0\,.$$

Finally, using the properties of the matrix P and possibly a new constant K gives

$$|y(t)|^2 \le K|y(0)|^2 e^{-\gamma_0 t} \to 0 \quad \text{as } t \to \infty\,. \tag{8.40}$$

Now consider write the first component of the linearized equation (8.38) as

$$\dot{\eta} = \varepsilon\gamma(\varepsilon,t)\eta + \varepsilon b(t)\chi_\varepsilon(t)^{m-1}\sum_{j=2}^{d} d_{1,j}(\varepsilon)y_j(t)\,,$$

where

$$\varepsilon\gamma(\varepsilon,t) := \lambda_1(\varepsilon) + m\varepsilon\alpha(\varepsilon)b(t)\chi_\varepsilon(t)^{m-1}\,.$$

Note that $\gamma(\varepsilon,t) \neq 0$, so this is either always positive or always negative. Multiplying by 2η, integrating and using the Cauchy inequality and the uniform boundedness of $b(t)\chi_\varepsilon(t)$ yields

$$|\eta(t)|^2 \le |\eta(0)|^2 e^{2\varepsilon\int_0^t \gamma(\varepsilon,s)\,\mathrm{d}s} + \varepsilon K\int_0^t e^{2\varepsilon\int_s^t \gamma(\varepsilon,\tau)\,\mathrm{d}\tau}|y(s)|^2\,\mathrm{d}s\,. \tag{8.41}$$

Suppose now that $\gamma(\varepsilon,t) < 0$. Then there exists $0 < \nu_0(\varepsilon) \to 0$ as $\varepsilon\searrow 0$ such that

$$2\varepsilon\gamma(\varepsilon,t) \le -\nu_0(\varepsilon)\gamma_0 \quad \text{for all } t \ge 0,$$

provided that $\varepsilon > 0$ is small enough (which is possible by the uniform boundedness and negativity of $\gamma(\varepsilon, t)$). Using the y-component estimate (8.40) and a generic constant K, the inequality (8.41) becomes

$$\begin{aligned}|\eta(t)|^2 &\le |\eta(0)|^2 e^{2\varepsilon \int_0^t \gamma(\varepsilon,s)\,\mathrm{d}s} + \varepsilon K |y(0)|^2 e^{-\gamma_0 t} \int_0^t e^{2\varepsilon \int_s^t \gamma(\varepsilon,\tau)\,\mathrm{d}\tau}\,\mathrm{d}s \\ &\le |\eta(0)|^2 e^{2\varepsilon \int_0^t \gamma(\varepsilon,s)\,\mathrm{d}s} + \varepsilon K |y(0)|^2 e^{-\gamma_0 t} \int_0^t e^{-\nu_0(\varepsilon)\gamma_0 (t-s)}\,\mathrm{d}s \\ &\le |\eta(0)|^2 e^{-\nu_0(\varepsilon)\gamma_0 t} + \varepsilon K |y(0)|^2 \frac{1}{\nu_0(\varepsilon)\gamma_0} e^{-\gamma_0 t}\left(1 - e^{-\nu_0(\varepsilon)\gamma_0 t}\right) \to 0\end{aligned}$$

as $t \to \infty$. Thus, if $\gamma(\varepsilon, t) < 0$, then the equilibrium solution $\eta = 0$ is asymptotically stable.

On the other hand, if $\gamma(\varepsilon, t) > 0$, then consider the solution of the linearized equation (8.38)–(8.39) with initial condition $\eta(0) \neq 0$ and $y(0) = 0$. Then $y(t) \equiv 0$, and the first component of the linearized equation (8.38) reduces to

$$\dot{\eta} = \varepsilon \gamma(\varepsilon, t)\eta$$

for which the zero solution is obviously asymptotically unstable.

The zero solution of the full linearized equations is thus asymptotically stable when $\gamma(\varepsilon, t) < 0$ and unstable if $\gamma(\varepsilon, t) > 0$. This gives the (local) stability assertions of the almost periodic solutions in Theorem 8.7 and completes the sketch proof.

4. Further developments

Bifurcation theory for vector-valued nonautonomous systems is fragmentary. Some further developments are briefly indicated in this section.

4.1. Nonautonomous linear part. The Duffing–van der Pol system arises in many physical and technological problems. With a time-dependent linear term, it has the form

$$\begin{pmatrix} \dot{x} \\ \dot{y} \end{pmatrix} = \begin{pmatrix} 0 & 1 \\ -\alpha + \sigma b(t) & \beta \end{pmatrix} \begin{pmatrix} x \\ y \end{pmatrix} - \begin{pmatrix} 0 \\ xy + x^3 \end{pmatrix}, \tag{8.42}$$

where $b(t)$ is either an almost periodic function or a bounded random process such as $\cos W_t$ with a Wiener process W_t (see Chapter 14).

In contrast to the nonautonomous differential equation (8.24) in the previous section, the Duffing–van der Pol system (8.42) has the form

$$\dot{x} = A(t)x + N(x), \quad \text{where } x \in \mathbb{R}^2 \text{ and } t \in \mathbb{R}, \tag{8.43}$$

with the linear part now time-dependent. A parameter can be introduced into either the linear or nonlinear term.

This structure of the nonautonomous differential equation (8.43) has a profound effect on the nature of the bifurcation. From the numerical simulations in Figure 8.1, this appears to occur in two steps. A possible explanation given in JOHNSON, KLOEDEN & PAVANI [**100**] suggests that this is due to an interval in the Sacker–Sell spectrum crossing the imaginary axis, see SACKER & SELL [**209**]. For a similar explanation in a simpler context, see the pitchfork bifurcations in Section 2.

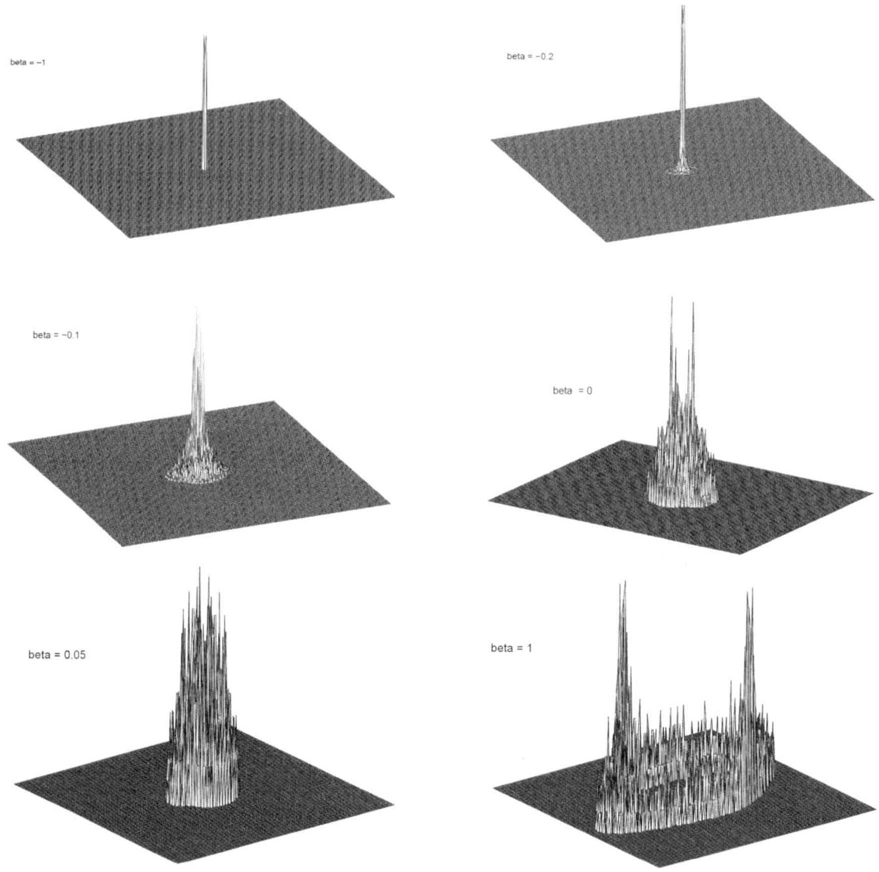

FIGURE 8.1. Two-step bifurcation in the nonautonomous Duffing–van der Pol system; [**100**] reprinted with permission of *World Scientific Publishing Company.*

4.2. Center manifold reduction. The reduction of the dimension of the system is central for the simplification of dynamical systems. The so-called center manifold reduction yields a reduced system, which describes the behavior on a center manifold. This reduction principle plays an important role in bifurcation theory, since the interesting dynamical behavior takes place on center manifolds. The center manifold reduction allows to study bifurcations of high-dimensional systems via low-dimensional bifurcation patterns. In this context, the Taylor approximation of invariant manifolds as developed in Section 4 of Chapter 6, is crucial, since such bifurcation patterns rely on Taylor coefficients. The nonautonomous center manifold reduction technique will be illustrated now by an application to a perturbed Lorenz system.

Consider a perturbed version of the well-known Lorenz equations (see, e.g., LORENZ [**159**]), given by the three-dimensional system

$$\begin{aligned}\dot{x}_1 &= \sigma_\nu(t)(x_2 - x_1)\,,\\ \dot{x}_2 &= \rho_\nu(t)x_1 - x_2 - x_1x_3\,,\\ \dot{x}_3 &= -\beta_\nu(t)x_3 + x_1x_2\,,\end{aligned} \tag{8.44}$$

and let $f_\nu : \mathbb{R}\times\mathbb{R}^3 \to \mathbb{R}^3$ denote the right hand side of this equation. As discussed in [**159**], the Lorenz equations are a simplified toy model of Raleigh–Bernard thermal convection. To incorporate an external perturbation into this model, it is interesting when all three parameters $\sigma_\nu, \rho_\nu, \beta_\nu$ are perturbed by a nonautonomous forcing, i.e., assume that the functions $\sigma_\nu, \rho_\nu, \beta_\nu : \mathbb{R} \to (0, \infty)$ are given by

$$\sigma_\nu(t) = \sigma_0 + \nu\sigma(t)\,, \quad \rho_\nu(t) = 1 + \rho_0 + \nu\rho(t) \quad \text{and} \quad \beta_\nu(t) = \beta_0 + \nu\beta(t)$$

with real constants $\sigma_0, \beta_0 > 0$, $\rho_0 \in \mathbb{R}$, bounded C^3-functions σ, ρ, β and $\nu \in \mathbb{R}$, which will serve as bifurcation parameter. In the following, the stability of the equilibrium $x = 0$ of (8.44) is analyzed for different values of ν. From the linearization

$$D_1f_0(t, 0) = \begin{pmatrix} -\sigma_0 & \sigma_0 & 0 \\ 1+\rho_0 & -1 & 0 \\ 0 & 0 & -\beta_0 \end{pmatrix},$$

it is clear that in case $\nu = 0$, the origin is attractive for $\rho_0 \in \left[-\left(\frac{\sigma_0+1}{2\sigma_0}\right)^2, 0\right)$ and unstable for $\rho_0 > 0$. More interesting is the nonhyperbolic case $\rho_0 = 0$, where a pitchfork bifurcation occurs as ρ_0 passes through 0. To mimic this situation, assume $\rho_0 = 0$ from now on. Before proceeding, augment the original system (8.44) by considering the parameter ν as an additional state space variable satisfying $\dot{\nu} = 0$ and, to simplify the calculations, apply the linear transformation

$$\begin{pmatrix} y_1 \\ y_2 \\ y_3 \\ y_4 \end{pmatrix} := \begin{pmatrix} -\sigma_0 & 0 & 1 & 0 \\ 1 & 0 & 1 & 0 \\ 0 & 1 & 0 & 0 \\ 0 & 0 & 0 & 1 \end{pmatrix} \begin{pmatrix} x_1 \\ x_2 \\ x_3 \\ \nu \end{pmatrix}$$

to the resulting equation $\dot{x} = f_\nu(t, x)$, $\dot{\nu} = 0$. This implies the four-dimensional system

$$\dot{y} = Ay + F(t, y) \tag{8.45}$$

with $A := \operatorname{diag}(-\sigma_0 - 1, -\beta_0, 0, 0)$ and the nonlinearity

$$F(t,y) := \begin{pmatrix} \frac{\sigma_0}{\sigma_0+1}y_1y_2 - \frac{\sigma(t)+\sigma_0(\sigma(t)+\rho(t))}{\sigma_0+1}y_1y_4 - \frac{1}{\sigma_0+1}y_2y_3 + \frac{\rho(t)}{\sigma_0+1}y_3y_4 \\ -\sigma_0y_1^2 + (1-\sigma_0)y_1y_3 - \beta(t)y_2y_4 + y_3^2 \\ \frac{\sigma_0^2}{\sigma_0+1}y_1y_2 + \frac{\sigma(t)+\sigma_0(\sigma(t)-\sigma_0\rho(t))}{\sigma_0+1}y_1y_4 - \frac{\sigma_0}{\sigma_0+1}y_2y_3 + \frac{\sigma_0\rho(t)}{\sigma_0+1}y_3y_4 \\ 0 \end{pmatrix}.$$

Thus, Theorem 6.10 can be applied to (8.45) to show that there exists a center-unstable manifold $\mathcal{W}^- \subseteq \mathbb{R} \times \mathbb{R}^3$ with two-dimensional fibers. The ansatz

$$w^-(t, y_3, y_4) = \sum_{i=0}^{2} y_3^{2-i}y_4^i \begin{pmatrix} w_{2-i,i}^1(t) \\ w_{2-i,i}^2(t) \end{pmatrix} + O\big((y_3^2 + y_4^2)^{3/2}\big)$$

yields that the equation (8.45) reduced to the center-unstable manifold $\mathcal{W}^-$ is given by

$$\dot{y}_3 = \frac{\sigma_0}{\sigma_0 + 1}\big(\nu\rho(t)y_3 - w^2_{2,0}(t)y_3^3\big) + O(\nu y_3^2, \nu^2 y_3, y_3^4)\,.$$

Using Theorem 6.19, one obtains $w^2_{2,0}(t) \equiv \frac{1}{\beta_0}$, and consequently, the bifurcation equation is given by

$$\dot{y}_3 = \frac{\sigma_0}{\sigma_0 + 1}\big(\nu\rho(t)y_3 - \tfrac{1}{\beta_0}y_3^3\big) + r(y_3, t, \nu)\,, \tag{8.46}$$

where the remainder r satisfies the three limit relations

$$\lim_{y_3 \to 0} \sup_{\nu \in (-|y_3|^3, |y_3|^3)} \sup_{t \in \mathbb{R}} \frac{|r(y_3, t, \nu)|}{|y_3|^3} = 0\,,$$

$$\lim_{\nu \to 0} \sup_{y_3 \in (-|\nu|, |\nu|)} \sup_{t \in \mathbb{R}} \frac{|r(y_3, t, \nu)|}{|\nu|^2} = 0\,,$$

$$\lim_{\nu \to 0} \frac{1}{\nu} \limsup_{y_3 \to 0} \sup_{t \in \mathbb{R}} \frac{|r(y_3, t, \nu)|}{|y_3|} = 0\,.$$

This implies that Theorem 8.4, which describes the nonautonomous pitchfork bifurcation, is applicable, if one assumes that

$$\liminf_{t \to -\infty} |\rho(t)| > 0\,.$$

In particular, this means that for the bifurcation equation (8.46), the following two statements hold:

(i) If $\liminf_{t \to -\infty} \rho(t) > 0$, then there exists a $\hat{\nu} > 0$ such that for all $\nu \in (0, \hat{\nu})$, there exists a nontrivial local pullback attractor $\mathcal{A}^\nu$ of (8.46) such that

$$\lim_{\nu \searrow 0} \operatorname{dist}(A^\nu_t, \{0\}) = 0 \quad \text{for all } t \in \mathbb{R}\,.$$

Furthermore, $\mathbb{R} \times \{0\}$ is a local pullback attractor of (8.46) for $\nu \in (-\hat{\nu}, 0]$.

(ii) If $\limsup_{t \to -\infty} \rho(t) < 0$, then there exists a $\hat{\nu} > 0$ such that for all $\nu \in (-\hat{\nu}, 0)$, there exists a nontrivial local pullback attractor $\mathcal{A}^\nu$ of (8.46) such that

$$\lim_{\nu \nearrow 0} \operatorname{dist}(A^\nu_t, \{0\}) = 0 \quad \text{for all } t \in \mathbb{R}\,.$$

Furthermore, $\mathbb{R} \times \{0\}$ is a local pullback attractor of (8.46) for $\nu \in [0, \hat{\nu})$.

Not only the reduced equation (8.46) admits a nonautonomous bifurcation of this type but also the Lorenz equation (8.44) itself. This is due to an asymptotic phase property of the center-unstable manifold (cf. AULBACH [**10**, Theorem 4] for a global version), i.e., every solution of (8.44) in a neighborhood of the manifold approaches exponentially a solution on the center-unstable manifold in forward time. Therefore, for small $\nu > 0$, there also exists a local pullback attractor of (8.44) shrinking down to $\{0\}$ for $\nu \searrow 0$.

Endnotes. Section 1 on bifurcations in nonautonomous Bernoulli differential equations is based on well-known results in the literature, see, e.g., Caraballo, Langa & Robinson [**33**] and Langa, Robinson & Suárez [**154, 155**] as well as Kloeden & Siegmund [**144**]. Section 2 is taken from Rasmussen [**196**]. The results in Section 3 are a special case of those in Kloeden [**119**], which is based on methods in Krasnosel'skiĭ, Burd & Kolesov [**150**] who investigated bifurcations of the zero solution in vector-valued ordinary differential equations with almost periodic coefficients (see also Hamaya [**91**] for bifurcations of almost periodic solutions in the discrete case). Johnson, Kloeden & Pavani [**100**] explore possible explanations for the two-phase behavior of many nonautonomous bifurcations in higher dimensions, and see also Pötzsche [**189, 183**]. See Rasmussen [**194**] for a systematic analysis of attractivity and bifurcations in nonautonomous systems and Pötzsche [**189, 183**] for abstract analytical bifurcation scenarios in higher dimensions. Bifurcations in nonautonomous difference equations are discussed in Rasmussen [**193**], and they are also well-studied for asymptotically autonomous system (see Rasmussen [**198**]). In Fabbri, Johnson & Mantellini [**74, 75, 101**], attractor-repeller bifurcations are analyzed for one-dimensional nonautonomous differential equations with strictly ergodic time dependence, and finally, also one-dimensional bifurcations of bounded solutions are considered in Núñez & Obaya [**169**] in this context.

CHAPTER 9

Set-valued nonautonomous dynamical systems

Set-valued dynamical systems arise as the solution mappings of set-valued differential equations, which are often called *differential inclusions*, and as the solution mappings of set-valued difference equations or *difference inclusions* in the discrete time case. Control systems are an important field of application of such systems.

Let (X, d_X) be a complete metric space, and let $\mathcal{K}(X)$ be the collection of nonempty compact subsets of X, which is a complete metric space with respect to the Hausdorff metric h. Recall that the Hausdorff metric is defined by $h(A, B) := \max\{\operatorname{dist}(A, B), \operatorname{dist}(B, A)\}$, where

$$\operatorname{dist}(A, B) := \max_{a \in A} \operatorname{dist}(a, B) := \max_{a \in A} \min_{b \in B} d_X(a, b)$$

is the Hausdorff semi-distance on $\mathcal{K}(X)$, and that a mapping $F : X \to \mathcal{K}(X)$ is said to be *upper semi-continuous* at x_0 if

$$\lim_{x_n \to x_0} \operatorname{dist}(F(x_n), F(x_0)) = 0$$

for all sequences $x_n \to x_0$ in X and *continuous* at x_0 if

$$\lim_{x_n \to x_0} h(F(x_n), F(x_0)) = 0$$

for all sequences $x_n \to x_0$ in X.

For any $A \in \mathcal{K}(X)$, define $F(A) := \bigcup_{a \in A} F(a)$, and define the set composition of two mappings $F, G : X \to \mathcal{K}(X)$ as $(F \circ G)(x) := F(G(x))$ for all $x \in X$. Denote the space of all upper semi-continuous mappings $F : X \to \mathcal{K}(X)$ by $C^{\mathrm{us}}(X, \mathcal{K}(X))$. Note that $F \circ G \in C^{\mathrm{us}}(X, \mathcal{K}(X))$ if $F, G \in C^{\mathrm{us}}(X, \mathcal{K}(X))$.

Set-valued autonomous dynamical systems or semi-dynamical systems have been extensively investigated. They are often called general control systems or general (semi-)dynamical systems and satisfy a set-valued version of the group or semi-group evolution property. Recall that the time set $\mathbb{T}$ is given by $\mathbb{T} = \mathbb{R}$ in the continuous time case and by $\mathbb{T} = \mathbb{Z}$ in the discrete time case.

DEFINITION 9.1 (Set-valued semi-dynamical system). A *set-valued semi-dynamical system* on X is given by a set-valued mapping $\mathbb{T}_0^+ \times X \ni (t, x_0) \mapsto \Phi(t, x_0) \in \mathcal{K}(X)$ such that

(i) $\Phi(0, x_0) = \{x_0\}$ for all $x_0 \in X$,
(ii) $\Phi(s + t, x_0) = \Phi(s, \Phi(t, x_0))$ for all $s, t \geq 0$ and all $x_0 \in X$,
(iii) $(t, x_0) \mapsto \Phi(t, x_0)$ is upper semi-continuous in $(t, x_0) \in \mathbb{T}_0^+ \times X$ with respect to the Hausdorff semi-metric, i.e.,

$$\operatorname{dist}(\Phi(s, y), \Phi(t, x)) \to 0 \quad \text{as } (s, y) \to (t, x) \in \mathbb{T}_0^+ \times X\,,$$

(iv) $t \mapsto \Phi(t, x_0)$ is continuous in $t \in \mathbb{T}_0^+$ with respect to the Hausdorff metric uniformly in x_0 in compact subsets of X.

EXERCISE 9.2. Determine explicitly the set-valued semi-dynamical system generated by the scalar ordinary differential equation

$$\dot{x} = \frac{3}{2} x^{\frac{2}{3}},$$

which does not have unique solutions at $x_0 = 0$. Show that property (iii) on the upper semi-continuity in the initial value cannot be strengthened to continuity here.

Analogously to their single-valued counterparts, set-valued nonautonomous dynamical systems can be formulated as processes or as skew product flows. Both will be considered here.

1. Set-valued processes

The definition of a set-valued process is a generalization of the single-valued process considered first in Chapter 2.

DEFINITION 9.3 (Set-valued process). A *set-valued process* on X is given by a set-valued mapping $(t, t_0, x_0) \mapsto \Phi(t, t_0, x_0) \in \mathcal{K}(X)$, where $t \geq t_0 \in \mathbb{T}$ and $x_0 \in X$, such that

(i) $\Phi(t_0, t_0, x_0) = \{x_0\}$ for all $x_0 \in X$ and all $t_0 \in \mathbb{T}$,
(ii) $\Phi(t_2, t_0, x_0) = \Phi(t_2, t_1, \Phi(t_1, t_0, x_0))$ for all $t_0 \leq t_1 \leq t_2$ in $\mathbb{T}$ and all $x_0 \in X$,
(iii) $(t, t_0, x_0) \to \Phi(t, t_0, x_0)$ is upper semi-continuous with respect to the Hausdorff semi-distance, i.e.,

$$\operatorname{dist}(\Phi(s, s_0, y_0), \Phi(t, t_0, x_0)) \to 0 \quad \text{as } (s, s_0, y_0) \to (t, t_0, x_0),$$

(iv) $t \to \Phi(t, t_0, x_0)$ is continuous in $t \in \mathbb{T}$ with respect to the Hausdorff metric uniformly in (t_0, x_0) in compact subsets of $\mathbb{T} \times X$, i.e.,

$$\sup_{(t_0, x_0) \in K} h(\Phi(s, t_0, x_0), \Phi(t, t_0, x_0)) \to 0 \quad \text{as } s \to t \in \mathbb{T}$$

for each $K \in \mathcal{K}(\mathbb{T} \times X)$.

Examples based on difference and differential inclusions, i.e., set-valued difference and differential equations, will be given in the next section in terms of set-valued skew product flows.

When investigating set-valued processes, it is often convenient to consider their single-valued trajectories.

DEFINITION 9.4 (Trajectory). A *trajectory* of a set-valued process is a single-valued continuous function $\phi : [\tau_0, \tau_1] \cap \mathbb{T} \to X$ for some $\tau_0 \leq \tau_1$ in $\mathbb{T}$ such that

$$\phi(t) \in \Phi(t, s, \phi(s)) \quad \text{for all } \tau_0 \leq s \leq t \leq \tau_1 .$$

The existence of trajectories joining points x_0 and $x_1 \in \Phi(t_1, t_0, x_0)$ was first proved by Barbashin (see CARABALLO, KLOEDEN & MARÍN-RUBIO [**29**] and ROXIN [**205**]). This theorem also says that the set of all trajectories emanating from a compact set of initial points is compact uniformly on compact time intervals.

THEOREM 9.5 (Barbashin's Theorem). *Let Φ be a set-valued process on a complete metric space (X, d_X). Then the following statements hold:*

(i) *There exists a trajectory from x_0 to $x_1 \in \Phi(t_1, t_0, x_0)$ for each $t_1 \geq t_0 \in \mathbb{T}$ and $x_0 \in X$.*
(ii) *The set of all trajectories of $\Phi(t_1, t_0, x_0)$ with $x_0 \in K_0$ is compact in $C([t_0, t_1], X)$ for all $t_1 \geq t_0$ and nonempty compact subset K_0 of X.*

There are two useful types of invariance concepts for set-valued processes, depending on whether the full sets are involved or just certain trajectories. These are called, respectively, *strong* and *weak.*

DEFINITION 9.6 (Strong and weak invariance). A nonempty nonautonomous set $\mathcal{A} \subset \mathbb{T} \times X$ is called *strongly invariant* for a set-valued process Φ if

$$\Phi(t, t_0, A_{t_0}) = A_t \quad \text{for all } t \geq t_0 . \tag{9.1}$$

If the equality in (9.1) is replace by "$\subset$" or "$\supset$", then $\mathcal{A}$ is called *strongly positively invariant* or *strongly negatively invariant*, respectively. $\mathcal{A}$ is called *weakly invariant* if for every $t_0 \in \mathbb{T}$ and $x_0 \in A_{t_0}$, there exists a trajectory $\phi : [t_0, \infty) \cap \mathbb{T} \to X$ with $\phi(t_0) = x_0$ such that

$$\phi(t) \in A_t \quad \text{for all } t \geq t_0 .$$

Usually, strong invariance will be meant when the qualifier strong is not explicitly mentioned. Important examples of strongly invariant families are pullback and forward attractors, which are defined as obvious generalizations of the single-valued cases. Invariant sets also provided useful information about the dynamics of the system such as the existence of entire trajectories. The proof of the following lemma is almost the same as its counterpart Lemma 2.15 in Chapter 2, except it needs Barbashin's Theorem (see Theorem 9.5 above).

LEMMA 9.7. *Let $\mathcal{A}$ be a nonempty nonautonomous set which is strongly invariant for a set-valued process Φ. Then for any $t_0 \in \mathbb{T}$ and $x_0 \in A_{t_0}$, there exists an entire solution ξ through x_0 which is contained in $\mathcal{A}$, i.e., $\xi(t_0) = x_0$ and $\xi(t) \in A_t$ for all $t \in \mathbb{T}$.*

The counterpart of Lemma 2.20 on the existence of a strongly invariant subfamily in a strongly positively invariant family of compact subsets also holds for set-valued processes. The proof can be found in KLOEDEN & MARÍN-RUBIO [**132**].

LEMMA 9.8. *Let Φ be a set-valued process on a complete metric space (X, d_X) for the time set $\mathbb{T}$ and let $\mathcal{A}$ be a nonempty compact nonautonomous set, which is Φ-strongly positively invariant. Then there exists a nonempty compact nonautonomous set $\mathcal{A}^\infty$ contained in $\mathcal{A}$, which is strongly invariant.*

The negative invariant case is more complicated by the fact that trajectories need not stay in the given sets. Moreover, the resulting invariant family need not be strongly invariant, but it is weakly invariant for the set-valued process.

THEOREM 9.9 (Existence of nonautonomous invariant sets). *Let Φ be a set-valued process on a complete metric space (X, d_X) for the time $\mathbb{T}$ and let $\mathcal{A}$ be a nonempty compact nonautonomous set, which is Φ-strongly negatively invariant. Then there* *exists a nonempty compact nonautonomous set $\mathcal{A}^\infty \subset \mathcal{A}$, which is Φ-strongly negatively invariant and Φ-weakly invariant.*

The proof of Theorem 9.9 is rather detailed and will be given in Section 4 at the end of this chapter.

REMARK 9.10. The nonautonomous set $\mathcal{A}^\infty$ in Theorem 9.9 is Φ-strongly negatively invariant like the original family $\mathcal{A}$, but it is also Φ-weakly invariant, which the original family need not be. Both types of invariance are needed to ensure the existence of entire solutions in $\mathcal{A}^\infty$ in the sense that for any $\tau \in \mathbb{T}$ and any $x_\tau \in A_\tau^\infty$, there exists an entire trajectory $\phi : \mathbb{T} \to X$ with $\phi(\tau) = x_\tau$ and $\phi(t) \in A_t^\infty$ for all $t \in \mathbb{T}$. Specifically, the weak invariance gives the existence of a trajectory forwards in time and the strong negative invariance allows one to construct one backwards in time. Then concatenate both parts to obtain an entire trajectory.

1.1. A bifurcation in a set-valued process. One can use Lemma 9.8 and Theorem 9.9 to show, for example, that a set-valued system undergoes a bifurcation.

Let $b : \mathbb{R} \to \mathbb{R}$ be a continuous function with bounded values, specifically with $0 < \alpha \leq b(t) \leq \beta$ for all $t \in \mathbb{R}$. Then the scalar ordinary differential equation

$$\dot{x} = \nu x - b(t)x^3 \tag{9.2}$$

generates a process, which undergoes a supercritical bifurcation as shown in Chapter 8.

This process has the zero solution for all ν, which is asymptotically stable for $\nu < 0$ and unstable for $\nu > 0$. There are no other steady state solutions when the function b does not depend on time. The differential equation (9.2) is, however, a Bernoulli equation and can be solved explicitly. Following the analysis in Chapter 8, one can show that there exist locally asymptotically stable entire solutions $\pm\bar{\phi}_\nu$, which are given explicitly by (8.4), i.e., the system has local nonautonomous attractors $\mathcal{A}_\infty^\pm$ in both the pullback and forward senses consisting of the singleton fibers $\{\xi_\nu^\pm(t)\}$ for all $t \in \mathbb{R}$.

This detailed structure cannot, however, be obtained using the method below. Nevertheless, such an analysis gives useful information when no alternative finer analysis is possible, for example in the following set-valued modification of the differential equation (9.2). Let $G : \mathbb{R} \to \mathcal{K}([1, 2])$ be continuous in the Hausdorff metric, and consider the nonautonomous differential inclusion

$$\dot{x} \in \nu x - b(t)x^3 + \varepsilon x G(x)\,,$$

where $\varepsilon > 0$ is very small, which generates a set-valued process Φ_ν (parameterized by ν) with $\Phi_\nu(t, t_0, 0) = \{0\}$ for all $t, t_0 \in \mathbb{R}_0^+$. As in the single-valued case, it follows that

$$2(\nu + \varepsilon)x^2 - 2\beta x^4 \leq \frac{\mathrm{d}}{\mathrm{d}t}x^2 \leq 2(\nu + 2\varepsilon)x^2 - 2\alpha x^4\,,$$

so the intervals

$$A^- = \left[-\sqrt{\frac{\nu}{\alpha}}, -\sqrt{\frac{\nu}{\beta}}\right] \quad \text{and} \quad A^+ = \left[\sqrt{\frac{\nu}{\beta}}, \sqrt{\frac{\nu}{\alpha}}\right]$$

are Φ_ν-positively invariant. Hence, each interval contains a strongly invariant nonautonomous set $\mathcal{A}_\infty^\pm$, which one can show are local pullback attractors for the set-valued process. The zero solution loses stability for some $\nu = \nu(\varepsilon) \in (-\varepsilon, 0)$. The families $\mathcal{A}_\infty^\pm$ are thus candidates for the counterparts of the bifurcating solutions in the single-valued case and provide an indication that some kind of supercritical bifurcation has occurred in the nonautonomous set-valued system. Theorem 9.9 is not needed here to show that there is an entire solution in a negatively invariant neighborhood of the origin, since the zero solution is already such a solution. However, it is needed for the differential inclusion

$$\dot{x} \in \nu x - b(t)x^3 + \varepsilon G(x)\,, \tag{9.3}$$

since there is no explicitly known solution in a neighborhood of the origin.

EXERCISE 9.11. Use Lemma 9.8 and Theorem 9.9 to establish the existence of entire solutions of the differential inclusion (9.3) for different values on the parameter ν.

2. Set-valued skew product flows

The definition of a set-valued skew product flow is a generalization of the single-valued skew product flow in Chapter 2. It involves a driving system, which is a single-valued autonomous dynamical system θ on parameter space P, where (P, d_P) is a complete metric space, and a cocycle mapping, which is now set-valued.

DEFINITION 9.12 (Set-valued skew product flow). Let θ be an autonomous dynamical system on P. A mapping $\Phi : \mathbb{T}_0^+ \times P \times X \to \mathcal{K}(X)$ with the properties

(i) $\Phi(0, p, x) = \{x\}$ for all $(p, x) \in P \times X$,
(ii) $\Phi(s + t, p, x) = \Phi\big(s, \theta_t(p), \Phi(t, p, x)\big)$ for all $s, t \in \mathbb{T}_0^+$ and $(p, x) \in P \times X$,
(iii) $(t, p, x) \mapsto \Phi(t, p, x)$ is upper semi-continuous with respect to the Hausdorff semi-metric,
(iv) $t \mapsto \Phi(t, p, x)$ is continuous with respect to the Hausdorff metric uniformly in (p, x) in compact subsets of $P \times X$,

is called a *set-valued cocycle mapping* on X with respect to a driving system θ. The pair (θ, Φ) consisting of an autonomous dynamical system θ on P and a set-valued cocycle mapping Φ on X with respect to θ is called *set-valued skew product flow*.

Continuous and discrete time set-valued skew product flows are generated, for example, by nonautonomous differential inclusions such as

$$\dot{x} \in F\big(\theta_t(p), x\big)$$

or nonautonomous difference inclusions such as

$$x_{n+1} \in F\big(\theta_n(p), x_n\big),$$

where $F : P \times \mathbb{R}^d \to \mathcal{K}(\mathbb{R}^d)$ satisfies appropriate properties and θ is a continuous or discrete time driving system on P. The simplest examples involve right-hand sides of the form

$$F(p, x) = f(p, x) + B_1(0)\,,$$

where f is a single-valued mapping and $B_1(0)$ is the closed unit ball in $\mathbb{R}^d$ centered on the origin.

EXAMPLE 9.13. Consider the scalar nonautonomous differential inclusion

$$\dot{x} \in -x + e^{-t}[0,1],$$

which has the explicit solution

$$\Phi(t, t_0, x_0) = -x_0 e^{-(t-t_0)} + [0, 1-t_0]e^{-(t-t_0)}.$$

This set-valued process can be considered as a set-valued skew product flow with driving system on $P = \mathbb{R}$ given by $\theta_t(t_0) = t + t_0$.

EXAMPLE 9.14. Following Example 2.11 in Chapter 2, consider a parametrically dependent difference inclusion defined in terms of a set-valued mapping $F : \left[\frac{1}{2}, 1\right] \times \mathbb{R} \to \mathcal{K}(\mathbb{R})$, where

$$F(q,x) := [-f(x,q), f(x,q)],$$

with the continuous single-valued mapping $f : \left[\frac{1}{2}, 1\right] \times \mathbb{R} \mapsto \mathbb{R}$

$$f(q,x) = \frac{|x| + q^2}{1+q}.$$

Hence, $F(q,\cdot) \in C^{\mathrm{us}}(\mathbb{R}, \mathcal{H}(\mathbb{R}))$ for each $q \in \left[\frac{1}{2}, 1\right]$ is, in fact, continuous with respect to the Hausdorff metric, and $F(\cdot, x)$ is continuous for each $x \in \mathbb{R}$.

Let $P = \left[\frac{1}{2}, 1\right]^{\mathbb{Z}}$ be the space of bi-infinite sequences $p = (p_n)_{n\in\mathbb{Z}}$ taking values in $\left[\frac{1}{2}, 1\right]$. P is a compact metric space with the metric

$$d_P(p, p') = \sum_{n=-\infty}^{\infty} 2^{-|n|} \left|p_n - p'_n\right|,$$

and let $\{\theta_n : n \in \mathbb{Z}\}$ be the group generated by the left shift operator θ on this sequence space.

The family of mappings $\Phi(n, p, \cdot) \in C^{\mathrm{us}}(\mathbb{R}, \mathcal{K}(\mathbb{R}))$, defined by

$$\Phi(0,p,x) := \{x\} \quad \text{and} \quad \Phi(n,p,x) := F_{p_{n-1}} \circ \cdots \circ F_{p_0}(x) \quad \text{for all } n \in \mathbb{N}$$

and $x \in \mathbb{R}$, where $p = (p_n)_{n\in\mathbb{Z}}$, is a set-valued cocycle mapping on $\mathbb{R}$ with the driving system θ. Note that the mappings $\Phi(n, \cdot, \cdot)$, $n \in \mathbb{N}$, are continuous.

2.1. Pullback attractors for set-valued skew product flows. The definitions of attractors for single-valued skew product flows adapt easily to set-valued skew product flows.

DEFINITION 9.15 (Pullback attractor of a set-valued skew product flow). Let (θ, Φ) be a set-valued skew product flow. A nonempty and compact nonautonomous set $\mathcal{A}$ is called a *pullback attractor* of (θ, Φ) if it is Φ-invariant, i.e.,

$$\Phi\big(t, p, A_p\big) = A_{\theta_t(p)} \quad \text{for all } t \geq 0,$$

and pullback attracting, i.e.,

$$\lim_{t\to\infty} \mathrm{dist}_X \left(\Phi(t, \theta_{-t}(p), D), A_p\right) = 0$$

for any bounded subset D of X and each $p \in P$.

A compact subset B of X is called an *absorbing set* for (θ, Φ) if for each $p \in P$ and every bounded subset D of X, there exists an $T = T(p, D) > 0$ such that

$$\Phi\big(t, \theta_{-t}(p), D\big) \subseteq B \quad \text{for all } t \geq T \text{ and } p \in P.$$

The proof of the following theorem is similar to that of Theorem 3.20 in Chapter 3 and can be found in KLOEDEN & SCHMALFUSS [**143**]. There it involves a state space $X = \mathbb{R}^d$.

THEOREM 9.16 (Existence of pullback attractors of set-valued skew product flows). *Let(θ, Φ) be a set-valued skew product flow on $\mathbb{R}^d$ with an absorbing set B. Then there exists a pullback attractor $\mathcal{A}$, which is uniquely determined by*

$$A_p = \bigcap_{\tau \geq 0} \overline{\bigcup_{t \geq \tau} \Phi(t, \theta_{-t}(p), B)} \quad \textit{for all } p \in P\,.$$

Theorem 9.16 is slightly more general than Theorem 3.20, because it does not require the absorbing set B to be positively invariant w.r.t. Φ, i.e., to satisfy $\Phi(t, p, B) \subset B$ for all $t \geq 0$. Further generalizations to absorbing nonautonomous sets $\mathcal{B}$ and attracting universes are also possible, in which case pullback attractor component subsets are given by

$$A_p = \bigcap_{\tau \geq 0} \overline{\bigcup_{t \geq \tau} \Phi\big(t, \theta_{-t}(p), B_{\theta_{-t}(p)}\big)}\,.$$

EXAMPLE 9.17. The set-valued skew product flow in Example 9.14 has a positively invariant absorbing set $B = [-1, 1]$, and Theorem 9.16 can be applied. Its pullback attractor has component sets defined by

$$A_p = [-1, 1] \sum_{n=1}^{\infty} \frac{q_{-n}^2}{(1 + q_{-1})(1 + q_{-2}) \cdot \ldots \cdot (1 + q_{-n})}$$

where $p = (q_n)_{n \in \mathbb{Z}} \in \left[\frac{1}{2}, 1\right]^{\mathbb{Z}}$.

3. Backward extension of autonomous semi-dynamical systems

In Chapter 10, skew product flows with an autonomous semi-dynamical system as its driving system will be introduced. Such a driving system may not be uniquely extendable backwards in time, but it will be shown here that the backward extension defines a set-valued semi-dynamical system *backwards in time* when restricted to a compact invariant subset. In particular, consider an autonomous semi-dynamical system $\theta = (\theta_t)_{t \geq 0}$ on a complete metric space (P, d_P), and suppose that P^* is a nonempty compact subset of P, which is invariant under θ, i.e., satisfies $\theta_t(P^*) = P^*$ for all $t \geq 0$ (note that continuous time systems are considered here, but the results also apply to their discrete time counterparts). Let $\mathcal{K}(P^*)$ denote the space of nonempty compact subsets of P^*. It is a complete and compact metric space with the Hausdorff metric h_{P^*} on P^*.

DEFINITION 9.18 (Backward extension). The *backward extension* of the semi-dynamical system θ on a compact subset P^* of a complete metric space (P, d) is defined as

$$\Theta_t(p) := \big\{q \in P^* : \theta_{|t|}(q) = p\big\} \quad \text{for all } t \leq 0\,.$$

Note that Θ usually cannot be concatenated with θ to form a "natural" topological extension of the semi-dynamical system θ on P^* to a dynamical system on $\mathcal{K}(P^*)$.

To see this, firstly define $\Theta_t(K) := \bigcup_{p\in K}\Theta_t(p)$ for each $t \le 0$ and $K \in \mathcal{K}(P^*)$. Secondly, note that in general,

$$\theta_t\big(\Theta_{-t}(p)\big) = \Theta_0(p) = \{p\} \subset \Theta_{-t}\big(\theta_t(p)\big) \quad \text{for all } t \ge 0\,,$$

with possibly strict set inclusion, since there may exist points $q \in P^*$ with $q \neq p$ for which $\theta_t(q) = \theta_t(p)$. In particular, $\theta_t\big(\Theta_{-s-t}(p)\big) = \Theta_{-s}(p)$ and $\Theta_{-s-t}\big(\theta_t(p)\big) \subset \Theta_{-s}(p)$ for $s, t \ge 0$, with possible strict set inclusion.

However, the mappings Θ_t do form a semigroup of set-valued mappings on P^*, specifically, a set-valued semi-dynamical system as defined in Definition 9.1 above.

THEOREM 9.19. *The system $\Theta = (\Theta_t)_{t\le 0}$ is a set-valued semi-dynamical system backwards in time on the compact metric space P^*, i.e., the set-valued mapping F defined by $F(t,p) := \Theta_{-t}(p) \in \mathcal{K}(P^*)$ for $(t,p) \in \mathbb{R}_0^+ \times P^*$ is an autonomous set-valued semi-dynamical system on P^*, specifically, with*

(i) *$F(0,p) = \{p\}$ for all $p \in P^*$,*
(ii) *$F(s+t,p) = F\,(s, F(t,p))$ for all $s,t \ge 0$ and all $p \in P^*$,*
(iii) *$(t,p) \mapsto F(t,p)$ is upper semi-continuous in $(t,p) \in \mathbb{R}_0^+ \times P^*$ with respect to the Hausdorff semi-metric, i.e.,*

$$\lim_{(s,q)\to(t,p)} \operatorname{dist}_{P^*}\big(F(s,q), F(t,p)\big) = 0\,,$$

(iv) *$t \mapsto F(t,p)$ is continuous in $t \ge 0$ with respect to the Hausdorff metric uniformly in $p \in P^*$.*

The proof of this theorem will be divided into five lemmata. For notational convenience, the subscript P^* on the metrics will be omitted and they will be denoted simply by d, dist and h, respectively. In addition, $\Theta_t(p)$ will be used instead of $\Theta_t(\{p\})$.

LEMMA 9.20. *$\Theta_t(p)$ is nonempty and compact for every $t \le 0$ and $p \in P^*$.*

PROOF. The result is trivial for $t = 0$ because $\Theta_0(p) = \{p\}$, so let $t < 0$. Since $\theta_{|t|}(P^*) = P^*$, there exists at least one $q \in P^*$ with $\theta_{|t|}(q) = p$, that is, $q \in \Theta_t(p)$. Thus, $\Theta_t(p)$ nonempty. Now let $q_n \in \Theta_t(p)$ with $q_n \to q \in P^*$ as $n \to \infty$. Then $p = \theta_{|t|}(q_n)$ for each n. But $\theta_{|t|}(q_n) \to \theta_{|t|}(q)$ as $n \to \infty$ by the continuity of $\theta_{|t|}(\cdot)$, so $\theta_{|t|}(q) = p$, which means $q \in \Theta_t(p)$. Thus $\Theta_t(p)$ is a closed subset of P^*, which is compact. Hence, $\Theta_t(p)$ is itself compact. □

LEMMA 9.21. *$\Theta_{s+t}(p) = \Theta_s\left(\Theta_t(p)\right)$ for all $s, t \le 0$ and $p \in P^*$.*

PROOF. The result is obvious if either $s = 0$ or $t = 0$, so assume that both $s < 0$ and $t < 0$.
Let $q \in \Theta_s\left(\Theta_t(p)\right)$. Then there exists a $\bar{q} \in \Theta_t(p)$ with $q \in \Theta_s(\bar{q})$. Thus, $\theta_{|s|}(q) = \bar{q}$ and $\theta_{|t|}(\bar{q}) = p$. The (forwards) semi-group property of θ gives

$$p = \theta_{|t|}(\bar{q}) = \theta_{|t|}\big(\theta_{|s|}(q)\big) = \theta_{|t|+|s|}(q) = \theta_{|s+t|}(q)\,,$$

since $|t| + |s| = |s+t|$ when $s, t < 0$. Thus, $q \in \Theta_{s+t}(p)$. But q was otherwise arbitrary, which means that $\Theta_s\left(\Theta_t(p)\right) \subset \Theta_{s+t}(p)$.
Let $q \in \Theta_{s+t}(p)$. Then $\theta_{|s+t|}(q) = p$. Define $\bar{q} = \theta_{|s|}(q)$. Then

$$\theta_{|t|}(\bar{q}) = \theta_{|t|}\big(\theta_{|s|}(q)\big) = \theta_{|t|+|s|}(q) = \theta_{|s+t|}(q) = p\,,$$

i.e., $\bar{q} \in \Theta_t(p)$. But $\bar{q} = \theta_{|s|}(q)$, so $q \in \Theta_s(\bar{q})$, and consequently, $q \in \Theta_s(\bar{q}) \subset \Theta_s(\Theta_t(p))$, which means $\Theta_{s+t}(p) \subset \Theta_s(\Theta_t(p))$.
Combining the above cases gives $\Theta_{s+t}(p) = \Theta_s\left(\Theta_t(p)\right)$ for each $p \in P^*$. $\square$

LEMMA 9.22. *The mapping* $(t,p) \mapsto \Theta_t(p)$ *is upper semi-continuous in* $t \leq 0$ *and* $p \in P^*$, *i.e.,*

$$\lim_{(s,q)\to(t,p)} \operatorname{dist}\big(\Theta_s(q), \Theta_t(p)\big) = 0 \quad \textit{for all } t \leq 0 \textit{ and } p \in P^*.$$

PROOF. Suppose first that $(t,p) \mapsto \Theta_t(p)$ is not upper semi-continuous. Then there exists an $\varepsilon_0 > 0$ and sequences $t_n \to t$ in $\mathbb{R}_0^-$ and $p_n \to p$ in P^* such that

$$\operatorname{dist}\big(\Theta_{t_n}(p_n), \Theta_t(p)\big) \geq \varepsilon_0 \quad \text{for all } n \in \mathbb{N}.$$

The set $\Theta_{t_n}(p_n)$ is compact, so there exists $q_n \in \Theta_{t_n}(p_n)$, i.e, with $\theta_{|t_n|}(q_n) = p_n$, such that $\operatorname{dist}\big(q_n, \Theta_t(p)\big) = \operatorname{dist}\big(\Theta_{t_n}(p_n), \Theta_t(p)\big)$, and hence, such that

$$\operatorname{dist}\big(q_n, \Theta_t(p)\big) \geq \varepsilon_0 \quad \text{for all } n \in \mathbb{N}.$$

Now $q_n \in \Theta_{t_n}(p_n) \subset P^*$ and P^* is compact. Hence, there exists a convergent subsequence $q_{n'} \to \bar{q} \in P^*$. Moreover, $\theta_{|t_{n'}|}(q_{n'}) = p_{n'} \to p$, and by continuity, $\theta_{|t_{n'}|}(q_{n'}) \to \theta_{|t|}(\bar{q})$ as $n' \to \infty$. Thus, $\theta_{|t|}(\bar{q}) = p$, that is $\bar{q} \in \Theta_t(p)$. This means that

$$\varepsilon_0 \leq \operatorname{dist}\left(q_{n'}, \Theta_t(p)\right) \leq d\left(q_{n'}, \bar{q}\right) \to 0 \quad \text{as} \quad n' \to \infty,$$

which is impossible. Thus, the mapping $(t,p) \mapsto \Theta_t(p)$ is upper semi-continuous. $\square$

LEMMA 9.23. $\Theta_t(K) \in \mathcal{K}(P^*)$ *for all* $t < 0$ *and* $K \in \mathcal{K}(P^*)$.

PROOF. This follows since K is compact and the mapping $p \to \Theta_t(p)$ is upper semi-continuous. See AUBIN & FRANKOWSKA [**9**] and ROXIN [**205**]. $\square$

LEMMA 9.24. *The mapping* $t \mapsto \Theta_t(p)$ *is continuous in* $t \leq 0$ *uniformly* $p \in P^*$, *i.e.,*

$$\lim_{s\to t} \sup_{p\in P^*} h\big(\Theta_s(p), \Theta_t(p)\big) = 0 \quad \textit{for all } t \leq 0.$$

PROOF. Apply Lemma 9.22 to see that $t \mapsto \Theta_t(p)$ is upper semi-continuous for each fixed $p \in P^*$. This convergence is in fact uniform in $p \in P^*$. Suppose now that $t \mapsto \Theta_t(p)$ is not lower semi-continuous. Then there exist $p \in P^*$, $\varepsilon_0 > 0$ and a sequence $t_n \to t$ in $\mathbb{R}_0^-$ such that

$$\operatorname{dist}\big(\Theta_t(p), \Theta_{t_n}(p)\big) \geq \varepsilon_0 \quad \text{for all } n \in \mathbb{N}.$$

Since $\Theta_t(p)$ is compact, there exist $q_n \in \Theta_t(p)$ with $\operatorname{dist}\big(\Theta_t(p), \Theta_{t_n}(p)\big) = \operatorname{dist}\big(q_n, \Theta_{t_n}(p)\big)$. Thus,

$$\varepsilon_0 \leq \operatorname{dist}\left(q_n, \Theta_{t_n}(p)\right) \leq d\left(q_n, \tilde{q}_n\right) \quad \text{for all } n \in \mathbb{N} \text{ and } \tilde{q}_n \in \Theta_t(p_n),$$

since $\operatorname{dist}(a, B) \leq d(a,b)$ for all $b \in B$. By the compactness of $\Theta_t(p)$, there is a convergent subsequence $q_{n'} \to \bar{q} \in \Theta_t(p)$. There also exists a subsequence $t_{n'} \to t \leq 0$ with either $t_{n'} \leq t$ or $t_{n'} \geq t$ for all n'.
Consider the case $t_{n'} \leq t$ for all n'. Then $t_{n'} - t \leq 0$ as well as $t \leq 0$, so $\Theta_{t_{n'}}(p) = \Theta_{(t_{n'}-t)+t}(p) = \Theta_{t_{n'}-t}\big(\Theta_t(p)\big)$. Pick $\bar{q}_{n'} \in \Theta_{t_{n'}}(p)$ with $\bar{q}_{n'} \in \Theta_{t_{n'}-t}(q_{n'})$. Hence, $\theta_{t-t_{n'}}(\bar{q}_{n'}) = q_{n'} \to \bar{q} \in \Theta_t(p)$. But $\bar{q}_{n'} \in P^*$, which is compact, so there exists a

convergent subsubsequence $\bar{q}_{n''} \to q^*$. Hence the continuity of θ in both variables gives $\theta_{t-t_{n''}}(\bar{q}_{n''}) \to \theta_0(q^*) = q^*$. But $\theta_{t-t_{n''}}(\bar{q}_{n''}) \to \bar{q}$, so $\bar{q} = q^*$. Thus,

$$\varepsilon_0 \leq \operatorname{dist}\left(q_{n''}, \Theta_{t_{n''}}(p)\right) \leq d(q_{n''}, \bar{q}_{n''}) \to 0 \quad \text{as } n'' \to \infty\,,$$

which is impossible.
Consider the case $t_{n'} \geq t$ for all n'. Then $\theta_{|t|}(q_{n'}) = p$. Define $\bar{q}_{n'} = \theta_{|t|-|t_{n'}|}(q_{n'})$. Then $\theta_{|t_{n'}|}(\bar{q}_{n'}) = \theta_{|t_{n'}|}\left(\theta_{|t|-|t_{n'}|}(\bar{q}_{n'})\right) = \theta_{|t|}(\bar{q}_{n'}) = p$, i.e., $\bar{q}_{n'} \in \Theta_{t_{n'}}(p)$. From above, $q_{n'} \to \bar{q} \in \Theta_t(p)$, so by continuity of θ in both variables, one has $\bar{q}_{n'} = \theta_{|t|-|t_{n'}|}(q_{n'}) \to \theta_{|t|-|t|}(\bar{q}) = \theta_0(\bar{q}) = \bar{q}$. Thus,

$$\varepsilon_0 \leq \operatorname{dist}\left(q_{n''}, \Theta_{t_{n''}}(p)\right) \leq d\left(q_{n''}, \bar{q}_{n''}\right) \to 0 \quad \text{as} \quad n'' \to \infty\,,$$

which is impossible.
These contradictions show that $t \mapsto \Theta_t(p)$ must be lower semi-continuous. □

4. Proof of existence of nonautonomous invariant sets

The proof of Theorem 9.9 is given here in several steps, first the easier autonomous case, for both discrete continuous time, and then the nonautonomous case, also for discrete and then continuous times. In the autonomous cases the families consist just of a single set.

4.1. Proof of Theorem 9.9. Discrete time autonomous systems. Let $\Phi : X \to \mathcal{K}(X)$ be upper semi-continuous and consider the discrete time autonomous set-valued system with $\Phi(n, x) := \Phi^n(x)$ for $n \in \mathbb{N}$ and $\Phi(0, x) = \{x\}$. In addition, let A be a nonempty compact subset of X such that $A \subset \Phi(A)$.
Define $A_0 = A$, and let A_{-1} be the maximal subset of A_0 such that for each $a_0 \in A_0$ there exists an $a_{-1} \in A_{-1}$ with $a_0 \in \Phi(a_{-1})$. Then $\Phi(a_{-1}) \cap A_0 \neq \emptyset$ for all $a_{-1} \in A_{-1}$ and $A_0 \subset \Phi(A_{-1})$. Moreover, A_{-1} is compact by the compactness of A_0 and the upper semi-continuity of Φ. To see this, consider sequences $a_k \in A_{-1}$ and $b_k \in \Phi(a_k) \cap A_0$. Since A_0 is compact, $b_k \in A_0$ and $a_k \in A_{-1} \subset A_0$, there are convergent subsequences $a_{k_j} \to \bar{a} \in A_0$ and $b_{k_j} \to \bar{b} \in A_0$. Then $b_{k_j} \in \Phi(a_{k_j})$ and by the upper semi-continuity, one has $\Phi(a_{k_j}) \to \Phi(\bar{a})$ in dist_X, so $\bar{b} \in \Phi(\bar{a}) \cap A_0$, i.e., $\bar{a} \in A_{-1}$, which means that A_{-1} is compact.
Repeating this procedure gives a nested sequence of nonempty compact subsets $(A_{-n})_{n \geq 0}$, such that for each $a_{-n} \in A_{-n}$ there exists an $a_{-n-1} \in A_{-n-1}$ with $a_{-n} \in \Phi(a_{-n-1})$. Moreover, $A_{-n-1} \subset A_{-n} \subset \Phi(A_{-n-1})$.
Hence the set defined by

$$A^\infty = \bigcap_{n \geq 0} A_{-n}$$

is a nonempty compact subset of A.
Moreover, A^∞ is Φ-weakly invariant since if $\bar{a} \in A^\infty$, then $\bar{a} \in A_{-n-1}$. Thus, there exist $b_n \in \Phi(\bar{a}) \cap A_{-n}$. Now $\Phi(\bar{a})$ is compact, so there is convergent subsequence $b_{n_j} \to \bar{b} \in \Phi(\bar{a})$. But $b_{n_j} \in A_{-n_j}$, and by Lemma 1.27,

$$\operatorname{dist}_X\left(\bar{b}, A^\infty\right) \leq d_X\left(\bar{b}, b_{n_j}\right) + \operatorname{dist}_X\left(A_{-n_j}, A^\infty\right) \to 0\,,$$

so $\bar{b} \in A^\infty$ and hence, $\bar{b} \in A^\infty \cap \Phi(\bar{a})$, which gives the desired weak invariance.
The set A^∞ is also Φ-strongly negatively invariant since if $\bar{a} \in A^\infty$, then $\bar{a} \in A_{-n}$ for all $n \in \mathbb{N}$, so there exist $b_n \in A_{-n}$ such that $\bar{a} \in \Phi(b_n)$. Now the $b_n \in A_0$,

which is compact. Hence, there exists a convergent subsequence $b_{n_j} \to \bar{b}$ in A_0 and as above $\bar{b} \in A^\infty$. Moreover, by upper semi-continuity $\bar{a} \in \Phi(b_{n_j}) \to \Phi(\bar{b})$ in dist, so $\bar{a} \in \Phi(\bar{b})$, which means that $A^\infty \subset \Phi(A^\infty)$. □

4.2. Proof of Theorem 9.9. Continuous time autonomous systems. Apply the result just proved to the discrete time system formed by the time-1 mapping $\Phi(1,\cdot) : X \to \mathcal{K}(X)$. This gives us a nonempty compact subset A_∞^1 of A which is the maximal $\Phi(1,\cdot)$-weakly invariant subset of A which is also $\Phi(1,\cdot)$-strongly negatively invariant. In particular, $\Phi(1, a_\infty^1) \cap A_\infty^1 \neq \emptyset$ for all $a_\infty^1 \in A_\infty^1$ and $A_\infty^1 \subset \Phi(1, A_\infty^1)$.

This means that there is a trajectory $\phi^1 : \mathbb{R}_0^+ \to X$ with $\phi^1(0) = a_\infty^1$ with $\phi^1(1) \in A_\infty^1$ in particular and $\phi^1(n) \in A_\infty^1$ for all $n \in \mathbb{N}$ in general (note that trajectories of $\Phi(1,\cdot)$ are defined only for non-negative integer times, but can be filled in for intermediate times using the continuous time system). The problem is that such a trajectory may not be contained in A_∞^1 for all intermediate times $t \in (0,1)$, etc.

Therefore, repeat the procedure for the discrete time system formed by the time-$\frac{1}{2}$ mapping $\Phi(\frac{1}{2},\cdot) : X \to \mathcal{K}(X)$ and obtain a nonempty compact subset A_∞^2 of A which is the maximal $\Phi(\frac{1}{2},\cdot)$-weakly invariant subset of A, i.e., for every $a_\infty^2 \in A_\infty^2$ there exists a trajectory $\phi^2 : \mathbb{R}_0^+ \to X$ with $\phi^2(0) = a_\infty^2$ with $\phi^2(\frac{1}{2}) \in A_\infty^2$ and $\phi^2(\frac{1}{2}n) \in A_\infty^2$ for all $n \in \mathbb{N}$. In particular, $\phi^2(1) = \phi^2(\frac{1}{2}\cdot 2) \in A_\infty^2$, so A_∞^2 is also a $\Phi(1,\cdot)$-strongly invariant subset of A. But A_∞^1 is the maximal compact $\Phi(1,\cdot)$-weakly invariant subset of A, so $A_\infty^2 \subset A_\infty^1$.

Repeat this procedure with the the discrete time system formed by the time-2^{-n} mapping $\Phi(2^{-n},\cdot) : X \to \mathcal{K}(X)$ and obtain a nonempty compact subset A_∞^n of A which is the maximal $\Phi(2^{-n},\cdot)$-weakly invariant subset of A, and thus also $\Phi(2^{-n+1},\cdot)$-weakly invariant. Hence $A_\infty^n \subset A_\infty^{n-1}$ for all $n \in \mathbb{N}$. This is a nested family of nonempty compact subsets, so the set defined by

$$A^\infty = \bigcap_{n\geq 1} A_\infty^n$$

is a nonempty compact subset of A. Moreover, by the discrete time case considered above, A^∞ is $\Phi(2^{-n},\cdot)$-weakly invariant for each $n \in \mathbb{N}$. To see this, take an arbitrary point $\bar{a} \in A^\infty$. Then $\bar{a} \in A_\infty^n$ for each $n \in \mathbb{N}$, and there exists a trajectory ϕ^n with $\phi^n(0) = \bar{a}$ and $\phi^n(2^{-n}j) \in A_\infty^n$ for each $j \in \{0, 1, \ldots, 2^n\}$. By Barbashin's Theorem 9.5, the set of all trajectories restricted to the time interval $[0,1]$ of the continuous time system Φ with initial conditions in the compact set A^∞ is compact in $C([0,1], X)$, so there is a convergent subsequence ϕ^{n_j} converging uniformly on $[0,1]$ to trajectory $\bar{\phi}$. Clearly, $\bar{\phi}(0) = \bar{a}$ and $\bar{\phi}(t_d) \in A^\infty$ for all dyadic $t_d \in [0,1]$. It follows then by the continuity of $\bar{\phi}$ that $\bar{\phi}(t) \in A^\infty$ for all $t \in [0,1]$. This argument can be repeated for all larger times. Thus, A^∞ is Φ-weakly invariant.

Finally, a similar argument to that for the discrete time case applied to the set-valued mapping $\Phi(t,\cdot)$ for fixed $t > 0$ shows that A^∞ is also strongly negative invariant. □

4.3. Proof of Theorem 9.9. Discrete time processes. Fix $n \in \mathbb{Z}$, define $A_0^j \equiv A$ for all $j \in \mathbb{Z}$ and let A_{-1}^n be the maximal subset of A_0^{n-1} such that for each $a_0^n \in A_0^n$ there exists an $a_{-1}^n \in A_{-1}^n$ for which $a_0^n \in \Phi(n, n-1, a_{-1}^n)$. Hence,

$\Phi(n, n-1, a^n_{-1}) \cap A^n_0 \neq \emptyset$ for each $a^n_{-1} \in A^n_{-1}$. Since A is compact and $\Phi(n, n-1, \cdot)$ is upper semi-continuous, the set A^n_{-1} is compact by a similar argument to the autonomous case above. Note that $A^n_{-1} \subset A^{n-1}_0$ and $A^n_0 \subset \Phi(n, n-1, A^n_{-1})$.
Repeating this procedure gives a sequence of nonempty compact subsets $(A^n_{-j})_{j\in\mathbb{N}_0}$ such that for each $a^n_{-j} \in A^n_{-j}$, there exists an $a^n_{-j-1} \in A^n_{-j-1}$ with $a^n_{-j} \in \Phi(n-j, n-j-1, a^n_{-j-1})$. Hence, $\Phi\big(n-j, n-j-1, a^n_{-j-1}\big) \cap A^n_{-j} \neq \emptyset$ and

$$\Phi\big(n, n-j-1, a^n_{-j-1}\big) \cap A^n_0 \neq \emptyset \quad \text{for all } j \in \mathbb{N}_0 \text{ and } a^n_{-j-1} \in A^n_{-j-1}\,.$$

Moreover, $A^n_{-j} \subset \Phi\big(n-j, n-j-1, A^n_{-j-1}\big)$.
Similarly, define nonempty compact subsets A^{n+k}_{-j} for all $j, k \in \mathbb{N}$. Note that $A^{n+k+1}_{-k-1} \subset A^{n+k}_{-k}$ for each $k \in \mathbb{N}$ and fixed $n \in \mathbb{Z}$, i.e., a nested family of nonempty compact subsets of A. To see this consider the case $k = 1$, recall that $A^{n+1}_{-1} \subset A \subset \phi\big(n+1, n, A^{n+1}_{-1}\big)$ and note that

$$\begin{aligned} A \subset \Phi\big(n+2, n, A^{n+2}_{-2}\big) &\subset \Phi(n+2, n+1, A) \\ &\subset \Phi\big(n+2, n+1, \Phi(n+1, n, A^{n+1}_{-1})\big) \\ &= \Phi\big(n+2, n, A^{n+1}_{-1}\big)\,, \end{aligned}$$

where the two-parameter semi-group property has been used for the last equality. Hence, the set defined by

$$A^\infty(n) = \bigcap_{k\geq 0} A^{n+k}_{-k} \quad \text{for } n \in \mathbb{Z}$$

is a nonempty compact subset of A.
Moreover, the family of nonempty compact subsets $\mathcal{A}^\infty = (A^\infty(n))_{n\in\mathbb{Z}}$ of A is Φ-weakly invariant, since if $\bar{a} \in A^\infty(n)$, then $\bar{a} \in A^{n+k}_{-k}$, which is the set of points a_{-k} in A for which $\Phi(n+1, n, a_{-k}) \cap A^{n+k}_{-k+1} \neq \emptyset$. Thus, there exist $b_k \in \Phi(n+1, n, \bar{a}) \cap A^{n+k}_{-k+1}$. Now $\Phi(n+1, n, \bar{a})$ is compact, so there is convergent subsequence $b_{k_j} \to \bar{b} \in \Phi(n+1, n, \bar{a})$. But $b_{k_j} \in A^{n+k_j}_{-k_j+1} = A^{n+1+k_j-1}_{-k_j+1} = A^{n+1+l_j}_{-l_j}$ with $l_j = k_j - 1$, and by Lemma 1.27,

$$\operatorname{dist}_X\big(\bar{b}, A^\infty\big) \leq d_X\big(\bar{b}, b_{k_j}\big) + \operatorname{dist}_X\Big(A^{n+1+l_j}_{-l_j}, A^\infty(n+1)\Big) \to 0 \quad \text{as } j \to \infty\,,$$

so $\bar{b} \in A^\infty(n+1)$, and hence, $\bar{b} \in A^\infty(n+1) \cap \Phi(n+1, n, \bar{a})$, which gives the desired weak invariance.
The family of nonempty compact subsets $\mathcal{A}^\infty = (A^\infty(n))_{n\in\mathbb{Z}}$ of A is also Φ-strongly negatively invariant. If $\bar{a} \in A^\infty(n)$, then $\bar{a} \in A^{n+l}_{-l}$ for all $l \geq 0$. But

$$A^{n+l}_{-l} = A^{n-1+l+1}_{-l} = A^{n-1+k}_{-k+1} \quad \text{for all } k = l+1 \geq 1\,.$$

Thus, there exist $b_k \in A^{n-1+k}_{-k}$ such that $\bar{a} \in \Phi(n, n-1, b_k)$. Now the $b_k \in A$, which is compact, so there exists a convergent subsequence $b_{k_j} \to \bar{b}$ in A. In fact, $\bar{b} \in A^\infty(n-1)$, since

$$\operatorname{dist}_X\big(\bar{b}, A^\infty(n-1)\big) \leq d_X\big(\bar{b}, b_{k_j}\big) + \operatorname{dist}_X\Big(A^{n-1+k_j}_{-k_j}, A^\infty(n-1)\Big) \to 0$$

in the limit $j \to \infty$. In addition, one has $\bar{a} \in \Phi\big(n, n-1, \bar{b}\big)$, since $\bar{a} \in \Phi\,(n, n-1, b_k) \to \Phi\big(n, n-1, \bar{b}\big)$ with respect to dist by upper semi-continuity, which implies that $A^\infty(n) \subset \phi\big(n, n-1, A^\infty(n-1)\big)$. □

4.4. Proof of Theorem 9.9. Continuous time processes. First consider the process restricted to the dyadic numbers in $\mathbb{R}$, and define $\mathbb{T}_0 = \mathbb{Z}$ and $\mathbb{D}_n = \{j \cdot 2^{-n} : j \in \{0, 1, \ldots, 2^n\}\}$. In addition, set

$$\mathbb{T}_n := \mathbb{Z} + \mathbb{D}_n = \left\{k + t_j^n : k \in \mathbb{Z} \text{ and } t_j^n \in \mathbb{D}_n\right\} \quad \text{for all } n \in \mathbb{N}.$$

Then apply the result above to the discrete time set-valued process formed by the restriction $\Phi|_{\mathbb{T}_0}$ of the set-valued mapping Φ to the time set $\mathbb{T}_0$. This gives us a family $\mathcal{A}_\infty^0 = \left(A_\infty^0(t)\right)_{t\in\mathbb{T}_0}$ of nonempty compact subsets of A which is the maximal $\Phi|_{\mathbb{T}_0}$-weakly invariant family of subsets of A, which is also $\Phi|_{\mathbb{T}_0}$-strongly negatively invariant, i.e., for all $a_{n+1}^0 \in A_\infty^0(n+1)$, there exists an $a_n^0 \in A_\infty^0(n)$ for which $a_{n+1}^0 \in \Phi(n+1, n, a_n^0)$, i.e., $A_\infty^0(n+1) \subset \Phi\big(n+1, n, A_\infty^0(n)\big)$ for any $n \in \mathbb{Z}$.
We will henceforth restrict attention to the case $n = 0$ and the time interval $[0, 1]$ in the following discussion. From above, there is a trajectory $\phi^0 : [0, 1] \to X$ of the set-valued process Φ with $\phi^0(0) = a_0^0 \in A_\infty^0(0)$ and $\phi^0(1) \in A_\infty^0(1)$. The problem is that such a trajectory may not be contained in A for all intermediate times $t \in (0, 1)$.
Therefore we repeat this procedure for the discrete time system formed by the restriction $\Phi|_{\mathbb{T}_1}$ of the set-valued mapping Φ to the time set $\mathbb{T}_1$ and obtain a family $\mathcal{A}_\infty^1 = \left(A_\infty^1(t)\right)_{t\in\mathbb{T}_1}$ of nonempty compact subsets of A which is the maximal $\Phi|_{\mathbb{T}_1}$-weakly invariant family of subsets of A as well as being $\Phi|_{\mathbb{T}_1}$-strongly negatively invariant. In particular, for each $a_1^1 \in A_\infty^1(1)$, there exists a trajectory $\phi^1 : [0, 1] \to X$ of Φ with $\phi^1(1) = a_1^1$ and $\phi^1(\frac{1}{2}) \in A_\infty^1(\frac{1}{2})$. Thus, $A_\infty^1(t_{j+1}) \subset \Phi\big(t_{j+1}, t_j, A_\infty^1(t_j)\big)$ for $t_{j+1} - t_j = \frac{1}{2}$.
By this and the semi-group property, one has

$$A_\infty^1(1) \subset \Phi\big(1, \tfrac{1}{2}, A_\infty^1(\tfrac{1}{2})\big) \subset \Phi\big(1, \tfrac{1}{2}, \Phi(\tfrac{1}{2}, 0, A_\infty^1(0))\big) = \Phi\big(1, 0, A_\infty^1(0)\big).$$

This holds on all intervals $[n, n+1]$, so $\mathcal{A}_\infty^1$ is also a $\Phi|_{\mathbb{T}_0}$-weakly invariant and strongly negative family of compact subsets of A. But $\mathcal{A}_\infty^0$ is the maximal $\Phi|_{\mathbb{T}_0}$-weakly invariant family of compact subset of A, so $A_\infty^1(t) \subset A_\infty^0(t)$ for all $t \in \mathbb{T}_0 \cap \mathbb{T}_0 = \mathbb{T}_0$.
We repeat this procedure with the discrete time system formed by the restriction $\Phi|_{\mathbb{T}_n}$ of the set-valued-mapping Φ to the time set $\mathbb{T}_n$ and obtain a family $\mathcal{A}_\infty^n = (A_\infty^n(t))_{t\in\mathbb{T}_n}$ of nonempty compact subsets of A, which is the maximal $\Phi|_{\mathbb{T}_n}$-weakly invariant and strongly negative family of subsets of A, and thus also $\Phi|_{\mathbb{T}_{n-1}}$-weakly invariant and strongly negative. Hence, $A_\infty^n(t) \subset A_\infty^{n-1}(t)$ for all $n \in \mathbb{N}$ and $t \in \mathbb{T}_{n-1} \cap \mathbb{T}_n = \mathbb{T}_{n-1}$.
Thus, for each $t_l \in \mathbb{T}_l$ for an arbitrary $l \in \mathbb{N}$, the subsets $A_\infty^n(t_l), n \geq l$, are nonempty, compact and nested. Hence, the set defined by

$$A^\infty(t_l) = \bigcap_{n\geq l} A_\infty^n(t_l)$$

is a nonempty compact subset of A. In this way we obtain a family $\mathcal{A}_\infty^{\text{dyadic}} = (A^\infty(t_d))_{t_d\in\bigcup_{l\geq 0}\mathbb{T}_l}$ of nonempty compact subsets of A.
By the discrete time case above, the family $\mathcal{A}_\infty^{\text{dyadic}}$ is also $\Phi|_{\mathbb{T}_n}$-strongly negatively

invariant for each $n \in \mathbb{N}$, i.e., one has

$$A^{\infty}(t_{j+1}^n) \subset \Phi\big(t_{j+1}^n, t_j^n, A^{\infty}(t_j^n)\big)$$

for all $t_{j+1}^n, t_{j+1}^n \in \mathbb{T}_n$ with $t_{j+1}^n - t_j^n = 2^{-n}$. From this and the semi-group property, it follows that $A^{\infty}(t_1) \subset \Phi\big(t_1, t_0, A^{\infty}(t_0)\big)$ for all dyadic numbers $t_0 \leq t_1$ in $[0,1]$. Finally, for non-dyadic t, we define $A^{\infty}(t)$ to be the cluster points of the sets $A^{\infty}(t_d)$ for dyadic $t_d \to t$ with $t_d < t$. It follows that $A^{\infty}(t)$ is a nonempty compact subset of A. Moreover, $h_X(A^{\infty}(t_d), A^{\infty}(t)) \to 0$ as $t_d \to t$ with $t_d < t$. The proof follows by contradiction. There are two cases.

Case 1. Suppose that there are for dyadic $t_d \to t$ with $t_d < t$ and an $\varepsilon_0 > 0$ such that

$$\varepsilon_0 \leq \operatorname{dist}(A^{\infty}(t_d), A^{\infty}(t)).$$

Then there exist $a_d \in A^{\infty}(t_d) \subset A$ such that

$$\varepsilon_0 \leq \operatorname{dist}(a_d, A^{\infty}(t)) = \operatorname{dist}_X(A^{\infty}(t_d), A^{\infty}(t)).$$

Since A is compact there is a convergent subsequence $a'_d \to \bar{a}$ in A. By definition $\bar{a} \in A^{\infty}(t)$. But

$$\varepsilon_0 \leq \operatorname{dist}\left(a'_d, A^{\infty}(t)\right) \leq d_X(a'_d, \bar{a}),$$

which is a contradiction.

Case 2. Suppose that there are dyadic $t_d \to t$ with $t_d < t$ and an $\varepsilon_0 > 0$ such that

$$\varepsilon_0 \leq \operatorname{dist}\left(A^{\infty}(t), A^{\infty}(t_d)\right).$$

Then there exist $a_d \in A^{\infty}(t) \subset A$ such that

$$\varepsilon_0 \leq \operatorname{dist}\left(a_d, A^{\infty}(t_d)\right) = \operatorname{dist}_X\left(A^{\infty}(t), A^{\infty}(t_d)\right).$$

Since $A^{\infty}(t)$ is compact, there exists a convergent subsequence $a'_d \to \bar{a}$ in $A^{\infty}(t)$. In addition, by definition of $A^{\infty}(t)$, there is a convergent subsequence $b'_d \to \bar{a}$ with $b'_d \in A^{\infty}(t'_d)$. But

$$\varepsilon_0 \leq \operatorname{dist}\left(a'_d, A^{\infty}(t_d)\right) \leq d_X(a'_d, \bar{a}) + d_X(\bar{a}, b'_d),$$

which is a contradiction.

Moreover, $A^{\infty}(t) \subset \Phi\big(t, s_d, A^{\infty}(s_d)\big)$ with $t > s_d$ where s_d is dyadic, since this holds for a dyadic $t = t_d$ and

$$\begin{aligned} &\operatorname{dist}_X\left(A^{\infty}(t), \Phi\left(t, s_d, A^{\infty}(s_d)\right)\right) \\ &\leq \operatorname{dist}\left(A^{\infty}(t), A^{\infty}(t_d)\right) + \operatorname{dist}\left(\Phi\left(t_d, s_d, A^{\infty}(s_d)\right), \Phi\left(t, s_d, A^{\infty}(s_d)\right)\right) \end{aligned}$$

for dyadic $t_d \to t$ with $s_d < t_d < t$ by continuity of Φ in its first variable.

Define $\mathcal{A}^{\infty} = (A^{\infty}(t))_{t \in \mathbb{R}}$. It remains to show that $A^{\infty}(t) \subset \Phi\big(t, s, A^{\infty}(s)\big)$ for all $s < t$ in $\mathbb{R}$. From above, the only remaining case to show is for s non-dyadic. First consider t dyadic and dyadic $s_d \to s$ with $s_d < s$. Then

$$\begin{aligned} \operatorname{dist}_X\left(A^{\infty}(t), \Phi\left(t, s, A^{\infty}(s)\right)\right) &\leq \operatorname{dist}_X\left(A^{\infty}(t), \Phi\left(t, s_d, A^{\infty}(s_d)\right)\right) \\ &\quad + \operatorname{dist}\left(\Phi\left(t, s_d, A^{\infty}(s_d)\right), \Phi\left(t, s, A^{\infty}(s)\right)\right) \end{aligned}$$

converges to zero by the upper semi-continuity of Φ in its second and third variables. Then for a general t, using the semi-group property, one gets

$$A^{\infty}(t) \subset \Phi\big(t, t_0, A^{\infty}(t_0)\big) \subset \Phi\big(t, t_0, \Phi\left(t_0, s, A^{\infty}(s)\right)\big) = \Phi\big(t, s, A^{\infty}(s)\big)$$

where $t_0 < s$ is dyadic but otherwise arbitrary. Thus, $\mathcal{A}^{\infty}$ is Φ-strongly negatively invariant.

Moreover, by the discrete time case above, the family $\mathcal{A}_\infty^{\text{dyadic}}$ is $\Phi\big|_{\mathbb{T}_n}$-weakly invariant for all $n \in \mathbb{N}_0$. To see this, take an arbitrary point $\bar{a} \in A^\infty(1)$. Then $\bar{a} \in A_\infty^n(1)$ for each $n \in \mathbb{N}$, and there exists a trajectory ϕ^n with $\phi^n(1) = \bar{a}$ and $\phi^n(j2^{-n}) \in A_\infty^n(j2^{-n})$ for each $j \in \{0, 1, \ldots, 2^n\}$. By the Barbashin's Theorem 9.5, the set of all trajectories restricted to the time interval $[0,1]$ of the continuous time system Φ with initial conditions in the compact set A is compact in $C([0,1], X)$, so there is a convergent subsequence ϕ^{n_j} converging uniformly on $[0,1]$ to trajectory $\bar{\phi}$. Clearly, $\bar{\phi}(1) = \bar{a}$ and $\bar{\phi}(t_d) \in A^\infty(t_d)$ for all dyadic $t_d \in [0,1]$. It follows then by the continuity of $\bar{\phi}$ that $\bar{\phi}(t) \in A^\infty(t)$ for all $t \in [0,1]$. This argument can be repeated for other time intervals $[n, n+1]$. Thus, $\mathcal{A}^\infty$ is Φ-weakly invariant. □

This completes the proof of Theorem 9.9.

Endnotes. Set-valued dynamical systems have a long history, an early account of which for the autonomous case can be found in the 1969 monograph of Szegö & Treccani [**229**] and for the nonautonomous case in the 1978 review paper Kloeden [**111**], see also Cheban & Fakhih [**39**]. They are closely associated to inclusion equations and control systems, see Roxin [**204**]. Usually nonautonomous set-valued systems are considered with the the process formalism, see, e.g., [**110, 111, 205, 204**], and see Melnik & Valero [**167**] for set-valued processes generated by evolution equations without unique solutions. Caraballo, Kloeden & Marín-Rubio [**29**] and Kloeden & Schmalfuß [**143**] consider set-valued skew product flows. The result on the existence of a weakly invariant subset of a strongly negatively invariant set (Theorem 9.9) and that of entire solutions is from Kloeden & Marín-Rubio [**133**]. For bifurcations in set-valued dynamical systems, see Kloeden & Marín-Rubio [**133**] and Lamb, Rasmussen & Rodrigues [**152**]. Weak pullback attractors for set-valued nonautonomous systems, i.e., requiring only at least one trajectory through each initial point to satisfy a given property, are not discussed in this book. Results about them can be found in Kloeden & Marín-Rubio [**134**] and Kloeden & Valero [**146**]. Also not discussed here are the results on the continuous dependence of attractors of set-valued processes on parameters in Li Desheng & Kloeden [**69**]. The backward extension of a single valued semi-dynamical system in a compact invariant set as set-valued dynamical systems comes from Kloeden [**117**].

CHAPTER 10

Nonautonomous semi-dynamical systems

A skew product semi-flow (θ, φ) consists of a cocycle mapping φ on a state space X and an autonomous semi-dynamical system θ on a base or parameter space P as its driving system. Suppose that θ has a global attractor P^* in P. Then, by Theorem 9.19, the backward extension Θ of θ on P^*, defined by

$$\Theta_t(p) := \left\{q \in P^* : \theta_{|t|}(q) = p\right\} \quad \text{for all } t \leq 0 \text{ and } p \in P^* , \tag{10.1}$$

is a set-valued semi-dynamical system on P^* backwards in time, i.e., it has the properties

(i) $\Theta_0(p) = \{p\}$ for all $p \in P^*$,
(ii) $\Theta_{s+t}(p) = \Theta_s \circ \Theta_t(p)$ for all $s, t \leq 0$ and $p \in P^*$,
(iii) $(t, p) \mapsto \Theta_t(p)$ is upper semi-continuous in $(t, p) \in \mathbb{R}_0^- \times P^*$ with respect to the Hausdorff semi-distance, i.e.,

$$\operatorname{dist}_{P^*}\left(\Theta_s(q), \Theta_t(p)\right) \to 0 \quad \text{as } (s, q) \to (t, p) \text{ in } \mathbb{R}_0^- \times P^* ,$$

(iv) $t \mapsto \Theta_t(p)$ is continuous in $t \leq 0$ with respect to the Hausdorff metric uniformly in $p \in P^*$.

With this backward extension, it is possible to generalize the concepts of pullback convergence and pullback attractors to skew product semi-flows.

1. Attractors of skew product semi-flows

Let (θ, φ) be a skew product semi-flow consisting of a cocycle mapping φ on a state space X and an autonomous semi-dynamical driving system θ on a base space P, where (X, d_X) and (P, d_P) are complete metric spaces. Suppose that the associated autonomous semi-dynamical system $\pi = (\theta, \varphi)$ on $P \times X$ has a global autonomous attractor

$$\mathcal{A} = \bigcup_{p \in P^*} \{p\} \times A_p . \tag{10.2}$$

Then the set $P^* \subset P$ is the global attractor of the driving system θ on the base space P, and θ has a set-valued backward extension Θ on P^*. There is no loss of generality in restricting attention to the autonomous semi-dynamical system $\pi^* = (\theta, \varphi)$ on $P^* \times X$, since it then also has a global autonomous attractor $\mathcal{A}^*$ which coincides with $\mathcal{A}$. Obviously, the sets A_p, $p \in P^*$, in (10.2) are nonempty compact subsets of X.

PROPERTY 10.1. The set-valued mapping $p \mapsto A_p$ is upper semi-continuous with respect to the Hausdorff semi-distance on X.

PROOF. The graph of the set-valued mapping $p \mapsto A_p$, $p \in P^*$, is the compact subset $\mathcal{A}$ in $P^* \times X$. The closedness of the graph is equivalent to the upper semi-continuity of the mapping $p \mapsto A_p$, see AUBIN & FRANKOWSKA [**9**]. □

Unlike the skew product flow case, the mapping $t \mapsto A_{\theta_t(p)}$ here is only continuous from the right for each $p \in P$ and upper semi-continuous from the left.

PROPERTY 10.2. The set-valued mapping $t \mapsto A_{\theta_t(p)}$ is continuous from the right for each $p \in P$ and upper semi-continuous from the left.

PROOF. Firstly, note that $\theta_s(p) \to \theta_t(p)$ as $s \to t$, so by the upper semi-continuity in Property 10.1, $A_{\theta_s(p)} \to A_{\theta_t(p)}$ upper semi-continuously as $s \to t$. Now let $s = t + \delta$ with $\delta > 0$. Then

$$\begin{aligned}
\operatorname{dist}_X\left(A_{\theta_t(p)}, A_{\theta_{t+\delta}(p)}\right) &\leq \operatorname{dist}_X\left(A_{\theta_t(p)}, \varphi(\delta, \theta_t(p), A_{\theta_t(p)})\right) \\
&\quad + \operatorname{dist}_X\left(\varphi(\delta, \theta_t(p), A_{\theta_t(p)}), A_{\theta_{t+\delta}(p)}\right) \\
&= \operatorname{dist}_X\left(A_{\theta_t(p)}, \varphi(\delta, \theta_t(p), A_{\theta_t(p)})\right) + 0 \\
&\leq h_X\left(\varphi(\delta, \theta_t(p), A_{\theta_t(p)}), A_{\theta_t(p)}\right) \to 0 \quad \text{as } \delta \to 0,
\end{aligned}$$

since $\varphi(\delta, \theta_t(p), A_{\theta_t(p)}) \subset A_{\theta_{t+\delta}(p)}$, so $\operatorname{dist}_X\left(\varphi(\delta, \theta_t(p), A_{\theta_t(p)}), A_{\theta_{t+\delta}(p)}\right) = 0$, and the set-valued mapping $t \to \varphi(t, p, C)$ is continuous for all $p \in P^*$ and any nonempty compact subset C of X, see Exercise 2.12. This means that the set-valued mapping $t \mapsto A_{\theta_t(p)}$ is also lower semi-continuous from the right. Hence, it is continuous from the right. □

The global attractor $\mathcal{A}$ of (10.2) is *φ-positive invariant*, i.e.,

$$\varphi(t, p, A_p) \subset A_{\theta_t(p)} \quad \text{for all } t \geq 0 \text{ and } p \in P^*. \tag{10.3}$$

However, it need not be φ-invariant as in the case of a skew product flow, i.e., it need not satisfy (10.3) with equality, since the θ_t may not be invertible, in particular not one-to-one, on P^*.

PROPERTY 10.3. $\mathcal{A}$ is *accumulative invariant* in the sense that

$$\bigcup_{q \in \Theta_{-t}(p)} \varphi(t, q, A_q) = A_p \quad \text{for all } t \geq 0 \text{ and } p \in P^*. \tag{10.4}$$

PROOF. This result follows from π-invariance of the global attractor $\mathcal{A}$, i.e., $\pi(t, \mathcal{A}) = \mathcal{A}$ for each $t \geq 0$, and the definition (10.1) of the backward extension $\Theta_{-t}(p)$. From (10.2), one gets

$$\pi(t, (p, A_p)) = \left(\theta_t(p), \varphi(t, p, A_p)\right) \subset \mathcal{A},$$

which yields $\varphi(t, p, A_p) \subset A_{\theta_t(p)}$, i.e., the positive invariance property (10.3). From this, one concludes that $\varphi(t, q, A_q) \subset A_p$ for each $q \in \Theta_{-t}(p)$. Hence,

$$\bigcup_{q \in \Theta_{-t}(p)} \varphi(t, q, A_q) \subset A_p.$$

To obtain the opposite inclusion, let $x \in A_p$, i.e., $(p, x) \in \mathcal{A}$, and fix $t \geq 0$. From the π-invariance $\pi(t, \mathcal{A}) = \mathcal{A}$, there exists $(q, y) \in \mathcal{A}$, i.e., with $y \in A_q$, such that

$\pi(t,(q,y)) = (p,x)$. This reads componentwise as $\theta_t(q) = p$ and $\varphi(t,q,y) = x$. Thus, $q \in \Theta_{-t}(p)$ and $x \in \varphi(t,q,A_q)$. Taking the union over all $x \in A_p$ gives

$$A_p \subset \bigcup_{q\in\Theta_{-t}(p)} \varphi(t,q,A_q)\,,$$

which finishes the proof. □

If $\mathcal{B}$ is φ-positively invariant and pullback absorbing such that $p \mapsto B_p$ is upper semi-continuous (which is no restriction of generality, see Chapter 7), then the sets $\widetilde{A}_p$ defined by

$$\widetilde{A}_p = \bigcap_{t\geq 0} \bigcup_{q\in\Theta_{-t}(p)} \varphi(t,q,B_q) \tag{10.5}$$

are nonempty and compact for each $p \in P^*$.

PROPERTY 10.4. The sets $\widetilde{A}_p$, $p \in P^*$, in (10.5) are nonempty and compact.

PROOF. Since $q \mapsto B_q$ is upper semi-continuous and the mapping $(t,q,x) \mapsto \varphi(t,q,x)$ continuous, the set-valued mapping $q \mapsto \varphi(t,q,B_q)$ is upper semi-continuous. Moreover, $\varphi(t,q,B_q)$ is nonempty and compact, because B_q is nonempty and compact. Finally, since $\Theta_{-t}(p)$ is nonempty and compact, the set $\bigcup_{q\in\Theta_{-t}(p)} \varphi(t,q,B_q)$ is nonempty and compact. Now let $\delta > 0$ and $t \geq 0$. Consider a $\bar{q} \in \Theta_{-t-\delta}(p)$, i.e., $p = \theta_{t+\delta}(\bar{q})$, and write $q = \theta_\delta(\bar{q})$, so $q \in \Theta_{-t}(p)$. Then

$$\begin{aligned}\varphi(t+\delta,\bar{q},B_{\bar{q}}) &= \varphi\big(t,\theta_\delta(\bar{q}),\varphi(\delta,\bar{q},B_{\bar{q}})\big)\\ &\subset \varphi\big(t,\theta_\delta(\bar{q}),B_{\theta_\delta(\bar{q})}\big) = \varphi(t,q,B_q)\end{aligned}$$

by the cocycle property and the positive invariance of $\mathcal{B}$, so

$$\bigcup_{\bar{q}\in\Theta_{-t-\delta}(p)} \varphi(t,\bar{q},B_{\bar{q}}) \subset \bigcup_{q\in\Theta_{-t}(p)} \varphi(t,q,B_q)\,.$$

Thus, $\widetilde{A}_p$ is nonempty and compact as the intersection of a family of nested nonempty compact sets. □

Moreover, if $\mathcal{B}$ corresponds to an absorbing set $\mathfrak{B}^* = \bigcup_{p\in P^*}\{p\} \times B_p$ of the global attractor $\mathcal{A}^*$, then the sets $\widetilde{A}_p$ coincide with the A_p in the component sets $\{p\} \times A_p$ of $\mathcal{A}^*$.

PROPERTY 10.5. Under the above condition, $\widetilde{A}_p = A_p$ for each $p \in P^*$.

PROOF. Let $\mathfrak{B}^* = \bigcup_{p\in P^*}\{p\} \times B^*_p$ be an absorbing set of the global attractor $\mathcal{A}^*$ in $P^* \times X$. Thus, $(p,a) \in \mathcal{A}^*$, i.e., $a \in A^*_p$, if and only if there exists a sequence $(p_n,t_n) \in \mathfrak{B}^*$ with $t_n \to \infty$, $\theta_{t_n}(p_n) \to p$ and $\varphi(t_n,p_n,b_n) \to a$ as $n \to \infty$.
Now let $a \in \widetilde{A}_p$. Then there exist sequences $t_n \to \infty$, $q_n \in \Theta_{-t_n}(p)$ and $b_n \in B^*_{q_n}$ such that $\varphi(t_n,q_n,b_n) \to a$ as $n \to \infty$. Now $(t_n,q_n) \in \mathcal{B}^*$ and $\theta_{t_n}(q_n) = p$. Obviously, $\theta_{t_n}(q_n) \to p$ as $n \to \infty$, so $(p,a) \in \mathcal{A}^*$, i.e., $a \in A^*_p$. Since a was otherwise arbitrary, this proves that $\widetilde{A}_p \subset A^*_p$.
Conversely, let $a \in A^*_p \equiv A_p$ and write $a = a_0$. Now

$$\bigcup_{q\in\Theta_{-1}(p)} \varphi(1,q,A_q) = A_p$$

by the accumulative invariance property with $t = 1$, so there exists $q_{-1} \in \Theta_{-1}(p)$ and $a_{-1} \subset A_{q_{-1}}$ such that $\varphi(1, q_{-1}, a_{-1}) = a_0$. Repeating this argument inductively with $\bigcup_{q\in\Theta_{-1}(q_{-n})} \varphi(1, q, A_q) = A_{q_{-n}}$, there exist $q_{-n-1} \in \Theta_{-1}(q_{-n}) \subset \Theta_{-n-1}(p)$ and $a_{-n-1} \subset A_{q_{-n-1}}$ such that $\varphi(1, q_{-n-1}, a_{-n-1}) = a_{-n}$. Thus, $\varphi(n, q_{-n}, a_{-n}) = a_0$ for each $n \in \mathbb{N}_0$. Since $a_{-n} \subset A_{q_{-n}} \subset B^*_{q_{-n}}$ for each $n \in \mathbb{N}_0$, it follows that $a_0 \in \varphi\big(n, q_{-n}, B^*_{q_{-n}}\big)$, and hence,

$$a = a_0 \in \bigcap_{n\geq 0} \varphi(n, q_{-n}, B^*_{q_{-n}}) \subset \bigcap_{n\geq 0} \bigcup_{q\in\Theta_{-n}(p)} \varphi(n, q, B^*_q)$$
$$\subset \bigcap_{t\geq 0} \bigcup_{q\in\Theta_{-t}(p)} \varphi(t, q, B^*_q) = \widetilde{A}_p\,.$$

Thus, $a \in \widetilde{A}_p$, which means $A^*_p \subset \widetilde{A}_p$. Combining the two results gives the desired equality $A^*_p = \widetilde{A}_p$. □

Finally, Definition 3.11 of pullback attraction for skew product flows in Chapter 3 is no longer applicable, but can be generalized in terms of the set-valued backward extension Θ defined by (10.1) for all $t \leq 0$, namely

$$\lim_{t\to\infty} \sup_{q\in\Theta_{-t}(p)} \operatorname{dist}_X \big(\varphi(t, q, D_q), A_p\big) = 0 \tag{10.6}$$

for any nonempty and compact (or bounded) nonautonomous set $\mathcal{D}$.

PROPERTY 10.6. $\mathcal{A}$ is pullback attracting in the sense of (10.6).

PROOF. It suffices to prove (10.6) for the absorbing set $\mathcal{B}$, since it pullback absorbs compact (or bounded) nonautonomous set. Suppose that there is an $\varepsilon_0 > 0$ and sequences $t_n \to \infty$ as $n \to \infty$ and $q_n \in \Theta_{-t_n}(p)$ such that

$$\operatorname{dist}_X \big(\varphi(t_n, q_n, B_{q_n}), A_p\big) \geq \varepsilon_0 \quad \text{for all } \ n \in \mathbb{N}\,.$$

By compactness, there exists a sequence $b_n \in B_{q_n}$ such that

$$\operatorname{dist}_X(a_n, A_p) = \operatorname{dist}_X \big(\varphi(t_n, q_n, B_{q_n}), A_p\big) \geq \varepsilon_0 \quad \text{for all } \ n \geq 0\,, \tag{10.7}$$

where $a_n = \varphi\,(t_n, q_n, b_n)$. But $\mathcal{B}^* = \bigcup_{p\in P^*} \{p\} \times B_p$ is π-positive invariant and $\theta_{t_n}(q_n) = p$, so $(p, a_n) = \big(\theta_{t_n}(q_n), \varphi(t_n, q_n, b_n)\big) \in \mathcal{B}^*$. Hence $a_n \in B_p$, which is compact, so there exists a convergent subsequence $a_{n_j} \to a \in B_p$ as $j \to \infty$. Consider the sequences $s_j = t_{n_j}$, $q'_j = q_{n_j} \in \Theta_{-t_{n_j}}(p) = \Theta_{-s_j}(p)$ and $b'_j = b_{n_j} \in B_{q_{n_j}} = B_{q'_j}$. Thus, $s_j \to \infty$ as $j \to \infty$, $\theta_t(q'_j) = p \to p$ (trivially) as $j \to \infty$ and $\varphi\big(s_j, q'_j, b'_j\big) \to a$ as $j \to \infty$. This means that $(p, a) \in \mathcal{A}^*$, i.e., $a \in A^*_p = A_p$, and hence,

$$\operatorname{dist}_X \big(\varphi(s_j, q'_j, b'_j), A_p\big) \leq \frac{1}{2}\varepsilon_0 \quad \text{for all } \ j \text{ large enough.}$$

This, however, contradicts the inequality (10.7), so the above assumption is false. □

2. The twisted horseshoe mapping

The *twisted horseshoe mapping* $f = (f_1, f_2)$ on the unit square $[0,1]\times[0,1]$ is defined by

$$f_1(x,y) = \begin{cases} 2x & : \ 0 \le x \le \frac{1}{2}\,, \\ 2-2x & : \ \frac{1}{2} < x \le 1\,, \end{cases} \qquad \text{and} \quad f_2(x,y) = \frac{x}{2} + \frac{y}{10} + \frac{1}{4}\,.$$

It was introduced and investigated by GUCKENHEIMER, OSTER & IPAKTCHI [**87**] in the 1970s as a biomathematical model, which has a saddle fixed point $(\bar{x}, \bar{y}) = \left(\frac{2}{3}, \frac{35}{54}\right)$ with the eigenvalues -2 and $\frac{1}{10}$. All the conditions of a theorem in KLOEDEN [**113**] which generalizes Marotto's snap-back repeller result to saddle points are thus satisfied, so the twisted horseshoe mapping behaves chaotically in a slight generalization of the sense of Li and Yorke. In fact, the twisted-horseshoe mapping also satisfies the Sharkovsky cycle co-existence ordering, due to a generalization in KLOEDEN [**112**] of Sharkovsky's theorem to triangular mappings.

The autonomous difference equation generated by the twisted horseshoe mapping can be interpreted as a nonautonomous difference equation

$$x_{n+1} = \frac{x_n}{10} + \frac{p_n}{2} + \frac{1}{4} \tag{10.8}$$

driven by the semi-dynamical system on $[0,1]$

$$p_{n+1} = \begin{cases} 2p_n & : \ 0 \le p_n \le \frac{1}{2}\,, \\ 2-2p_n & : \ \frac{1}{2} < p_n \le 1\,. \end{cases} \tag{10.9}$$

Note that any trajectory of the driving system can be extended backwards indefinitely, though typically not uniquely. Let $p = (p_n)_{n\in\mathbb{Z}}$ be such an entire trajectory and hold it fixed. Then the explicit solution of (10.8) given by

$$x_n = 10^{-n+n_0} x_{n_0} + \frac{1}{2} \sum_{j=n_0}^{n-1} 10^{j+1-n} p_j + \frac{1}{4} \sum_{j=n_0}^{n-1} 10^{j+1-n}. \tag{10.10}$$

EXERCISE 10.7. Verify (10.10).

Let n fixed, and let the x_{n_0} take values in a fixed bounded subset of $\mathbb{R}$. Then, taking the pullback limit as $n_0 \to -\infty$ gives

$$\bar{x}_n = \frac{1}{2} \sum_{j=-\infty}^{n-1} 10^{j+1-n} p_j + \frac{1}{4} \sum_{j=-\infty}^{n-1} 10^{j+1-n}\,.$$

This means that

$$\bar{x}_n = \frac{1}{2} \sum_{j=-\infty}^{n-1} 10^{j+1-n} p_j + \frac{10}{36}\,.$$

In particular, for two different initial values x_{n_0} and x'_{n_0} with the same driving sequence p, it is easily shown that

$$|x_n - x'_n| = |x_{n_0} - x'_{n_0}| 10^{-n+n_0} \to 0 \quad \text{as } n \to \infty\,,$$

and hence, with $x'_n \equiv \bar{x}_n$, one gets

$$|x_n - \bar{x}_n| \to 0 \quad \text{as } n \to \infty\,.$$

Thus, all solutions of (10.8) with different starting points converge to the same limit if they correspond to the same driving sequence.

Define $\bar{x}(p) := \bar{x}_0$, and let θ be the shift operator on the space of sequences $P = [0,1]^{\mathbb{Z}}$. Then $\bar{x}_n = \bar{x}(\theta_n(p))$ and the singleton subsets $A_p := \{\bar{x}(p)\}$ form the pullback attractor (which is also forwards attracting here) of the skew product semi-flow generated by the twisted horseshoe system.

Let X_p be the space $\mathbb{R}$ with the origin shifted to $\bar{x}(p)$. With the change of coordinates $z_n := x_n - \bar{x}(p)$, the mapping (10.8) simplifies to

$$z_{n+1} = \frac{1}{10} z_n \,, \tag{10.11}$$

with $z_n \in X_{\theta^n(p)}$ and $z_{n+1} \in X_{\theta^{n+1}(p)}$. This makes the behavior of the system transparent and provides a simple example of a generalization of skew product flows for which the state space is parameterized by the base space variable, i.e., is given by a fibration of the state space of the associated autonomous semi-dynamical system, see, e.g., CHEBAN [**38**], CHEBAN, KLOEDEN & SCHMALFUSS [**41**] and PÖTZSCHE [**188**].

Endnotes. This chapter is based mainly on the paper Kloeden [**117**], with the example of the twisted horseshoe mapping taken from Diamond, Kloeden & Kozyakin [**71**]. The twisted horseshoe mapping was introduced by Guckenheimer, Oster & Ipaktachi [**87**]. Kloeden & Li Zhong [**130**] reviews of chaos, in particular higher dimensional generalizations of the Li & Yorke result. The generalization of Sharkovsky's theorem to triangular mappings can be found in Kloeden [**112**]. For attractors in pinched skew product flows, see Glendinning [**83**].

CHAPTER 11

Approximation and perturbation of attractors

In this chapter, three aspects of the effects of approximation and perturbation on attractors are considered. In the first section, an autonomous differential equation is subjected to a nonautonomous perturbation; in the second section, a numerical scheme is applied to a nonautonomous equation with a uniform nonautonomous attractor; and in the third section, the perturbation of the driving system of a skew product flow is considered.

1. Nonautonomous perturbations of an autonomous system

A global attractor of an autonomous system is uniformly asymptotically stable and hence *totally stable*, i.e., nearby perturbations of the system are also dissipative. By Theorem 7.9, there exists a Lyapunov function which characterizes the uniform asymptotic stability of such an attractor. Such a Lyapunov function is used in the proof of the next theorem to construct an absorbing set for a nonautonomous perturbation of an autonomous system, from which it follows that the nonautonomous system has a pullback attractor.

THEOREM 11.1 (Nonautonomous perturbations of attractors). *Suppose that $f : \mathbb{R}^d \to \mathbb{R}^d$ is uniformly Lipschitz and that the autonomous system*

$$\dot{x} = f(x) \tag{11.1}$$

has a global attractor A. Furthermore, suppose that $g : \mathbb{R} \times \mathbb{R}^d \to \mathbb{R}^d$ is continuous in $t \in \mathbb{R}$, uniformly Lipschitz continuous in $x \in \mathbb{R}^d$ and uniformly bounded on $\mathbb{R} \times \mathbb{R}^d$, specifically

$$\sup_{(t,x)\in\mathbb{R}\times\mathbb{R}^d} \|g(t,x)\| \le K < \infty .$$

Then the nonautonomous system

$$\dot{x} = f(x) + \varepsilon g(t,x) \tag{11.2}$$

generates a process $\phi^\varepsilon(t, t_0, x_0)$ which has a pullback attractor $\mathcal{A}$ such that

$$\lim_{\varepsilon\searrow 0} \operatorname{dist}(A_t^\varepsilon, A) = 0 \quad \textit{for all } t \in \mathbb{R} . \tag{11.3}$$

In addition, the component sets A_t^ε have constant Hausdorff dimension

$$\dim_H A_{t_0}^\varepsilon = \dim_H A_{t_1}^\varepsilon \quad \textit{for all } t_0, t_1 \in \mathbb{R} .$$

PROOF. Since the attractor A is globally uniformly asymptotically stable, by Theorem 7.9, there exists a Lyapunov function V on $\mathbb{R}^d$, which will be used to construct an absorbing set for the nonautonomous system. It follows from the definition of the upper left Dini derivative (see (7.9)) and the Lipschitz property of V that for any $t \in \mathbb{R}$ and $x \in \mathbb{R}^d$, one has

$$\begin{aligned} D^+_{(11.2)}V(x) &= \limsup_{h\searrow 0} \frac{V\big(x + h(f(x) + \varepsilon g(t,x))\big) - V(x)}{h} \\ &= \limsup_{h\searrow 0} \Bigg(\frac{V\big(x + h(f(x) + \varepsilon g(t,x))\big) - V(x + hf(x))}{h} \\ &\qquad\qquad + \frac{V(x + hf(x)) - V(x)}{h} \Bigg) \\ &\le \limsup_{h\searrow 0} \left(\frac{Lh\varepsilon\|g(t,x)\|}{h} + \frac{V(x + hf(x)) - V(x)}{h} \right) \\ &\le L\varepsilon\|g(t,x)\| + \limsup_{h\searrow 0} \frac{V(x + hF(x)) - V(x)}{h} \\ &\le LK\varepsilon + D^+_{(11.1)}V(x) \\ &\le LK\varepsilon - cV(x)\,. \end{aligned}$$

Hence,

$$D^+_{(11.2)}V(x) \le -LK\varepsilon \quad \text{for all } x \notin B_\varepsilon := \left\{ x \in \mathbb{R}^d : V(x) \le 2LK\varepsilon/c \right\},$$

from which it follows that

$$V\left(x(t)\right) \le V\left(x_0\right) - LK\varepsilon\left(t - t_0\right)$$

for all $t \ge t_0$ such that $x(\tau) \notin B_\varepsilon$ for $t_0 \le \tau \le t$. Thus, the set B_ε is an absorbing set for the solutions of the nonautonomous system (11.2), which consequently exist for all future and initial times $t \ge t_0 > -\infty$. The process $\phi^\varepsilon(t, t_0, x_0)$ thus exists and has a positively invariant uniform absorbing set B_ε. Applying Theorem 3.18 to this process and absorbing set gives the existence of a pullback attractor $\mathcal{A}^\varepsilon$ inside B_ε.

For $x \in B_\varepsilon$, one has

$$\alpha(\operatorname{dist}(x, A)) \le V(x) \le 2LK\varepsilon/c\,,$$

so $\operatorname{dist}(x, A) \le \alpha^{-1}(2LK\varepsilon/c)$. Hence,

$$\operatorname{dist}(B_\varepsilon, A) \le \alpha^{-1}(2LK\varepsilon/c)$$

and, since $A^\varepsilon(t) \subset B_\varepsilon$, it follows that

$$\operatorname{dist}(A^\varepsilon(t), A) \le \alpha^{-1}(2LK\varepsilon/c)\searrow 0 \quad \text{as } \varepsilon\searrow 0\,.$$

This proves (11.3).

To show that the Hausdorff dimension is constant, note that $\phi^\varepsilon(t_1, t_0, \cdot)$ for any $t_0 \le t_1 \in \mathbb{R}$ is a bi-Lipschitz mapping by Lemma 11.2 below. Thus, by FALCONER [**76**, Corollary 2.4], the sets $A^\varepsilon(t_0)$ and $A^\varepsilon(t_1)$ have the same Hausdorff dimension. This completes the proof of Theorem 11.1. □

LEMMA 11.2. *Let $A = \left(1 + M^{-1}\right)e^{M|t_1 - t_0|}$, where $M > 0$ is a Lipschitz constant of $G(t,x) := f(x) + \varepsilon g(t,x)$ in x uniformly in $t \in \mathbb{R}$. Then one has*

$$A^{-1}\|x - y\| \le \left\|\phi^\varepsilon(t_1, t_0, x) - \phi^\varepsilon(t_1, t_0, y)\right\| \le A\|x - y\|\,.$$

PROOF. Write $\Delta\phi^{\varepsilon}_{t,t_0} := \phi^{\varepsilon}(t,t_0,x) - \phi^{\varepsilon}(t,t_0,y)$ for $0 \le t \le t_1 - t_0$. Then

$$\Delta\phi^{\varepsilon}_{t,t_0} = (x-y) + \int_{t_0}^{t+t_0} [G(s,\phi^{\varepsilon}(s,t_0,x)) - G(s,\phi^{\varepsilon}(s,t_0,y))]\, ds\,,$$

so using the Lipschitz condition

$$\left\|\Delta\phi^{\varepsilon}_{t,t_0}\right\| \le \|x-y\| + M\int_{t_0}^{t+t_0} \left\|\Delta\phi^{\varepsilon}_{s,t_0}\right\| ds\,,$$

and the right hand inequality then follows by the Gronwall inequality. To obtain the left hand inequality, repeat the argument starting at t_1 and integrate backwards to t_0 from the previously obtained endpoint at t_1. □

2. Numerical approximation of uniform attractors

Consider a nonautonomous differential equation

$$\dot{x} = f(\theta_t(p), x) \tag{11.4}$$

on $\mathbb{R}^d$ for each $p \in P$, where $\theta = (\theta_t)_{t\in\mathbb{R}}$ is an autonomous dynamical system on a compact metric space P, and suppose that $f : P \times \mathbb{R}^d \to \mathbb{R}^d$ satisfies the assumptions of Theorem 7.10, which means that (11.4) generates a skew product flow (θ,φ) on $P \times \mathbb{R}^d$. Suppose, in addition, as in Corollary 7.12, that the skew product flow (θ,φ) has a global uniform attractor $\mathcal{A}$, i.e., $\mathcal{A}$ is a uniformly pullback and forward attractor.

Consider also a one-step explicit numerical scheme corresponding to the differential equation (11.4), such as the Euler scheme,

$$x_{n+1} = x_n + h_n F\left(h_n, \theta_{t_n}p, x_n\right) \tag{11.5}$$

with variable time step sizes $h_n = t_{n+1} - t_n \in (0,1]$. It will be assumed that the increment function $F : [0,1] \times P \times \mathbb{R}^d \to \mathbb{R}^d$ is continuous in all of its variables and that the numerical scheme (11.5) satisfies a local discretization error estimate of the form

$$|\varphi(h,p,x_0) - x_1| \le h\mu_R(h) \quad \text{for all } |x_0| \le R\,, \tag{11.6}$$

for each $R > 0$, where $\mu_R : [0,1] \to \mathbb{R}^+_0$ is a strictly increasing function with $\mu_R(h) > 0$ for $h > 0$ and $\mu_R(0) = 0$.

REMARK 11.3. For one-step order schemes such as the Euler and Runge–Kutta schemes, $\mu(h)$ is typically of the form $K_R h^{\nu}$ for some integer $\nu \ge 1$. This would require the differentiability of F in p as well as x and of θ_t in t, which is too restrictive for certain applications.

The numerical scheme (11.5) with variable time steps is a discrete time nonautonomous dynamical system, which generates a skew product flow provided the choice of admissible step size sequences is restricted. For fixed $\delta > 0$, define $\mathcal{H}^{\delta}$ to be the set of all two sided sequences $h = (h_n)_{n\in\mathbb{Z}}$ satisfying

$$\frac{1}{2}\delta \le h_n \le \delta \quad \text{for all } n \in \mathbb{Z}$$

(the particular factor $\frac{1}{2}$ here is chosen just for convenience).

EXERCISE 11.4. Show that the set $\mathcal{H}^\delta$ is compact metric space with the metric

$$\rho_{\mathcal{H}^\delta}\left(h, \bar{h}\right) = \sum_{n=-\infty}^{\infty} 2^{-|n|} \left|h_n - \bar{h}_n\right|.$$

Consider the shift operator $\tilde{\theta} : \mathcal{H}^\delta \to \mathcal{H}^\delta$ defined by

$$\tilde{\theta}(h) = \tilde{\theta}((h_n)_{n\in\mathbb{Z}}) := (h_{n+1})_{n\in\mathbb{Z}}.$$

This is a homeomorphism on the compact metric space $\left(\mathcal{H}^\delta, \rho_{\mathcal{H}^\delta}\right)$ and its iterates $(\tilde{\theta}_n)_{n\in\mathbb{Z}}$ form a discrete time group on $\mathcal{H}^\delta$.

For a given sequence $h = (h_n)_{n\in\mathbb{Z}}$, set $t_0 = 0$ and define $t_n = t_n(h) := \sum_{j=0}^{n-1} h_j$ and $t_{-n} = t_{-n}(h) := -\sum_{j=1}^{n} h_{-j}$ for all $n \in \mathbb{N}$. In addition, let $\Theta = \left(\Theta_n\right)_{n\in\mathbb{Z}}$ be the group of continuous mappings on $\mathcal{Q}^\delta := \mathcal{H}^\delta \times P$ with $\Theta_n : \mathcal{Q}^\delta \to \mathcal{Q}^\delta$ for $n \in \mathbb{Z}$ defined by iteration of the basic mappings

$$\Theta_0 := \mathrm{id}_{\mathcal{Q}^\delta}, \quad \Theta_1(h,p) := \left(\tilde{\theta}h, \theta_{h_0}p\right) \quad \text{and} \quad \Theta_{-1}(h,p) := \left(\tilde{\theta}_{-1}h, \theta_{-h_{-1}}p\right),$$

where $\theta_{-1} := \theta^{-1}$ (inverse mapping), and define the map $\psi : \mathbb{Z}^+ \times \mathcal{Q}^\delta \times \mathbb{R}^d \to \mathbb{R}^d$ by

$$\psi(0,q,x_0) := x_0 \quad \text{and} \quad \psi(n,q,x_0) = \psi(n,(h,p),x_0) := x_n \quad \text{for all } n \in \mathbb{N},$$

where x_n is the n-th iterate of the numerical scheme (11.5) with initial value $x_0 \in \mathbb{R}^d$, initial parameter $p \in P$ and step size sequence $h \in \mathcal{H}^\delta$. Then (Θ, ψ) is a discrete time skew product flow on $\mathcal{Q}^\delta \times \mathbb{R}^d$.

EXERCISE 11.5. Show that the above mappings Θ_n and ψ are continuous in their variables and satisfy the group and cocycle properties, respectively.

How one actually chooses or should choose the step sizes is a very important issue for actual computations. Whatever mechanism one uses to select an admissible step size sequence, once one has chosen such a sequence the subsequent dynamics are included in the skew product flow formalism proposed here.

THEOREM 11.6. *Suppose that the vector field $f : P \times \mathbb{R}^d \to \mathbb{R}^d$ of the differential equation* (11.4) *satisfies the assumptions of Theorem 7.10 and Corollary 7.12 and that the continuous time skew product flow (θ, φ) generated by* (11.4) *has a global uniform pullback attractor $\mathcal{A}$. Assume, in addition, that the increment function $F : [0,1] \times P \times \mathbb{R}^d \to \mathbb{R}^d$ in the numerical scheme* (11.5) *is continuous in all of its variables and that* (11.5) *satisfies the local discretization error estimate* (11.6). *Then the discrete time skew product flow (Θ, ψ) generated by the numerical scheme* (11.5) *has a global uniform pullback attractor $\mathcal{A}^\delta$, provided the maximal step size δ is sufficiently small, such that*

$$\limsup_{\delta\searrow 0} \sup_{p\in P} \sup_{h\in\mathcal{H}^\delta} \operatorname{dist}\left(A^\delta_{(h),p}, A_p\right) = 0.$$

The assumption that the vector field f in the differential equation (11.4) satisfies a global Lipschitz property is not a major limitation, since the essential dynamics that are being approximated occur in a (possibly very large) compact subset of $\mathbb{R}^d$ on which a local Lipschitz property could be used. Such a condition is fulfilled, e.g., by C^1-mappings with local Lipschitz constants being uniform in p.

2.1. Proof of Theorem 11.6. The proof of Theorem 11.6 is based on the proof of a similar result of KLOEDEN & LORENZ [**131**] for autonomous systems. It uses a Lyapunov function $V : P \times \mathbb{R}^d \to \mathbb{R}_0^+$ that characterizes the uniform pullback attractor. The existence of such a Lyapunov function is provided by Theorem 7.10 and Corollary 7.12.

Note that for a global pullback attractor, the φ-positively invariant nonautonomous set $\mathcal{N}$ in the forwards convergence property (P5) in Theorem 7.10) and the pullback absorbing family $\mathcal{B} = (B_p)_{p\in P}$ in Lemma 7.14 can be constructed to be arbitrarily large, i.e., with $B_R(A_p) \subset N_p \subset B_p$ for each $p \in P$ given an arbitrary $R > 0$.

The key tool in the proof is provided by the following Lyapunov inequality in which the function $\mu = \mu_R$ is from the local discretization error estimate (11.6) of the numerical scheme (11.5) with R chosen so large such that $\bigcup_{p\in P} N_p \subset B_R(0)$.

LEMMA 11.7. *One has*

$$V\left(\theta_h(p), x_1(h,p)\right) \le e^{-ch}\, V\left(p, x_0\right) + h\mu(h)\,, \tag{11.7}$$

where $x_1(h,p)$ is the first iteration of the numerical scheme (11.5) *with any step size $h > 0$ and initial state $x_0 \in N_p$ for the parameter p.*

PROOF. From the Lipschitz property of V with Lipschitz constant $L = 1$, one gets

$$V\big(\theta_h(p), x_1(h,p)\big) \le V\big(\theta_h(p), \varphi(h,p,x_0)\big) + \|x_1(h,p) - \varphi(h,p,x_0)\|.$$

Applying the forwards exponential decay in equality (P5) for the Lyapunov function V in Theorem 7.10 to the first term on the right side, and the local discretization error estimate (11.6) to the second, then gives (11.7) for all $h \ge 0$. □

2.1.1. *Construction of an absorbing set for the discretized system.* Fix $\delta > 0$ and define

$$\eta = \eta(\delta) := \frac{2\delta\mu(\delta)}{1 - e^{-c\delta/2}}\,.$$

LEMMA 11.8. *The set $\Lambda_p^\delta := \left\{x \in \mathbb{R}^d : V(p,x) \le \eta(\delta)\right\}$ is a nonempty compact subset of $\mathbb{R}^d$ with*

$$\operatorname{dist}\left(\Lambda_p^\delta, A_p\right) \le \alpha^{-1}(\eta)\,. \tag{11.8}$$

PROOF. Λ_p^δ is nonempty, since it contains $A_p = \left\{x \in \mathbb{R}^d : V(p,x) = 0\right\}$, and it is closed in view of the continuity of V in x. Its boundedness, hence compactness, and the inequality (11.8) follow from the inequality

$$\alpha(\operatorname{dist}(x, A_p)) \le V(p,x)$$

in the uniform lower bound property (P2') in Corollary 7.12 and the definition of Λ_p^δ. Hence,

$$\operatorname{dist}(x, A_p) \le \alpha^{-1}\left(V(p,x)\right) \le \alpha^{-1}\left(\eta\right) \quad \text{for all } x \in \Lambda_p(\eta)\,.$$

This finishes the proof of this lemma. □

The following lemma says that the family $\Lambda^\delta = \left(\Lambda_p^\delta\right)_{p\in P}$ is positively invariant with respect to the discrete time cocycle mapping ψ formed by iterating the numerical scheme.

LEMMA 11.9. *One has* $\psi\big(1,(h,p),\Lambda_p^\delta\big) \subseteq \Lambda_{\theta_h p}^\delta$.

PROOF. It suffices to consider any single iterate $x_1 = x_1(h,p)$ for an arbitrary $x_0 \in \Lambda_p^\delta$, which means that $V(p,x_0) \leq \eta = \eta(\delta)$. Then the key inequality (11.7) with $h \in \left[\frac{1}{2}\delta, \delta\right]$ and the definition of $\eta(\delta)$ give

$$\begin{aligned} V\big(\theta_h(p), x_1(h,p)\big) &\leq e^{-ch} V(p,x_0) + h\mu(h) \\ &\leq e^{-c\delta/2} V(p,x_0) + \delta\mu(\delta) \\ &\leq e^{-c\delta/2}\eta(\delta) + \frac{1}{2}\big(1 - e^{-c\delta/2}\big)\eta(\delta) \\ &= \frac{1}{2}\big(1 + e^{-c\delta/2}\big)\eta(\delta) \leq \eta(\delta)\,, \end{aligned}$$

so $x_1(h,p) \in \Lambda_{\theta_h p}^\delta$. □

The family $\Lambda^\delta = \big(\Lambda_p^\delta\big)_{p\in P}$ is, in fact, forwards absorbing under ψ uniformly in p, provided δ is chosen small enough.

LEMMA 11.10. *There exists a* $\delta^* > 0$ *such that for each family* $\mathcal{D} = (D_p)_{p\in P}$ *of compact subsets with* $D_p \subset N_p$ *and* $\delta \in (0,\delta^*)$, *there exists an integer* $N = N(\mathcal{D},\delta) \geq 1$ *for which*

$$\psi\left(n,(h,p),D_p\right) \subseteq \Lambda_{\theta_{t_n} p}^\delta$$

for all $n \geq N$ *uniformly in* $p \in P$.

PROOF. As in KLOEDEN & LORENZ [**131**, Lemma 3.4], there exists a $\gamma > 0$ such that

$$1 + e^{-c\gamma/2} = 2e^{-c\gamma/4} \quad \text{and} \quad 1 + e^{-c\delta/2} < 2e^{-c\delta/4}$$

for all $0 < \delta < \delta^* := \gamma$.

Consider $x_1(h_0,p) = \psi\big(1,(h,p),x_0\big)$ with $x_0 \notin \Lambda_p^\delta$. Then, by the Lyapunov inequality (11.7) and the definition of $\eta = \eta(\delta)$, one obtains

$$\begin{aligned} V\big(\theta_h(p), x_1(h_0,p)\big) &\leq e^{-c\delta/2} V(p,x_0) + \delta\mu(\delta) \\ &= e^{-c\delta/2} V(p,x_0) + \frac{1}{2}\big(1 - e^{-c\delta/2}\big)\eta(\delta) \\ &< \frac{1}{2}\big(1 + e^{-c\delta/2}\big) V(p,x_0) \\ &< e^{-c\delta/4} V(p,x_0)\,, \end{aligned}$$

since $V(p,x_0) > \eta(\delta)$ and $0 < \delta < \delta^*$. Repeating this argument,

$$V\left(\theta_{t_n}(p), x_n\right) < e^{-cn\delta/4} V(p,x_0)\,,$$

where $x_n = \psi\left(n,(h,p),x_0\right)$, as long as

$$\begin{aligned} &x_0 \notin \Lambda_p^\delta\,, \\ &x_1(h_0,p) \notin \Lambda_{\theta_{t_1}(p)}^\delta\,, \ \ldots\ , x_{n-1} = \psi\left(n-1,(h,p),x_0\right) \notin \Lambda_{\theta_{t_{n-1}}(p)}^\delta\,. \end{aligned}$$

Now

$$V(p,x_0) \leq \operatorname{dist}(x_0,A_p) \leq \operatorname{dist}(D_p,A_p) < \infty \quad \text{for all } x_0 \in D\,,$$

which means that

$$V\left(\theta_{t_n}(p), x_n\right) < e^{-cn\delta/4}\, h(D_p,A_p)$$

as long as

$$x_0 \notin \Lambda^\delta_p\,,$$
$$x_1(h_0,p) \notin \Lambda^\delta_{\theta_{t_1}(p)}\,, \ \ldots\ , x_{n-1} = \psi(n-1,(h,p),x_0) \notin \Lambda^\delta_{\theta_{t_{n-1}}(p)}\,.$$

Define $N = N(\mathcal{D},\delta)$ to be the smallest integer n for which

$$e^{-cn\delta/4}\operatorname{dist}(D_p,A_p) \le \eta(\delta) < e^{-c(n-1)\delta/4}\operatorname{dist}(D_p,A_p)\,.$$

Thus, for each $x_0 \in D_p$, there exists an integer $n \ge N$, possibly 0, such that $\psi(n,(h,p),x_0) \in \Lambda^\delta_{\theta_{t_n}(p)}$. By the positive invariance of the Λ^δ_p proved in Lemma 11.9, it follows that the j-th iterate of ψ then remains in $\Lambda^\delta_{\theta_{t_j}(p)}$ for all $j \ge n$. □

2.1.2. *Existence and convergence of the discretized pullback attractor.* Applying Theorem 3.20 to the numerical skew product flow (Θ,ψ) and the absorbing family $\Lambda^\delta = \left(\Lambda^\delta_p\right)_{p\in P}$ gives the existence of a uniform pullback attractor

$$\mathcal{A}^\delta = \left(A^\delta_q\right)_{q=(p,h)\in\mathcal{Q}^\delta}$$

for (Θ,ψ), i.e., with

$$A^\delta_q = \bigcap_{n\ge 0} \psi\left(|t_{-n}|\,,\Theta_{-n}q,\Lambda^\delta_{\theta_{t_{-n}}(p)}\right)$$

(note that $t_{-n} < 0$). Since $A_p \subseteq \Lambda_{(h,p)}(\eta)$ for all $p \in P$ and $h \in \mathcal{H}^\delta$, it follows that

$$\operatorname{dist}\left(A^\delta_{(h,p)},A_p\right) < \operatorname{dist}\left(\Lambda^\delta_p,A_p\right) \le \alpha^{-1}(\eta(\delta)) \to 0 \quad \text{as } \delta \to 0$$

uniformly in $p \in P$ and $h \in \mathcal{H}^\delta$. This completes the proof of Theorem 11.6. □

REMARK 11.11. Figures 15.1 and 15.2 in Chapter 15 provide an application of Theorem 11.6. They were obtained by a numerical scheme applied to the differential equations under consideration. Since the component subsets of the uniform attractor are singleton sets here, the convergence of the numerical attractor to the original one is in fact continuous, i.e., with respect to the Hausdorff metric.

3. Perturbation of the driving system

The influence of the driving system on a skew product flow generated by a triangular system of differential equations can be perturbed in two ways, directly by perturbing the vector field of the driving system component itself or indirectly by perturbing its input variable in the vector field of the coupled component. These perturbations have profoundly different effects on a nonautonomous attractor of the driven system.

Consider a triangular system of autonomous differential equations

$$\dot{x} = f(x,p)\,, \tag{11.9}$$
$$\dot{p} = g(p)\,, \tag{11.10}$$

which generates a skew product flow. It will be supposed that this skew product flow has an attractor of some kind. The questions we ask are as follows.

What happens when the driving component is slightly perturbed? What conditions on the equations (11.9)–(11.10) *ensure the persistence of such an attractor under perturbation of the driving component?*

The answers to these questions are not as obvious or easy as it might at first seem. Indeed, even an infinitesimal uniform perturbation of the vector field g in (11.10) may cause substantial changes to the individual trajectories of (11.10) and destroy the attractor. This effect seen shown in the following example.

EXAMPLE 11.12. Consider the system of differential equations

$$\dot{x} = p_1 x - \sqrt{x}\,, \tag{11.11}$$

$$\dot{p}_1 = -p_2\gamma(\lambda, p) \quad \text{and} \quad \dot{p}_2 = p_1\gamma(\lambda, p)\,, \tag{11.12}$$

where $x \in \mathbb{R}^+$, $p = (p_1, p_2) \in \mathbb{S}^1 = \{p \in \mathbb{R}^2 : p_1^2 + p_2^2 = 1\}$ and

$$\gamma(\lambda, p) := 1 + \frac{\lambda^2}{1+2\lambda} + \frac{\lambda(1+\lambda)}{1+2\lambda}\operatorname{sgn}(p_1)$$

depending on a parameter $\lambda \in \mathbb{R}$. The function $\gamma(\lambda, p)$ is close to 1 for small λ and provides the "perturbation" of the driving system (11.12) for λ different from 0. The discontinuity in $\gamma(\lambda, p)$ and hence in the vector field of equation (11.12) at $p_1 = 0$ can be avoided, but at the expense of the convenient explicit solutions used below.

Equations (11.11)–(11.12) determine a skew product flow on $\mathbb{S}^1 \times \mathbb{R}^+$ with the driving component being defined by (11.12) on $\mathbb{S}^1$. The p_1-component of the solution $p(t) = (p_1(t), p_2(t))$ is 2π-periodic for any small value of parameter λ, having the mean value

$$\bar{p}_{1,\lambda} := \lim_{t\to\infty} \frac{1}{t}\int_0^t p_1(s)\,\mathrm{d}s = \frac{1}{2\pi}\int_0^{2\pi} p_1(s)\,\mathrm{d}s = \frac{2}{\pi}\frac{\lambda}{1+\lambda}\,.$$

The solution $x(t) = \varphi(t, p_0, x_0)$ of the forced equation (11.11) is given explicitly by

$$x(t) = \begin{cases} e^{\int_0^t p_1(s)\,\mathrm{d}s} w^2(t) & : \ w(t) \geq 0\,, \\ 0 & : \ w(t) < 0\,, \end{cases}$$

where

$$w(t) := \sqrt{x_0} - \frac{1}{2}\int_0^t e^{-\frac{1}{2}\int_0^s p_1(u)\,\mathrm{d}u}\,\mathrm{d}s\,.$$

Now $\bar{p}_{1,\lambda} \leq 0$ for small $\lambda \leq 0$, and the zero solution $\bar{x}(t) \equiv 0$ of the driven system (11.11) is globally asymptotically stable uniformly in $p \in \mathbb{S}^1$ (see Figure 11.1 left). In fact, it has a nonautonomous, both forward and pullback, attractor $\mathcal{A}$ with fibers $A_p = \{0\}$ for all $p \in P = \mathbb{S}^1$, while the corresponding autonomous semi-dynamical system on $\mathbb{S}^1 \times \mathbb{R}_0^+$ has the global attractor $\mathcal{A} = \mathbb{S}^1 \times \{0\}$.

On the other hand, for small $\lambda > 0$, the behavior of the driven system (11.11) changes dramatically. Indeed, the solution $x(t)$ for sufficiently small initial value x_0 still tends to zero (thin solid lines in Figure 11.1 right), but for a large initial value x_0 it tends to infinity (dotted lines in Figure 11.1 right). These two kinds of solutions are separated by a unique periodic solution $x_{\text{per}}(t)$ (thick solid line in Figure 11.1 right) corresponding to the initial value $x_{\text{per}}(0)$ satisfying

$$\sqrt{x_{\text{per}}(0)} = \frac{1}{2}\int_0^\infty e^{-\frac{1}{2}\int_0^s p_1(u)\,\mathrm{d}u}\,\mathrm{d}s\,.$$

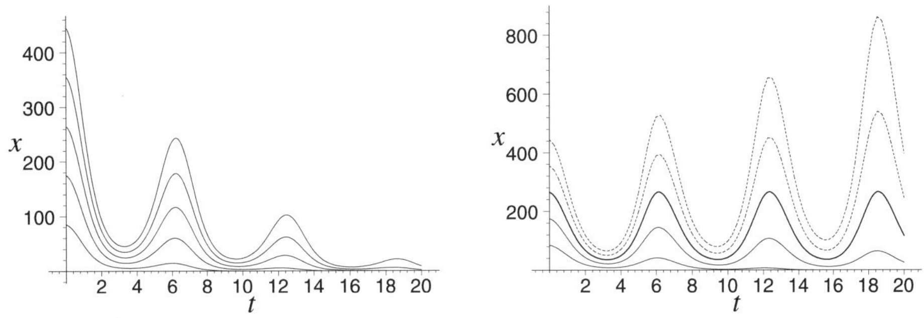

FIGURE 11.1. [**126**] reprinted with permission of the *American Institute of Mathematical Sciences.*

The periodic solution $x_{\text{per}}(t)$ *bifurcates from infinity* when $\lambda \to 0$ with $\lambda > 0$ and is obviously unstable. Specifically, it is the bifurcation from infinity of this unstable periodic solution that destroys the global attractor of the unperturbed system. See KRASNOSEL'SKIĬ & MAWHIN [**149**] for bifurcations at infinity.

The answers to the above questions depend on the nature of the perturbations of the driving component. This will be classified as *strong* and *weak* depending on whether the vector field g of the driving equation in (11.10) or only the p-component in the vector field f of the driven equation (11.9) is perturbed.

DEFINITION 11.13 (Strong perturbations). The system

$$\dot{x} = f(x,p)\,, \tag{11.13}$$

$$\dot{p} = g(p) + h(t,p)\,, \tag{11.14}$$

where $\|h(t,p)\| \leq \varepsilon$, is called a *strong perturbation* of (11.9)–(11.10).

DEFINITION 11.14 (Weak perturbations). The system

$$\dot{x} = f(x, p + q(t,p))\,, \tag{11.15}$$

$$\dot{p} = g(p)\,, \tag{11.16}$$

where $\|q(t,p)\| \leq \varepsilon$ is called a *weak perturbation* of (11.9)–(11.10).

Finding answers seems to be easier for weak perturbations than for strong perturbations and will be handled in terms of inflated attractors for the "inflation" of the vector field f of the driven system. The main difficulty in the strong case is that a trajectory of the perturbed driving system (11.14) may have very different long-term behavior to the corresponding trajectory of the unperturbed driving system (11.10), no matter how small perturbation term $h(t,p)$. However, this can be overcome by assuming that the unperturbed driving system (11.10) is shadowing. Then, for any trajectory of the perturbed driving system, there will exist a trajectory of the unperturbed driving system, which remains close to it for all time, provided the magnitude of the perturbation is sufficiently small.

3.1. Weak perturbations and inflated attractors. The effects of perturbations of the driving component variable p in the driven equation (11.15) can be investigated in terms of the "inflation" of the vector field of equation (11.15), see KLOEDEN & KOZYAKIN [**124, 125**], specifically by replacing the driven equation (11.9) by a differential inclusion

$$\dot{x} \in F_\varepsilon(x, p)\,, \tag{11.17}$$

while the driving system (11.10) remains unaltered. Here $F_\varepsilon(x,p)$ is a nonempty compact convex set, which is defined as the *internal ε-inflation*

$$F_\varepsilon(x,p) := \left\{ y \in \mathbb{R}^d : \|y - f(x, p+q)\| \le \rho_\varepsilon(x,p) \right\},$$

where

$$\rho_\varepsilon(x,p) := \sup\left\{ \|f(x,p+q) - f(x,p)\| : \|q\| \le \varepsilon \right\}.$$

This allows direct access to the well-developed theory of differential inclusions. The set-valued mapping $(\varepsilon, x, p) \mapsto F_\varepsilon(x,p)$ is continuous in all of its variables with compact convex values. Hence, for any initial value $x(0) = x_0$, the differential inclusion (11.17) has an absolutely continuous solution $x(t)$ satisfying

$$\dot{x}(t) \in F_\varepsilon\big(x(t), \theta_t(p_0)\big) \tag{11.18}$$

for almost all $t \in [0, T)$ for some maximal $T = T(\varepsilon, x_0, p_0) \le \infty$, where $\theta_t(p_0)$ denotes the solution $p(t, p_0)$ of the driving equation (11.10) with the initial condition $p(0) = p_0 \in P$.

Define $\Phi^\varepsilon(t, x_0, p_0)$ to be the set of all points $y \in \mathbb{R}^d$ for which there exists a solution $x(t)$ of the differential inclusion (11.18) for this $p_0 \in P$ with $x(0) = x_0$ and $x(t) = y$. The set-valued mapping Φ^ε will be called the *internal ε-inflation* of the single-valued cocycle mapping φ generated by unperturbed system (11.9)–(11.10).

It will be supposed that $T(\varepsilon, x_0, p_0) = \infty$ for all $x_0 \in \mathbb{R}^d$ and $p_0 \in P$. Then Φ^ε is a set-valued cocycle mapping driven by the autonomous system θ and $(\theta, \Phi^\varepsilon)$ is a set-valued skew product flow as in Chapter 9. An attractor (forward or pullback) of the internal ε-inflated set-valued skew product flow $(\theta, \Phi^\varepsilon)$ will be called the *internal ε-inflated attractor* (forward or pullback) of the unperturbed skew product flow (θ, φ). The proof of the following theorem is left to the reader.

THEOREM 11.15. *Suppose that the single-valued skew product flow (θ, φ) has an internal ε_0-inflated pullback attractor $\mathcal{A}^{\varepsilon_0}$ for some $\varepsilon_0 > 0$. Then (θ, φ) has an internal ε-inflated pullback attractor $\mathcal{A}^\varepsilon$ for every $\varepsilon \in [0, \varepsilon_0]$, and these are related through*

$$A_p^\varepsilon \subset A_p^{\varepsilon'} \quad \textit{and} \quad A_p^\varepsilon = \bigcap_{\varepsilon < \varepsilon'} A_p^{\varepsilon'}$$

for any $0 \le \varepsilon < \varepsilon' \le \varepsilon_0$ and each $p \in P$.

Dissipativity conditions on the vector field function f of (11.9) ensure the existence of the set-valued cocycle mapping Φ^{ε_0} and an inflated pullback attractor for some $\varepsilon_0 > 0$. It follows then from Theorem 11.15 that weak perturbations of the driving system variable in the vector field of the driven equation (11.9) give rise to a pullback attractor of the perturbed system, which is componentwise close to the unperturbed pullback attractor in the sense of upper semi-continuous convergence as the perturbation magnitude ε approaches zero.

3.2. Strong perturbations and shadowing. The autonomous dynamical system θ generated by equation (11.10) is said to have the *shadowing property* if for any $\varepsilon > 0$, there is a $\delta = \delta(\varepsilon) > 0$ such that for any absolutely continuous function $q : \mathbb{R} \to P$ satisfying

$$\|\dot{q}(t) - g(q(t))\| < \delta \quad \text{for almost all}\, t \in \mathbb{R}\,, \tag{11.19}$$

there exists a solution $p(t)$ of unperturbed equation (11.10) such that

$$\|q(t) - p(t)\| < \varepsilon \quad \text{for all } t \in \mathbb{R}\,. \tag{11.20}$$

Let φ^h be the cocycle solution mapping of system (11.9) corresponding to the solution mapping θ_t^h of the perturbed driving system (11.14) for a strong perturbation mapping $h(t,p)$.

LEMMA 11.16. *Suppose that the autonomous system* (11.10) *has the shadowing property and that the perturbation term $h(t,p)$ in* (11.14) *satisfies $\|h(t,p)\| < \delta$ for $\delta = \delta(\varepsilon)$ as given in the shadowing property. Then for any nonempty compact subset D of $\mathbb{R}^d$, one has*

$$\varphi^h(t,q,D) \subseteq \bigcup_{\|p-q\|\leq\varepsilon} \Phi^\varepsilon(t,p,D) \quad \textit{for all } t \in \mathbb{R}_0^+ \tag{11.21}$$

and

$$\varphi^h\left(t,\theta_{-t}^h(q),D\right) \subseteq \bigcup_{\|p-q\|\leq\varepsilon} \Phi^\varepsilon\left(t,\theta_{-t}(p),D\right) \quad \textit{for all } t \in \mathbb{R}_0^+\,, \tag{11.22}$$

where Φ^ε is the internal ε-inflated cocycle solution mapping of (11.9)–(11.10).

PROOF. Fix $\varepsilon > 0$, and let $\|h(t,p)\| < \delta$ with $\delta = \delta(\varepsilon)$ as given in the shadowing property for the equation (11.10). Moreover, let $q(t)$ be a solution of the perturbed driving equation (11.14), i.e.,

$$\dot{q}(t) = g(q(t)) + h(t,q(t)) \quad \text{for all } t \in \mathbb{R}\,.$$

The inequality (11.19) thus holds for $q(t)$, so by the shadowing property there exists a solution $p(t)$ of the unperturbed equation (11.10) such that the inequality (11.20) holds.
Let $x(t) := \varphi^h(t,q_0,x_0)$ denote the solution of the driven equation of

$$\dot{x}(t) = f(x(t),q(t))\,, \qquad x(0) = x_0\,,$$

i.e., with $q(t) = \theta_t^h(q_0)$ here. Then

$$\dot{x}(t) = f\big(x(t),p(t) + (q(t) - p(t))\big)\,, \qquad x(0) = x_0\,,$$

and by the shadowing inequality (11.20),

$$\dot{x}(t) \in F_\varepsilon(x(t),p(t))\,, \qquad x(0) = x_0\,.$$

It then follows from the definition of the solution mapping for the internally inflated system that

$$\varphi^h(t,q_0,x_0) = x(t) \in \Phi^\varepsilon(t,p_0,x_0)\,,$$

and the required forwards inclusion (11.21) is an immediate consequence of the fact that $\|p(0) - q_0)\| < \varepsilon$. The backwards inclusion (11.22) is proved analogously. $\square$

In spite of its simplicity, Lemma 11.16 is important because it clearly demonstrates the differing influences of the weak and strong perturbations on the behavior of the system, the former manifesting itself through the second ε in (11.21) and (11.22), i.e., in the Φ^{ε} term, and the latter through the first ε in (11.21) and (11.22), i.e., under the set union symbol.

THEOREM 11.17. *Suppose that skew product flow (θ, φ) generated by the system (11.9)–(11.10) has a uniform internal ε-inflated attractor $\mathcal{A}^{\varepsilon}$ for each $\varepsilon \in [0, \varepsilon_0]$ for some $\varepsilon_0 > 0$, and that the driving system (11.10) has the shadowing property. In addition, suppose that the perturbation term $h(t,p)$ in (11.14) satisfies $\|h(t,p)\| < \delta$ for $\delta = \delta(\varepsilon)$ as given in the shadowing property for some $\varepsilon \in (0, \varepsilon_0]$. Then, the strongly perturbed system (11.13)–(11.14) has a pullback attractor $\mathcal{A}^h$ such that*

$$A_q^h \subseteq \bigcup_{\substack{p \in P: \\ \|q-p\| \leq \varepsilon}} A_p^{\varepsilon} \quad \text{for all } q \in P. \tag{11.23}$$

PROOF. Let $\varepsilon \in (0, \varepsilon_0]$ correspond to the $\delta(\varepsilon)$ that bounds the perturbation h. Then by the uniformity assumption on the internal ε-inflated attractor $\mathcal{A}^{\varepsilon}$, for any $\sigma > 0$ and nonempty compact subset D of $\mathbb{R}^d$, there exists a $T = T(\varepsilon, \sigma, D) \geq 0$ such that

$$\Phi^{\varepsilon}(t, p, D) \subset B_{\sigma}\big(A_{\theta_t(p)}^{\varepsilon}\big) \quad \text{for all } t \geq T \text{ and } p \in P.$$

It follows immediately from this inclusion and from Lemma 11.16 that

$$\varphi^h(t, q, D) \subseteq B_{\sigma}\Big(\bigcup_{p \in P: \|p-q\| \leq \varepsilon} A_{\theta_t(p)}^{\varepsilon}\Big) \tag{11.24}$$

and

$$\varphi^h\big(t, \theta_{-t}^h(q), D\big) \subseteq B_{\sigma}\Big(\bigcup_{p \in P: \|p-q\| \leq \varepsilon} A_p^{\varepsilon}\Big) \tag{11.25}$$

for all $t \geq T$ and $q \in P$. Now fix an arbitrary $\sigma > 0$, and define

$$C_{\sigma} := B_{\sigma}\Big(\bigcup_{p \in P} A_p^{\varepsilon_0}\Big),$$

where $\mathcal{A}^{\varepsilon_0}$ is the uniform internally ε_0-inflated attractor. The set C_{σ} is compact, since P and the sets $A_p^{\varepsilon_0}$ are compact and the mapping $p \mapsto A_p^{\varepsilon_0}$ is upper semi-continuous. Moreover, by Theorem 11.15, C_{σ} contains any set A_p^{ε} with $\varepsilon \in [0, \varepsilon_0)$ and $p \in P$. Hence, by (11.24), one gets

$$\varphi^h(t, q, D) \subseteq B_{\sigma}\Big(\bigcup_{p \in P: \|p-q\| \leq \varepsilon} A_{\theta_t(p)}^{\varepsilon}\Big) \subseteq C_{\sigma}$$

and, by (11.25),

$$\varphi^h\big(t, \theta_{-t}^h(q), D\big) \subseteq B_{\sigma}\Big(\bigcup_{p \in P: \|p-q\| \leq \varepsilon} A_p^{\varepsilon}\Big) \subseteq C_{\sigma} \tag{11.26}$$

for all $t \geq T$ and $q \in P$.

The existence of an attractor (forwards and pullback) $\mathcal{A}^h$ of the strongly perturbed system (11.13)–(11.14) follows then from the above inclusions by Theorem 3.18 and its generalizations. In particular, the definition of the pullback attractor component sets (3.9) for (θ^h, φ^h) instead of (θ, φ) with the pullback absorbing system consisting of the same subset C_{σ} gives

$$A_q^h = \bigcap_{\tau \geq 0} \overline{\bigcup_{t \geq \tau} \varphi^h\big(t, \theta_{-t}^h(q), C_{\sigma}\big)} \subseteq B_{\sigma}\Big(\bigcup_{p \in P: \|p-q\| \leq \varepsilon} A_p^{\varepsilon}\Big) \quad \text{for all } q \in P,$$

where the set inclusion follows from (11.26) with $D = C_\sigma$. The desired inclusion (11.23) then follows since $\sigma > 0$ can be chosen arbitrarily small. □

REMARK 11.18. Although the system (11.9)–(11.10) is supposed to have a *uniform* attractor, it is not clear whether the attractor $\mathcal{A}^h$ of the perturbed system (11.13)–(11.14) will be uniform or not.

Endnotes. Section 1 is based on Kloeden & Stonier [**145**] and Section 3 on Kloeden & Kozyakin [**126, 128**], see also Johnson & Kloeden [**99**] for a digitized driving system. Other papers on the numerical approximation of nonautonomous attractors are Kloeden & Schmalfuß [**140**], Kloeden [**142**], Cheban, Kloeden & Schmalfuß [**40**] and Aulbach, Rasmussen & Siegmund [**11**]. For a review article, see Grüne & Kloeden [**84**].

CHAPTER 12

Infinite-dimensional systems

It was assumed in the previous chapters that the state space is a complete metric space and that the absorbing sets are compact (see Theorem 3.20 and Theorem 3.27), which, with the continuity of the flow, suffices to ensure the existence of an attractor. This is no restriction in the state space $\mathbb{R}^d$, where closed and bounded equivalently means compact, but this is not true for a general state space. As mentioned in Remark 3.28, compact subsets of infinite-dimensional spaces are "thin" and atypical, and it is much easier to determine an absorbing property for a closed and bounded subset, such as a unit ball, rather than a compact subset. Counterparts of Theorem 3.20 and Theorem 3.27 then hold if it is additionally assumed that the flow is compact (i.e., maps bounded subsets into precompact subsets) or, more generally, asymptotically compact.

Such generalizations will be considered in this chapter, first in terms of autonomous systems and then for nonautonomous systems. The results are applicable to systems based on evolution equations such as parabolic and hyperbolic partial differential equations and delay differential equations, for which the state spaces are infinite-dimensional function spaces. The attractors of such systems are often finite-dimensional, but need not be. Readers are referred to the monograph of CARVALHO, LANGA & ROBINSON **[35]** for a comprehensive treatment of the infinite-dimensional case.

1. Squeezing and flattening properties: the autonomous case

Some important concepts and results in the investigation of infinite-dimensional dynamical systems are illustrated here through simple examples of autonomous flows. The infinite-dimensional space $\mathbb{R}^\infty$ will be endowed with the supremum norm, so is essentially the space ℓ_∞.

EXAMPLE 12.1. The semi-dynamical system generated by the infinite-dimensional differential equation

$$\dot{x}_1 = x_1(1 - x_1^2), \quad \dot{x}_i = -x_i \quad \text{for all } i \in \{2, 3, 4, \dots\}$$

in $\mathbb{R}^\infty$ has the global attractor

$$A = [-1, 1] \times \prod_{i \geq 2} \{0\},$$

which is a compact subset of $\mathbb{R}^\infty$ of dimension 1.

This example satisfies the *squeezing property*, which was introduced by FOIAS & TEMAM [**81**] in the context of the Navier–Stokes equations. It enures that the global attractor is finite-dimensional.

DEFINITION 12.2 (Squeezing property). Suppose that a semi-flow ϕ on a Banach space $(X, \|\cdot\|)$ has a bounded absorbing set $B \subset X$, and let P be a projection onto a finite-dimensional subspace of X. Then the semi-flow ϕ is said to have the *squeezing property* if for $x, y \in B$, either

$$\left\|(\mathbb{1} - P)(\phi(1,x) - \phi(1,y))\right\| \leq \left\|P(\phi(1,x) - \phi(1,y))\right\|,$$

i.e., the higher modes are bounded by the lower modes, or

$$\left\|\phi(1,x) - \phi(1,y)\right\| \leq \delta\|x - y\| \quad \text{for some } \delta \in (0,1),$$

i.e., the solutions are squeezed together.

EXAMPLE 12.3. The semi-dynamical system generated by the infinite-dimensional differential equation

$$\dot{x}_i = x_i\left(\tfrac{1}{i^2} - x_i^2\right) \quad \text{for all } i \in \mathbb{N} \tag{12.1}$$

in $\mathbb{R}^\infty$ has the global attractor

$$A = \prod_{i\in\mathbb{N}} \left[-\tfrac{1}{i}, \tfrac{1}{i}\right]$$

which is a compact subset of $\mathbb{R}^\infty$ and is infinite-dimensional.

This example does not satisfy the squeezing property. To see this, compare two solutions with initial values x_0 with all components equal to zero except for one which takes the value of the equilibrium point on the boundary of the attractor, i.e.,

$$x_{0,i^*} = \frac{1}{i^*} \quad \text{and} \quad x_{0,i} = 0 \quad \text{for all } i \neq i^*$$

for two different values of $i^* > N$, for some given N (which would be the dimension of the finite-dimensional subspace if the squeezing property were to hold). The corresponding solution is

$$\phi(t, x_0) \equiv x_0 \quad \text{for all } t \geq 0\,.$$

The lower-dimensional components, i.e., modes, of the difference of two such solutions are identically zero, but the higher modes are not squeezing.

The system (12.1) does, however, satisfy a weaker property, which was called the *flattening property* in KLOEDEN & LANGA [**129**], and was introduced earlier under another name by WANG, ZHONG & ZHOU [**234**].

DEFINITION 12.4 (Flattening property). Suppose that a semi-flow ϕ on a Banach space $(X, \|\cdot\|)$ has a bounded absorbing set $B \subset X$. Then ϕ is said to have the *flattening property* if for any bounded set $D \subset X$ and $\varepsilon > 0$, there exist $T_\varepsilon(D) > 0$, a finite-dimensional subspace X_ε of X and a bounded projector $P_\varepsilon : X \to X_\varepsilon$ such that $\bigcup_{t \geq T_\varepsilon(D)} P_\varepsilon \phi(t, D)$ is bounded and

$$\left\|(\mathbb{1} - P_\varepsilon)\phi(t, x_0)\right\| < \varepsilon \quad \text{for all } t \geq T_\varepsilon(D) \text{ and } x_0 \in D\,.$$

It was shown in [**129**] that the squeezing property implies the flattening property under appropriate conditions.

2. Pullback asymptotic compactness

Nontrivial examples of compact flows are those generated by parabolic partial differential equations and delay differential equations. However, the flows generated by neutral delay equations and hyperbolic partial differential equations are, in general, not compact, but they are compact in an asymptotic sense.

A pullback version of flattening for skew product flows is introduced here and then two pullback versions of asymptotic compactness are treated, the formulation of which requires a measure of noncompactness. These three concepts will then be shown to be equivalent under a mild restriction on the state space.

2.1. Measure of noncompactness.

The concept of a measure of noncompactness indicates how much a subset of an infinite-dimensional space differs from a compact set. The following definition is due to Kuratowski. See DEIMLING [**66**] for further details.

DEFINITION 12.5 (Kuratowski measure of noncompactness). Let (X, d) be a metric space and D be a bounded subset of X. The *Kuratowski measure of noncompactness* of D is defined by

$$\gamma(D) := \inf \left\{ \delta > 0 : D \text{ admits a finite cover by sets of diameter less than } \delta \right\}.$$

The following lemma summarizes some of the basic properties of this measure of noncompactness (see, e.g., DEIMLING [**66**]).

LEMMA 12.6. *Let $(X, \|\cdot\|)$ be a Banach space and γ be a Kuratowski measure of noncompactness. Then the following statements hold:*

(i) $\gamma(D) = 0$ *if and only if* $\overline{D}$ *is compact,*
(ii) $\gamma(D_1 + D_2) \leq \gamma(D_1) + \gamma(D_2)$,
(iii) $\gamma(D_1) \leq \gamma(D_2)$ *for all* $D_1 \subset D_2$,
(iv) $\gamma(D_1 \cup D_2) \leq \max\{\gamma(D_1), \gamma(D_2)\}$,
(v) $\gamma(\overline{D}) = \gamma(D)$.

In addition, let X be an infinite-dimensional Banach space with a decomposition $X = X_1 \oplus X_2$, and let $P : X \to X_1$ and $Q : X \to X_2$ be projectors with $\dim X_1 < \infty$. *Then one has*

(vi) $\gamma(B_\varepsilon(0)) = 2\varepsilon$,
(vii) $\gamma(D) < \varepsilon$ *for any bounded subset D of X for which the diameter of QD is less than ε.*

The following result generalizes Lemma 1.27. It is due to WANG, ZHONG & ZHOU [**234**, Lemma 2.11].

LEMMA 12.7. *Let $(A_n)_{n\in\mathbb{N}}$ be a family of nested (i.e., $A_{n+1} \subset A_n$ for $n \in \mathbb{N}$) closed and bounded subsets of a Banach X with measures of noncompactness $\gamma(A_n) < \frac{1}{n}$ for $n \in \mathbb{N}$. Then $A := \bigcap_{n\in\mathbb{N}} A_n$ is a nonempty compact subset of X.*

2.2. Equivalent asymptotic compactness concepts. Three asymptotic compactness concepts are now introduced for skew product flows (θ, φ) on $P \times X$, where (P, d_P) is a complete metric space and $(X, \|\cdot\|)$ is a Banach space. The topological properties of P are, in fact, not important here.

DEFINITION 12.8 (Pullback flattening). A skew product flow (θ, φ) on a Banach space $(X, \|\cdot\|)$ is said to be *pullback flattening* if for every bounded nonautonomous set $\mathcal{B}$, $\varepsilon > 0$ and $p \in P$, there exists a $T_0 = T_0(\mathcal{B}, \varepsilon, p) > 0$ and a finite-dimensional subspace X_ε of X such that

$$\bigcup_{t \geq T_0} P_\varepsilon \varphi\big(t, \theta_{-t}(p), B_{\theta_{-t}(p)}\big) \text{ is bounded}$$

and

$$\Big\|(\mathbb{1} - P_\varepsilon) \bigcup_{t \geq T_0} \varphi\big(t, \theta_{-t}(p), B_{\theta_{-t}(p)}\big)\Big\|_X < \varepsilon\,, \tag{12.2}$$

where $P_\varepsilon : X \to X_\varepsilon$ is a bounded projection and (12.2) is understood in the sense that $\big\|(\mathbb{1} - P_\varepsilon)\varphi(t, \theta_{-t}(p), x_0)\big\|_X < \varepsilon$ for all $x_0 \in B_{\theta_{-t}(p)}$ and $t \geq T_0$.

DEFINITION 12.9 (Pullback limit-set compact). A skew product flow (θ, φ) on a Banach space $(X, \|\cdot\|)$ is said to be *pullback limit-set compact* if for every bounded nonautonomous set $\mathcal{B}$, $\varepsilon > 0$ and $p \in P$, there exists a $T_1 = T_1(\mathcal{B}, \varepsilon, p) > 0$ such that

$$\gamma\Big(\bigcup_{t \geq T_1} \varphi\big(t, \theta_{-t}(p), B_{\theta_{-t}(p)}\big)\Big) < \varepsilon\,,$$

where γ is a measure of noncompactness defined on the subsets of X.

DEFINITION 12.10 (Pullback asymptotically compact). A skew product flow (θ, φ) on a Banach space $(X, \|\cdot\|)$ is said to be *pullback asymptotically compact* in X if for every bounded nonautonomous set $\mathcal{B}$, each $p \in P$ and sequences $t_k \to \infty$ and $x_k \in B_{\theta_{-t_k}(p)}$, $k \in \mathbb{N}$, the set $\big\{\varphi\big(t_k, \theta_{-t_k}(p), x_k\big) : k \in \mathbb{N}\big\}$ is precompact in X.

It will be shown in the following that these three concepts are equivalent, provided the state space is a uniformly convex.

DEFINITION 12.11 (Uniformly convex). A Banach space $(X, \|\cdot\|)$ is said to be *uniformly convex* if for all $\varepsilon > 0$, there exists $\delta > 0$ such that

$$\frac{\|x + y\|}{2} < 1 - \delta$$

for all $x, y \in X$ with $\|x\|, \|y\| \leq 1$ and $\|x - y\| > \varepsilon$.

Requiring a space to be uniformly convex is not a severe restriction in applications, since this property is satisfied by all Hilbert spaces, the L^p spaces with $1 < p < \infty$ and most Sobolev spaces $W^{k,p}$ with $1 < p < \infty$, see BRÉZIS [**25**, Section III.7].

THEOREM 12.12. *Let (θ, φ) be a skew product flow on $P \times X$, where (P, d_P) is a metric space and $(X, \|\cdot\|)$ is a uniformly convex Banach space. Then the following three statements are equivalent:*

(i) *(θ, φ) is pullback flattening,*
(ii) *(θ, φ) is pullback limit-set compact,*
(iii) *(θ, φ) is pullback asymptotically compact.*

PROOF. It will be shown that (i) $\Rightarrow$ (ii), (ii) $\Rightarrow$ (iii) and (iii) $\Rightarrow$ (i).
Part 1 (Pullback flattening $\Rightarrow$ pullback limit-set compact). Suppose that the skew product flow is pullback flattening and consider an arbitrary bounded nonautonomous set $\mathcal{B}$. Then for each $p \in P$, one has

$$\begin{aligned}\gamma\Big(\bigcup_{t\geq T_1}\varphi\big(t,\theta_{-t}(p),B_{\theta_{-t}(p)}\big)\Big) &\leq \gamma\Big(P_\varepsilon \bigcup_{t\geq T_1}\varphi\big(t,\theta_{-t}(p),B_{\theta_{-t}(p)}\big)\Big)\\ &\quad + \gamma\Big((\mathbb{1}-P_\varepsilon)\bigcup_{t\geq T_1}\varphi\big(t,\theta_{-t}(p),B_{\theta_{-t}(p)}\big)\Big)\\ &\leq 0 + \gamma(B_\varepsilon(0)) = 2\varepsilon\,.\end{aligned}$$

Hence, the skew product flow is pullback limit-set compact.
Part 2 (Pullback limit-set compact $\Rightarrow$ pullback asymptotically compact). Suppose that the skew product flow is pullback limit-set compact, and let $\mathcal{B}$ be a bounded nonautonomous set such that for each $p \in P$ and $\varepsilon > 0$, there exists a $T_1 = T_1(\mathcal{B},\varepsilon,p) > 0$ with

$$\gamma\Big(\bigcup_{t\geq T_1}\varphi\big(t,\theta_{-t}(p),B_{\theta_{-t}(p)}\big)\Big) < \varepsilon\,.$$

Choose $\varepsilon_n := \frac{1}{n}$ and define $\tau_n := T_1\big(\mathcal{B},\frac{1}{n},p\big)$ for $n \in \mathbb{N}$ such that one has $\tau_i < \tau_{i+1}$ for all $i \in \mathbb{N}$. Then

$$\gamma\Big(\bigcup_{t\geq \tau_n}\varphi\big(t,\theta_{-t}(p),B_{\theta_{-t}(p)}\big)\Big) < \frac{1}{n} \quad \text{for all } n \in \mathbb{N}\,,$$

and it follows from the properties of the measure of noncompactness that

$$\gamma\Big(\overline{\bigcup_{t\geq \tau_n}\varphi\big(t,\theta_{-t}(p),B_{\theta_{-t}(p)}\big)}\Big) < \frac{1}{n} \quad \text{for all } n \in \mathbb{N}\,.$$

Now the bounded sets $A_n := \overline{\bigcup_{t\geq \tau_n}\varphi\big(t,\theta_{-t}(p),B_{\theta_{-t}(p)}\big)}$ are nested, i.e., with $A_{n+1} \subset A_n$ for $n \in \mathbb{N}$, so by Lemma 12.7, their intersection is a nonempty compact subset of X:

$$\emptyset \neq A_\infty = \bigcap_{n\geq 1} A_n = \bigcap_{n\geq 1}\overline{\bigcup_{t\geq \tau_n}\varphi\big(t,\theta_{-t}(p),B_{\theta_{-t}(p)}\big)}\,. \tag{12.3}$$

Now consider arbitrary sequences $t_k \to \infty$ and $x_k \in B_{\theta_{-t_k}(p)}$, $k \in \mathbb{N}$, let $F_j(p) := \big\{\varphi(t_k,\theta_{-t_k}(p),x_k) : k \geq j\big\}$ and (discarding a finite number of k if necessary) define

$$n_j := \max\big\{n \in \mathbb{N} \;:\; \tau_n \leq t_j\big\}\,,$$

so $\lim_{j\to\infty} n_j = \infty$. Since for all $j \in \mathbb{N}$ and $k \geq j$, one has

$$\varphi\big(t_k,\theta_{-t_k}(p),x_k\big) \in \varphi\big(t_k,\theta_{-t_k}(p),B_{\theta_{-t_k}(p)}\big) \subset A_{n_j}\,,$$

it follows that $F_j(p) \subset \overline{F_j(p)} \subset A_{n_j}$ for all $k \geq j$ and $j \in \mathbb{N}$, and hence,

$$\gamma\big(\overline{F_j(p)}\big) < \frac{1}{n_j} \to 0 \quad \text{as } j \to \infty\,.$$

But $F_{j+1}(p) \subset F_j(p)$ for $j \in \mathbb{N}$, i.e., the sets are nested, so their intersection is nonempty and compact with

$$\emptyset \neq \overline{F}(p) := \bigcap_{j\geq 1}\overline{F_j(p)} \subset A_\infty\,.$$

However, the set $F_1(p) := \big\{\varphi\big(t_k,\theta_{-t_k}(p),x_k\big) : k \geq 1\big\}$ is precompact, which means that the skew product flow is pullback asymptotically compact.
Part 3 (Pullback asymptotically compact $\Rightarrow$ pullback flattening). Suppose that skew

product flow is pullback asymptotically compact, and let $\mathcal{B}$ be a bounded nonautonomous set. Moreover, let $p \in P$ be arbitrary but fixed, and consider the set

$$A_\infty = \bigcap_{n\geq 1} A_n = \bigcap_{n\geq 1} \overline{\bigcup_{t\geq \tau_n} \varphi\big(t, \theta_{-t}(p), B_{\theta_{-t}(p)}\big)}\,.$$

It is clear that $a \in A_\infty$ if and only if there exist sequences $t_k \to \infty$ and $x_k \in B_{\theta_{-t_k}(p)}$, $k \in \mathbb{N}$, such that $\varphi(t, \theta_{-t}(p), a_k) \to a$ as $k \to \infty$. From the assumed pullback asymptotic compactness of the skew product flow, it is clear that A_∞ is nonempty. Now consider a sequence $a_k \in A_\infty$, $k \in \mathbb{N}$. Then for each $k \geq 1$ there exist sequences $t_k \geq k$ and $y_k \in \varphi\big(t, \theta_{-t_k}(p), B_{\theta_{-t_k}(p)}\big)$ such that $\|y_k - a_k\| \leq \frac{1}{k}$. By the pullback asymptotic compactness of the skew product flow, there is a convergent subsequence (which can be relabeled as the original one) such that $\lim_{k\to\infty} y_k = a \in A_\infty$. Thus, A_∞ is both nonempty and compact. Now suppose that A_∞ does not pullback attract $\mathcal{B}$. Then there exist an $\varepsilon_0 > 0$ and sequences $t_k \to \infty$ in $\mathbb{R}_0^+$ and $y_k \in \varphi\big(t_k, \theta_{-t_k}(p), B_{\theta_{-t_k}(p)}\big)$ such that

$$\operatorname{dist}_X(y_k, A_\infty) > \varepsilon_0 \quad \text{for all } k \in \mathbb{N}\,. \tag{12.4}$$

Thus, by pullback asymptotic compactness of the skew product flow, there is a convergent subsequence (which can be relabeled) such that $\lim_{k\to\infty} y_k = a \in A_\infty$. This, however, contradicts (12.4), so A_∞ does in fact pullback attract $\mathcal{B}$. In particular, for any $\varepsilon > 0$, there is a $T_\varepsilon = T_\varepsilon(\varepsilon, \mathcal{B}, p) > 0$ such that

$$\operatorname{dist}_X\big(\varphi\big(t, \theta_{-t}(p), B_{\theta_{-t}(p)}\big), A_\infty\big) < \frac{\varepsilon}{4} \quad \text{for all } t \geq T_\varepsilon\,.$$

This means in particular that

$$\varphi\big(t, \theta_{-t}(p), B_{\theta_{-t}(p)}\big) \subset B_{\varepsilon/4}(A_\infty)\,.$$

Since A_∞ is a compact subset of X, there exist N_ε and points $x_1, x_2, \dots, x_{N_\varepsilon}$ in X such that

$$A_\infty \subset \bigcup_{i=1}^{N_\varepsilon} B_{\varepsilon/4}(x_i)\,,$$

from which it follows that for $t \geq T_\varepsilon$, one has

$$\varphi\big(t, \theta_{-t}(p), B_{\theta_{-t}(p)}\big) \subset B_{\varepsilon/4}(A_\infty) \subset \bigcup_{i=1}^{N_\varepsilon} B_{\varepsilon/2}(x_i)\,, \tag{12.5}$$

which implies that

$$\bigcup_{t\geq T_\varepsilon} \varphi\big(t, \theta_{-t}(p), B_{\theta_{-t}(p)}\big) \subset \bigcup_{i=1}^{N_\varepsilon} B_{\varepsilon/2}(x_i)\,.$$

Now let $X_{N_\varepsilon} := \operatorname{span}\{x_1, x_2, \dots, x_{N_\varepsilon}\}$. Since X is uniformly convex, there exists a projection $P_{N_\varepsilon} : X \to X_{N_\varepsilon}$ such that

$$\|x - P_{N_\varepsilon} x\| = \operatorname{dist}_X\big(x, X_{N_\varepsilon}\big) \quad \text{for all } x \in X\,.$$

Then

$$\Big\|(\mathbb{1} - P_{N_\varepsilon}) \textstyle\bigcup_{t\geq T_\varepsilon} \varphi\big(t, \theta_{-t}(p), B_{\theta_{-t}(p)}\big)\Big\| \leq \dfrac{\varepsilon}{2} < \varepsilon\,.$$

Moreover, the set $P_{N_\varepsilon} \bigcup_{t\geq T_\varepsilon} \varphi\big(t, \theta_{-t}(p), B_{\theta_{-t}(p)}\big)$ is bounded in X, since $\bigcup_{t\geq T_\varepsilon} \varphi\big(t, \theta_{-t}(p), B_{\theta_{-t}(p)}\big)$ is bounded in X. Thus, the skew product flow is flattening. This completes the proof of this theorem. $\square$

Note that the implication (iii) $\Rightarrow$ (i) is also easy to prove, i.e., if an skew product flow is pullback asymptotically compact, then it is pullback limit-set compact. Indeed, from (12.5), one has

$$\gamma\big(\varphi(t,\theta_{-t}(p),B_{\theta_{-t}(p)})\big) \le \gamma\big(A_\infty(\mathcal{B},p)\big) \le \frac{\varepsilon}{4} \quad \text{for all } t \ge T_\varepsilon\,,$$

which means that the skew product flow is pullback limit-set compact.

REMARK 12.13. The uniform convexity of the Banach space was used only in the last implication (iii) $\Rightarrow$ (i) of the preceding proof. Thus, if X is a general Banach space, then the following holds:

- Pullback limit-set compact is equivalent to pullback asymptotically compact.
- Pullback flattening implies pullback asymptotically compact.

2.3. Existence of a pullback attractor. Pullback flattening and the existence of a pullback absorbing family of bounded subsets imply the existence of a pullback attractor.

THEOREM 12.14 (Existence of a pullback attractor). *Suppose that a skew product flow (θ,φ) is pullback flattening and that the bounded nonautonomous set $\mathcal{B}$ is pullback absorbing. Then, it has a unique pullback attractor $\mathcal{A}$ with $A_p \subset B_p$ for every $p \in P$.*

PROOF. Since the skew product flow has a pullback absorbing set $\mathcal{B}$, every bounded nonautonomous set $\mathcal{D}$ is pullback absorbed into $\mathcal{B}$ in a finite time. Hence, only the asymptotic behavior for the skew product flow starting in $\mathcal{B}$ need to be considered. In addition, the skew product flow is flattening, so by Theorem 12.12, it is pullback limit-set compact and from the proof of Theorem 12.12, there is a nonempty compact subset $A_\infty = A_\infty(\mathcal{B},p)$ of X for each $p \in P$ as defined in (12.3). Define $A_p := A_\infty(\mathcal{B},p)$ for each $p \in P$ and consider the so-generated nonautonomous set $\mathcal{A}$. This set is φ-invariant and pullback attracting by the same arguments as in the proof of Theorem 3.18. Thus, $\mathcal{A}$ is the desired pullback attractor. □

It also follows that pullback flattening is a necessary condition for the existence of pullback attractors in a uniformly convex Banach space, since a family of closed ε-neighborhoods of the component sets A_p of the pullback attractor $\mathcal{A}$ is pullback absorbing and (12.5) holds. This gives the following corollary.

COROLLARY 12.15. *Let (θ,φ) be a skew product flow on a uniformly convex Banach space such that there exists a pullback attractor. Then (θ,φ) is pullback flattening.*

Endnotes. The monographs by Carvalho, Langa & Robinson [**35**] and Chepyzhov & Vishik [**43**] deal mainly with infinite-dimensional nonautonomous dynamical systems generated by evolutionary partial differential equations. See also Hale [**90**], Ladyzhenskaya [**151**], Robinson [**201**], Sell & You [**220**], Temam [**230**] and Vishik [**233**]. Flattening was introduced in Kloeden & Langa [**129**] for nonautonomous and random dynamical systems, but had been used earlier under another name by Qingfeng Ma, Shouhong Wang & Chenkui Zhong [**163**] for autonomous systems as well as Yejuan Wang, Chenkui Zhong &

Shenfan Zhou [**234**] for nonautonomous systems. The squeezing property is due to Foias, Manley & Temam [**79**] and Foias & Temam [**81**]. See Deimling [**65**] for infinite-dimensional ordinary differential equations.

CHAPTER 13

Switching and control systems

Switching and control systems are structurally similar. They have been investigated intensively by control and systems engineers as well as mathematicians for many years. Both are intrinsically nonautonomous dynamical systems and can be formulated as skew product flows, the theory of which provides a broader understanding of their asymptotic behavior. Switching systems are also of considerable interest to the computing and scientific computing communities.

1. Switching systems

Consider a finite collection of ordinary differential equations

$$\dot{x} = f_i(x) \quad \text{for } i \in \{1, \ldots, N\}, \tag{13.1}$$

in $\mathbb{R}^d$. Given a sequence of indices $i_n \in \{1, \ldots, N\}$ for $n \in \mathbb{Z}$ and an increasing sequence of times $(\tau_n)_{n \in \mathbb{Z}}$, the system

$$\dot{x} = f_{i_n}(x) \quad \text{for all } \tau_n < t < \tau_{n+1} \,. \tag{13.2}$$

is said to be a *switching system*.

The engineering literature on switching systems has mainly concentrated on the asymptotic stability or stabilizability of a common equilibrium solution, which is usually taken to be the zero solution. In general, however, the un-switched systems may have different equilibria or even different attractors of more complicated form. Since the switching of such systems are nonautonomous, the theory of nonautonomous dynamical systems is tailor-made to investigate their asymptotic behavior.

1.1. Formulation of switching systems as skew product flows.

The switching system (13.2) can be reformulated as an affine control system

$$\dot{x} = \sum_{i=1}^{N} f_i(x) u_i \,, \tag{13.3}$$

with the *switching control* $u = (u_1, \ldots, u_N) : \mathbb{R} \to \{e_1, \ldots, e_N\}$, where e_j is the j-th unit vector in $\mathbb{R}^N$, satisfying $u(t) = e_{i_n}$ for $\tau_n \leq t < \tau_{n+1}$ for all $n \in \mathbb{Z}$. In particular, switching controls are piecewise constant functions, which are continuous from the right with a limit from the left.

Let $\mathcal{S}$ denote the collection of all such switching controls, with N considered fixed. In addition, denote by $\mathcal{S}_\tau$ the subset consisting of all switching controls with a *dwell time* $\tau > 0$, i.e., with successive switches separated by at least time τ. A metric d on $\mathcal{S}$ and $\mathcal{S}_\tau$ is given by

$$d(u,v) = \sum_{n=1}^{\infty} \tfrac{1}{2^n} \int_{-n}^{n} \|u(s) - v(s)\|_\infty \, \mathrm{d}s \tag{13.4}$$

where $\|\cdot\|_\infty$ is the sup-norm on $\mathbb{R}^N$. The metric space $(\mathcal{S}, d)$ is not complete. This is verified with the sequence

$$u^n = \begin{cases} 1 & : \quad (2j+1)^{-2} \le t < (2j)^{-2} \text{ and } j \in \{1, \dots, n\} \\ 0 & : \quad \text{otherwise} \end{cases}$$

which is a Cauchy sequence with $d(u^n, u^{n+k}) < (2n)^{-2}$ for all $k \in \mathbb{N}_0$, but its limiting function $\bar{u} : \mathbb{R} \to \{e_1, \dots, e_N\}$ is not continuous from the right at $t = 0$, so $\bar{u} \notin \mathcal{S}$. However, it was proved in KLOEDEN [**121**] that

THEOREM 13.1. *$(\mathcal{S}_\tau, d)$ is a compact metric space for any $\tau > 0$.*

Define the shift operator $\theta_t : \mathcal{S}_\tau \to \mathcal{S}_\tau$ for each $t \in \mathbb{R}$ by $\theta_t(u(\cdot)) := u(t + \cdot)$. Then $\theta = (\theta_t)_{t\in\mathbb{R}}$ is a group of operators with $\theta_0 = \mathrm{id}$ and $\theta_t \circ \theta_s = \theta_{t+s}$.

LEMMA 13.2. *The mapping $t \mapsto \theta_t(u)$ is uniformly continuous in the metric d on $\mathcal{S}_\tau$ with*

$$d\big(\theta_s(u), \theta_t(u)\big) = d\big(\theta_{s-t}(u), u\big) \le 4\frac{|s-t|}{\tau} \quad \textit{for all } |s-t| < \tau \textit{ and } u \in \mathcal{S}_\tau\,.$$

PROOF. The desired result follows from the inequality

$$d\big(\theta_t(u), u\big) \le 4\frac{|t|}{\tau} \quad \text{for all } |t| < \tau \text{ uniformly in } u \in \mathcal{S}_\tau\,.$$

To prove this, note that a switching control $u \in \mathcal{S}_\tau$ has at most one switch in an interval of length less than τ, so

$$\int_{t^*-\tau/2}^{t^*+\tau/2} \|u(t+s) - u(s)\|_\infty \, \mathrm{d}s \le |t| \quad \text{for all } |t| < \tau \text{ and } t^* \in \mathbb{R}\,.$$

Hence, the interval $[-n, n]$ contains at most $2n/\tau$ switches, which gives

$$\int_{-n}^{n} \|\theta_t(u(s)) - u(s)\|_\infty \, \mathrm{d}s \le \frac{2n}{\tau}|t| \quad \text{for all } |t| < \tau\,.$$

The result then follows from the fact that $\sum_{n=1}^{\infty} n2^{-n+1} = 4$. □

Assume now that the initial value problem

$$\dot{x} = \sum_{i=1}^{N} f_i(x) u_i\,, \qquad x(0) = x_0\,,$$

with a given switching control $u \in \mathcal{S}_\tau$ has a unique solution $x(t) = \varphi(t, u, x_0)$ which exists for all $t \ge 0$ (this holds, for example, if the functions f_i are all locally Lipschitz and satisfy a growth bound or a dissipativity condition such as (13.5) below). Then φ satisfies the initial condition

$$\varphi(0, u, x_0) = x_0$$

as well as the cocycle property

$$\varphi(s+t,u,x_0) = \varphi(t,\theta_s(u),\varphi(s,u,x_0)) \quad \text{for all} \;\; s,t \geq 0\,.$$

In addition, φ is continuous.

LEMMA 13.3. *The mapping $(t,u,x_0) \mapsto \varphi(t,u,x_0)$ is continuous in $(t,u,x_0) \in \mathbb{R}_0^+ \times \mathcal{S}_\mathcal{T} \times \mathbb{R}^d$.*

PROOF. Consider solutions $x(t) = x(t,u,x_0)$ and $y(t) = y(t,v,y_0)$ of the switching system. Then one has

$$\begin{aligned}
&\|x(\tau) - y(t)\| - \|x_0 - y_0\| \\
&\leq \sum_{i=1}^{N} \left\| \int_t^\tau f_i(x(s))u_i(s)\,\mathrm{d}s + \int_0^t \big(f_i(x(s))u_i(s) - f_i(y(s))v_i(s)\big)\,\mathrm{d}s \right\| \\
&\leq \sum_{i=1}^{N} \left(\left| \int_t^\tau \|f_i(x(s))u_i(s)\|\,\mathrm{d}s \right| + \int_0^t \|f_i(x(s))u_i(s) - f_i(y(s))v_i(s)\|\,\mathrm{d}s \right) \\
&\leq \sum_{i=1}^{N} \left| \int_t^\tau \|f_i(x(s))\|\|u_i(s)\|\,\mathrm{d}s \right| + \sum_{i=1}^{N} \int_0^t \|f_i(x(s))\|\|u_i(s) - v_i(s)\|\,\mathrm{d}s \\
&\quad + \sum_{i=1}^{N} \int_0^t \|f_i(x(s)) - f_i(y(s))\|\|v_i(s)\|\,\mathrm{d}s \\
&\leq NK \left| \int_t^\tau \mathrm{d}s \right| + NK \int_0^t \|u(s) - v(s)\|_\infty\,\mathrm{d}s + NL \int_0^t \|x(s) - y(s)\|\,\mathrm{d}s \\
&\leq NK|\tau - t| + NK2^T d(u,v) + NL \int_0^t \|x(s) - y(s)\|\,\mathrm{d}s
\end{aligned}$$

for all $0 \leq \tau$ and $t \leq T$, where L is a common (local) Lipschitz constant for the functions f_i and K a common bound for the $|f_i(x)|$ for $|x| \leq R$ for an appropriately large R. The Gronwall inequality then gives

$$\big\|x(t,u,x_0) - y(t,v,y_0)\big\| \leq \big(\|x_0 - y_0\| + NK|t-t| + NK2^T d(u,v)\big)e^{NLT}\,,$$

from which the asserted continuity follows. □

COROLLARY 13.4. *The switching system (θ, φ) is a skew product flow on $\mathcal{S}_\mathcal{T} \times \mathbb{R}^d$.*

1.2. Formulation of switching systems as processes. Sometimes one is interested in only a specific switching control $\bar{u}$ rather than in all switching controls in $\mathcal{S}_\mathcal{T}$ or those in the smaller hull $H(\bar{u})$ of a specific switching control $\bar{u}$, that is, one wants to investigate the asymptotic behavior of the nonautonomous differential equation

$$\dot{x} = \sum_{i=1}^{N} f_i(x)\bar{u}_i(t)\,.$$

Assuming future existence and uniqueness, the solution $x_{\bar{u}}(t,t_0,x_0)$ for an initial value x_0 at time t_0 forms a process. The process formalism may sometimes provide useful information or insights when the skew product formalism does not or cannot.

EXAMPLE 13.5. Consider two scalar differential equations with $f_1(x) = x$ and $f_2(x) = -x$, which have the zero solution as a common equilibrium, and the switching control $\bar{u}(t) = (1,0)$ for $t < 0$ and $\bar{u}(t) = (0,1)$ for $t \geq 0$. The general solution is given by

$$x_{\bar{u}}(t, t_0, x_0) = \begin{cases} e^{t-t_0}x_0 & : \quad t_0 \leq t \leq 0\,, \\ e^{-t+|t_0|}x_0 & : \quad \text{otherwise.} \end{cases}$$

The zero solution is obviously asymptotically Lyapunov stable (though not uniformly in t_0), and there is a forward attractor $\mathcal{A}$ with component sets $A_t = \{0\}$ for all $t \in \mathbb{R}$, but this is not a pullback attractor. If the roles of the systems are reversed, then $\mathcal{A}$ becomes a pullback attractor, but is not a forward attractor and the zero solution is, in fact, unstable.

1.3. Asymptotic behavior to switching systems. The asymptotic stability or stabilizability of a common equilibrium solution, which is usually taken to be the zero solution, is of major interest for engineers. Various sufficient conditions have been established, such as the existence of a common Lyapunov function (see LIBERZON [**157**, Theorem 2.2]), or equivalently, the strong asymptotic stability of the zero (equilibrium) solution of the differential inclusion

$$\dot{x} \in \operatorname{conv}\left\{f_1(x), \ldots, f_N(x)\right\}$$

consisting of all convex combinations of the vector fields of the individual systems (13.1) ([**157**, Corollary 2.3]) as well as certain commutativity and Lie algebra conditions ([**157**, Theorems 2.6 and 2.7]).

Finding a common Lyapunov function is not always easy, but one can often show that the individual systems have a common absorbing set, e.g., by means of a dissipativity condition such as

$$\langle x, f_i(x)\rangle \leq K - L\|x\|^2 \quad \text{for all } i \in \{1, \ldots, N\}\,. \tag{13.5}$$

LEMMA 13.6. *Suppose that the un-switched autonomous systems* (13.1) *have a common forward absorbing set, which is positively invariant for each of them. Then there is a common pullback absorbing set for any switching systems* (13.3), *which is uniform in the switching controls.*

PROOF. Let B be the common forward absorbing set, which can be assumed without loss of generality to be a closed ball $\overline{B_R(0)}$ about the origin with a suitably large radius $R > 0$. Each of the un-switched systems is thus equi-ultimately bounded as defined in YOSHIZAWA [**239**, Theorem 20.3], i.e., for each $i \in \{1, \ldots, N\}$, there exists a Lyapunov-like function $V_i(x)$ defined for $|x| \geq R$ such that

$$a_i(|x|) \leq V_i(x) \quad \text{and} \quad D_i^+V_i(x) \leq -c_iV_i(x) \quad \text{for all } |x| \geq R\,,$$

where $r \mapsto a_i(r)$ is a continuous, positive and strictly increasing function defined for $r \geq R$ and $c_i > 0$, and D_i^+ is the upper Dini derivative from the right with respect to the i-th un-switched system. Define

$$V(t,x) := \sum_{i=1}^{N} u_i(t)V_i(x)$$

where $u(t) = (u_1(t), \dots, u_N(t))$ is the switching control under consideration. Note that $a^*(r) := \min_{i=1,\dots,N} a_i(r)$ is continuous, positive and strictly increasing in $r \geq R$ with

$$V(t,x) = \sum_{i=1}^{N} u_i(t)V_i(x) \geq \sum_{i=1}^{N} u_i(t)a_i(|x|) \geq \sum_{i=1}^{N} u_i(t)a^*(|x|) \geq a^*(|x|),$$

and, since only one of the u_i is "active" at any one time,

$$\begin{aligned} D_u^+ V(t,x) = \sum_{i=1}^{N} u_i(t) D_i^+ V_i(x) &\leq -\sum_{i=1}^{N} u_i(t) c_i V_i(x) \\ &\leq -c^* \sum_{i=1}^{N} u_i(t) V_i(x) = -c^* V(t,x), \end{aligned}$$

where $c^* = \min_{i=1,\dots,N} c_i > 0$. By [**239**, Theorem 10.5], the switching system (13.3) with this switching control is equi-ultimately bounded in $\overline{B_R(0)}$, which is thus an absorbing set. Moreover, the absorption time is uniformly bounded over all switching controls since a^* and C^* are independent of the switching control, so $\overline{B_R(0)}$ is a uniform pullback absorbing set. □

The theory of nonautonomous dynamical systems applied to switching systems considered as skew product flows with the compact base space $P = \mathcal{S}_\tau$ gives the following well-known result.

THEOREM 13.7. *Suppose that the zero solution is asymptotically stable for each of the un-switched systems* (13.1) *and that these systems have a common positively invariant absorbing set. Suppose also that the zero solution is the only bounded entire solution of any switching system for a switching control in $\mathcal{S}_\tau$. Then the zero solution is asymptotically stable uniformly for all such switching systems.*

PROOF. Consider the switching system as a skew product flow (θ, φ) on $\mathcal{S}_\tau \times \mathbb{R}^d$. This skew product flow has a uniform forward absorbing set $B_u \equiv B$, which is thus also pullback absorbing. Hence, it has a pullback attractor $\mathcal{A}$ with fibers $A_u = \{0\}$ for all $u \in P$. Otherwise, there would exist a point $x_u \in A_u \setminus \{0\}$ for some $u \in \mathcal{S}_\tau$, and thus, a bounded entire solution $\bar{x}_u(\cdot)$ of the switching system (13.3) corresponding to this control with $\bar{x}_u(t) = x_u$ and $\bar{x}_u(t) \in A_{\theta_t(u)}$ for all $t \in \mathbb{R}$, in contradiction to the assumption that the zero solution is the only bounded entire solution. The set-valued mapping $u \mapsto A_u \equiv \{0\}$ is trivially continuous, and it follows from Theorem 3.41 that $\mathcal{A}$ is a uniform forward attractor, i.e., the zero solution is uniformly asymptotically stable for the switching systems corresponding to all switching controls in $\mathcal{S}_\tau$. □

Of course, it may be difficult in practice to show that the zero solution is the only bounded entire solution. However, this holds automatically if the un-switched systems are linear. Simple examples in [**157**] show that asymptotic stability may fail to hold when the un-switched linear systems do not have a common absorbing set.

COROLLARY 13.8. *Suppose that each of the un-switched systems* (13.1) *is linear, i.e., $f_i(x) = A_i x$, has asymptotically stable zero solution and a common positively* *invariant absorbing set. Then the zero solution of any switching system* (13.3) *is asymptotically stable uniformly for all switching controls.*

The proof follows by observing that for any $\rho > 0$, each linear system $\dot{x} = A_i x$ is invariant under the coordinate transformation $x \to z := \rho x$, i.e., resulting in the transformed system $\dot{z} = A_i z$, which is just the original system written in terms of another variable. Thus an invariant ball $B_R(0)$ in the x-system transforms to the invariant ball $B_R(0)$ in the z-system. Forgetting the transformation and using the fact the two systems are identical apart from the notation allows us to conclude that $B_R(0)$ is an invariant ball for the linear system, whatever the notation used. Moreover, this holds for each $\rho > 0$, no matter how small, from which it follows as above that $A_u \equiv \{0\}$ for all switching controls u.

1.4. Nontrivial pullback attractors. The skew product flow formalism is particularly advantageous for investigating switching systems for which the resulting pullback attractor $\mathcal{A}$ is nontrivial, i.e., does not correspond to a common zero solution as considered above. The component sets A_u need not in general be identical or single valued, but they nevertheless provide information about the forward asymptotic behavior of the switching systems. From results in Chapter 3, the best one can expect in the general case is forward convergence to the component sets of the parametrically inflated pullback attractor rather than to those of the pullback attractor itself, i.e.,

$$\lim_{t\to\infty} \operatorname{dist}\left(\varphi(t,u,D), A_{\theta_t(u)}[\varepsilon_0]\right) = 0$$

for any bounded subset D of $\mathbb{R}^d$, where $A_u[\varepsilon_0] = \bigcup_{d(v,u)\leq\varepsilon_0} A_v$. This holds for any $\varepsilon_0 > 0$. The convergence is also uniform in the switching controls $u \in P$, which could be the whole space $\mathcal{S}_\tau$ or the smaller hull $H(u)$ of a particular switching control u, in which case the above union is over the ε_0 neighborhood of u in a smaller space.

Alternatively, if the skew product flow has a pullback absorbing family consisting of a single set $B_u \equiv B$ for all $u \in \mathcal{S}_\tau$ for which the absorbing times are independent of u and if the set-valued mapping $u \mapsto A_u$ is continuous in the Hausdorff metric rather than just upper semi-continuous, then the pullback attractor $\mathcal{A}$ is also a uniform forward attractor by Theorem 3.41. Such continuity always holds when the component sets A_u are singleton sets and also when θ_t is nontrivially periodic as was seen in Lemma 3.45.

However, more can be said with additional information about the switching system and the pullback attractor. For example, from Theorem 3.41, the pullback attractor $\mathcal{A}$ is also a forward attractor if the set-valued mapping $p \mapsto A_p$ is continuous in the Hausdorff metric. Thus, the following can be concluded from Lemmata 3.45 and 13.6.

LEMMA 13.9. *If P is the hull of a periodic switching control u, then the pullback attractor $\mathcal{A}$ is also a forward attractor, and the set-valued mapping $t \mapsto A_{\theta_t(u)}$ is periodic.*

Similarly, from Lemma 13.6 above and WANG, LI & KLOEDEN [**237**, Theorems 3.3 and 3.5], one obtains the following.

LEMMA 13.10. *If P is the hull of an almost periodic switching control u and the set-valued mapping $u \mapsto A_u$ is continuous in the Hausdorff metric, then the pullback attractor $\mathcal{A}$ is also a forward attractor and the set-valued mapping $t \mapsto A_{\theta_t(u)}$ is almost periodic.*

From Theorem 3.47, the pullback attractors in Lemmata 13.9 and 13.10 are, in fact, uniformly asymptotically Lyapunov stable.

1.5. Systems with a one-sided dissipative Lipschitz condition. If the un-switched systems (13.1) have a common positively invariant absorbing set B, then each will have a global autonomous attractor A_i, $i \in \{1, \ldots, N\}$. These will be component sets of the pullback attractor for the space $P = \mathcal{S}_\tau$, corresponding to the (non)switching controls $u(t) \equiv e_i$, respectively, and are obviously forward attracting for each of these controls. If nontrivial switching between the systems occurs, then the corresponding pullback attractor component sets can be quite different. This is illustrated by the following situation where each un-switched system satisfies a one-sided dissipative Lipschitz condition and thus has an asymptotically stable equilibrium, which may differ from system to system. In particular, assume for all $x, y \in \mathbb{R}^d$ that

$$\langle x - y, f_i(x) - f_i(y) \rangle \leq -L_i \|x - y\|^2 \quad \text{for all } i \in \{1, \ldots, N\}, \tag{13.6}$$

and define $L_* := \min\{L_1, \ldots, L_N\} > 0$.

THEOREM 13.11. *If each of the un-switched systems (13.1) satisfies the one-sided dissipative Lipschitz condition (13.6), then the pullback attractor of the nonautonomous dynamical system induced by the switching systems (13.3), corresponding to all possible switching controls in $\mathcal{S}_\tau$, has singleton component sets and is uniformly asymptotically Lyapunov stable.*

PROOF. For any switching controls $u \in \mathcal{S}_\tau$, the switching system (13.3) satisfies a one-sided dissipative Lipschitz condition (13.6) with constant L_*. Its vector field is $f_u(t, x) = \sum_{i=1}^N f_i(x) u_i(t)$, so

$$\begin{aligned}
\langle x - y, f_u(t, x) - f_u(t, y) \rangle &= \Big\langle x - y, \sum_{i=1}^N f_i(x) u_i(t) - \sum_{i=1}^N f_i(y) u_i(t) \Big\rangle \\
&= \sum_{i=1}^N u_i(t) \langle x - y, f_i(x) - f_i(y) \rangle \\
&\leq - \sum_{i=1}^N u_i(t) L_i \|x - y\|^2 \\
&\leq -L_* \|x - y\|^2 \sum_{i=1}^N u_i(t) = -L_* \|x - y\|^2,
\end{aligned}$$

since $\sum_{i=1}^N u_i(t) \equiv 1$. The switching system is thus *extremely stable* in the terminology of YOSHIZAWA [**239**], which is known as *incrementally stable* by engineers (see ANGELI [**5**]), i.e., with

$$\|x_u(t) - y_u(t)\| \leq e^{-L_*(t - t_0)} \|x_u(t_0) - y_u(t_0)\| \to 0 \quad \text{as } t \to \infty \tag{13.7}$$

for any two solutions of the switching system. Moreover, this is uniform in the choice of switching control. It is thus uniformly asymptotically stable in the sense of Theorem 3.47, i.e., uniformly in $t_0 \in \mathbb{R}$.
The theory of pullback attractors can be used to determine to what these solutions converge. By taking $y = 0$ in (13.6) and defining

$$K_* := \max\left\{\|f_1(0)\|, \ldots, \|f_N(0)\|\right\} \geq 0,$$

it follows that the switching system satisfies the dissipativity condition

$$\left\langle x, f_u(t,x)\right\rangle \leq \frac{2K_*^2}{L_*} - \frac{L_*}{2}\|x\|^2$$

and thus has a positive invariant absorbing set $\overline{B_{R^*}(0)}$, the closed ball of radius $R^* := 1 + 4K_*^2/L_*^2$ about the origin, which is uniform in the switching control. The associated skew product flow thus has a pullback attractor $\mathcal{A}$. The fibers A_u of this pullback attractor are in fact singleton sets (this can be proved as in Theorem 3.21). Write $A_u := \{\bar{x}_u\}$, and define $\bar{x}_u(t) := \bar{x}_{\theta_t(u)}$ for all $t \in \mathbb{R}$ and $u \in \mathcal{S}_\tau$. It is easy to see that $\bar{x}_u(\cdot)$ is the entire solution of the switching system (13.3) with the initial value $\bar{x}_u(0)$ and switching control u. Using the above results on pullback attractors, one concludes that the pullback attractor here, given by the solutions $\bar{x}_u(\cdot)$, are asymptotically Lyapunov stable uniformly in the switching controls. Alternatively, one can show this directly through the inequality (13.7) taking $\bar{x}_u(\cdot)$ as one of the solutions, i.e.,

$$\|x_u(t) - \bar{x}_u(t)\| \leq e^{-L^*(t-t_0)}\|x_u(t_0) - \bar{x}_u(t_0)\| \to 0 \quad \text{as } t \to \infty.$$

This completes the proof of this theorem. □

EXAMPLE 13.12. Consider the switching of linear scalar systems

$$\dot{x} = -x + a_i \quad \text{for } i \in \{1, \ldots, N\},$$

which satisfy the one-sided dissipativity condition (13.6) with $L_i = 1$. Here a_i is the Lyapunov asymptotically stable steady state solution of the i-th system. Since $\sum_{i=1}^N u_i(t) \equiv 1$, a switching system here has the form

$$\dot{x} = -x + \sum_{i=1}^N u_i(t) a_i \quad \text{for } i \in \{1, \ldots, N\},$$

which is linear and thus solvable with an integrating factor to give the solution

$$x_u(t, t_0, x_0) = e^{-(t-t_0)} x_0 + e^{-t} \sum_{i=1}^N a_i \int_{t_0}^t e^s u_i(s)\, ds$$

for an initial value x_0 at time t_0 (this is the same as the solution starting at time 0 with the switching control $\theta_{t_0}(u(\cdot)) = u(\cdot + t_0)$). Taking $t_0 \to -\infty$ with t held fixed, one obtains the pullback limit

$$\bar{x}_u(t) := e^{-t} \sum_{i=1}^N a_i \int_{-\infty}^t e^s u_i(s)\, ds \quad \text{for all } t \in \mathbb{R}.$$

This bounded entire solution of the switching system is Lyapunov asymptotically stable. The pullback attractor here has singleton fibers $A_u = \{\bar{x}_u(0)\}$ and is also a forward attractor uniformly in the switching controls.

1.6. Different dwell times. The proof of the next lemma follows directly from the definitions.

LEMMA 13.13. *Let (θ, φ) be a skew product flow on $P \times \mathbb{R}^d$ with a forward (respectively, pullback) attractor $\mathcal{A}$, and let Q be a nonempty closed subset of P which is θ-invariant, i.e., $\theta_t(Q) = Q$ for all $t \in \mathbb{R}$. Then $\mathcal{A}^Q = \mathcal{A} \cap (Q \times \mathbb{R}^d)$ is a forward (respectively, pullback) attractor of the restriction of (θ, φ) to $Q \times \mathbb{R}^d$.*

It follows from this, for example, that Theorem 13.7 also holds for any θ-invariant closed subset Q of P such as $Q = H(\bar{u})$, the hull of a specific switching control $\bar{u}$ in $\mathcal{S}_\tau$. In addition, Lemma 13.13 allows one to consider pullback attractors for switching systems with different dwell times, since $\mathcal{S}_\tau$ is θ-invariant and $\mathcal{S}_\tau \subset \mathcal{S}_{\tau'}$ for $\tau' < \tau$.

THEOREM 13.14. *Suppose that the switching system (θ, φ) on $\mathcal{S}_{\tau'} \times \mathbb{R}^d$ has a pullback attractor $\mathcal{A}^{\tau'} = (A_u)_{u \in \mathcal{S}_{\tau'}}$. Then the restriction of (θ, φ) to $\mathcal{S}_\tau \times \mathbb{R}^d$ has a pullback attractor $\mathcal{A}^\tau = (A_u)_{u \in \mathcal{S}_\tau}$ for any $\tau > \tau'$.*

The set of switching controls for all dwell times, $\mathcal{S}^* = \bigcup_{\tau > 0} \mathcal{S}_\tau$, is not complete under the metric d defined by (13.4). However, if the un-switched systems (13.1) satisfy a common dissipativity condition such as (13.5), then the switching systems have a common absorbing set which is uniformly absorbing in the controls $u \in \mathcal{S}_\tau$ and the dwell times $\tau > 0$. Thus, the switching system (θ, φ) on $\mathcal{S}_\tau \times \mathbb{R}^d$ has a pullback attractor $\mathcal{A}^\tau$ for each $\tau > 0$ with $\mathcal{A}^\tau \subset \mathcal{A}^{\tau'}$ for all $\tau' < \tau$. One could then interpret $\mathcal{A}^* := \bigcup_{\tau > 0} \mathcal{A}^\tau$ to be the pullback attractor of the switching system (θ, φ) on $\mathcal{S}^* \times \mathbb{R}^d$, without restriction on the dwell time. However, some care is needed here with sequences of switching controls of the form $u_n \in \mathcal{S}_{\tau_n}$ with $\tau_n \to 0$. To handle switching controls in $\mathcal{S} \setminus \mathcal{S}^*$, one could use the weaker and less intuitive weak* topology on $\mathcal{S}$ and its corresponding pullback attractors. This topology is used on the space of open loop controls in the affine control systems, which will be considered in the next section.

1.7. How realistic is pullback convergence? An advantage of pullback convergence is that it provides a means for constructing the component sets of a pullback attractor. This obviously requires knowledge of the past history of the system. In some important subclasses of switching controls such as the periodic and almost periodic switching controls considered in Lemmata 13.9 and 13.10, this past history is known.

There are, however, many modeling situations for which the past history is not known. Instead the system is known – in fact, prescribed – on future time intervals of the form $[t_0, \infty)$. Forward attractors can be easily modified to these situations. Pullback attractors can also be used with an invented past history, but the pullback attractor will then depend on which "history" is chosen and it is not clear which history should be used.

The generalization of the theory of nonautonomous dynamical systems to nonautonomous semi-dynamical systems as proposed in Chapter 10 offers some insight here. The driving system θ on P is then assumed to be only a semi-dynamical system, but the corresponding skew product flow (θ, φ) is still an autonomous semi-dynamical system on $P \times \mathbb{R}^d$. Suppose that P is compact and that the skew product flow has a global autonomous attractor $\mathcal{A} \subset P \times \mathbb{R}^d$ with cross-section structure $\mathcal{A} = \bigcup_{p \in P} \{p\} \times A_p$. Then the family $\mathcal{A} = (A_p)_{p \in P}$ of nonempty compact subsets of $\mathbb{R}^d$ forms a pullback attractor with pullback convergence modified to

$$\lim_{t \to \infty} \text{dist}\left(\varphi(t, \Theta_{-t}(p), x_0), A_p\right) = 0,$$

where $\Theta_{-t}(p) := \{q \in P : \theta_t(q) = p\}$ for each $t \geq 0$ and $p \in P$. It was seen in Chapter 9 that $F(t, p) := \Theta_{-t}(p)$ for all $t \geq 0$ and $p \in P$ defines an autonomous set-valued dynamical system on P. Essentially, it contains all possible past histories of the driving system and the modified pullback attractor component sets are the accumulative effect of all of these past histories. In this sense, pullback attraction is still meaningful and useful for nonautonomous systems defined only for future time.

These observations can be applied to switching systems with $P = \mathcal{S}_\tau^+$ consisting of piecewise continuous switching controls $u : \mathbb{R}_0^+ \to \{e_1, \ldots, e_N\}$ with dwell time τ and the forward shift operator θ.

2. Affine control systems

Switching systems were formulated as special cases of affine control systems. The general form of an affine control system in $\mathbb{R}^d$ is

$$\dot{x} = f_0(x) + \sum_{i=1}^{m} u_i f_i(x), \qquad u(t) = (u_1(t), \ldots, u_m(t)) \in U, \tag{13.8}$$

where U is a compact convex subset of $\mathbb{R}^m$ with $0 \in \operatorname{int} U$, and the coefficient functions $f_0, \ldots, f_m : \mathbb{R}^d \to \mathbb{R}^d$, which may be nonlinear, are assumed to be regular enough to ensure existence and uniqueness of solutions globally both forwards and backwards in time. This happens for bilinear affine control systems, for which $f_i(x) = A_i x$ for $i \in \{1, \ldots, m\}$.

Denote the space of (time-varying) open loop controls with values in U by

$$\mathcal{U} = \{u : \mathbb{R} \to U : u \text{ is measurable}\} \subset L^\infty(\mathbb{R}, \mathbb{R}^m), \tag{13.9}$$

and write the solution of (13.8) with the initial value $x \in \mathbb{R}^d$ and control $u \in \mathcal{U}$ as $\varphi(t, u, x)$ for $t \in \mathbb{R}$, i.e., $\varphi(0, u, x) = x$.

The system (13.8) is nonautonomous due to the time variation of the controls, so does not itself define an autonomous dynamical system on $\mathbb{R}^d$. However, it does form a skew product flow when combined with the shift operator on the space of controls.

2.1. Affine control systems as skew product flows.

Consider the shift operator $\theta : \mathbb{R} \times \mathcal{U} \to \mathcal{U}$ on the space $\mathcal{U}$ of control functions, which is defined by $\theta(t, u(\cdot)) = u(t + \cdot)$ and will often be written $\theta_t(u(\cdot))$ or $\theta_t(u)$. The following result is proved in COLONIUS & KLIEMANN [**53**, Lemmata 4.2.1 and 4.2.4].

LEMMA 13.15. *Let $\mathcal{U}$ be the space of open loop controls defined by (13.9) for a compact convex subset U of $\mathbb{R}^m$. Then there exists a metric d on $\mathcal{U}$ such that*

(i) *the topology generated by this metric on $\mathcal{U}$ coincides with the restriction of the weak* topology on $L^\infty(\mathbb{R}, \mathbb{R}^m)$, which is dual to $L^1(\mathbb{R}, \mathbb{R}^m)$,*

(ii) $(\mathcal{U}, d)$ *is a compact metric space, and*
(iii) *the shift* $\theta : (t, u) \mapsto u(t + \cdot)$ *defines continuous autonomous dynamical system.*

The *control flow* $\Phi : \mathbb{R} \times \mathcal{U} \times \mathbb{R}^d \to \mathcal{U} \times \mathbb{R}^d$ is then defined by

$$\Phi(t, u, x) = \big(\theta_t(u), \varphi(t, u, x)\big)\,, \tag{13.10}$$

assuming that the solutions of the the affine control system (13.8) exist for all time.

THEOREM 13.16. *A control flow* Φ *is a time-reversible skew product flow on* $\mathcal{U} \times \mathbb{R}^d$.

EXERCISE 13.17. Show that the control flow Φ satisfies the cocycle property.

The main part of the proof of Theorem 13.16 is to show the continuity of the control flow Φ.

THEOREM 13.18 (Continuity of the control flow). *The mapping* $(t, u, x) \mapsto \Phi(t, u, x)$ *is continuous.*

PROOF. It is necessary to prove the following assertion for fixed $u_* \in \mathcal{U}$, $x_* \in \mathbb{R}^d$ and $t_* \in \mathbb{R}$. For all $\varepsilon > 0$, there exists a $\delta > 0$ such that

$$\|\varphi(t, x, u) - \varphi(t_*, x_*, u_*)\| < \varepsilon\,,$$

whenever $\max\big\{|t - t_*|, \|x - x_*\|, d(u, u_*)\big\} < \delta$. For all $t \in \mathbb{R}$, $x \in \mathbb{R}^d$ and $u \in \mathcal{U}$ the triangle inequality yields

$$\begin{aligned}&\|\varphi(t, x, u) - \varphi(t_*, x_*, u_*)\| \\ &\le \|\varphi(t, x, u) - \varphi(t, x_*, u_*)\| + \|\varphi(t, x_*, u_*) - \varphi(t_*, x_*, u_*)\|\,.\end{aligned}$$

The second summand tends to zero in the limit $t \to t_*$ by the continuity of $\varphi(\cdot, x_*, u_*)$, so it remains to prove that the first summand also becomes arbitrarily small if $\max\big\{|t - t_*|, \|x - x_*\|, d(u, u_*)\big\} < \delta$ for a sufficiently small δ. It can be assumed without loss of generality that $t_* > 0$. Since $\varphi\big([0, t_* + c], x_*, u_*\big)$ is compact for $c > 0$, there exits an open set $V \subset \mathbb{R}^d$ with compact closure such that

$$\varphi(t, x_*, u_*) \in V \quad \text{for all } t \in I := [0, t_* + c]\,.$$

Thus, there exists a C^1-function $\rho : \mathbb{R}^d \to [0, 1]$ with compact support, $\operatorname{supp}\rho$, such that

$$\rho(x) = 1 \quad \text{for all } x \in V\,.$$

Define C^1-vector fields $g_i : \mathbb{R}^d \to \mathbb{R}^d$ by $g_i(x) := \rho(x) f_i(x)$ for all $i \in \{0, 1, \ldots, m\}$, and set

$$g(x, u) := g_0(x) + \sum_{i=1}^{m} u_i g_i(x) = \theta(x) f(x, u)\,.$$

The functions $g_0, g_1, \ldots, g_m$ and $g(\cdot, u)$ are globally Lipschitz continuous in x (uniformly in u in the last case). Hence, all solutions $\psi(t, x, u)$ of the control system $\dot{x}(t) = g(x(t), u(t))$ are defined globally. Comparing

$$\xi(t) := \varphi(t, x_*, u_*) = \psi(t, x_*, u_*) \quad \text{for all } t \in [0, t_* + c]\,,$$

and $\eta_{x,u}(t) := \psi(t, x, u)$ for arbitrarily chosen $x \in \mathbb{R}^d$ and $u \in \mathcal{U}$ gives

$$\begin{aligned}
&\xi(t) - \eta_{x,u}(t) \\
&= (x_* - x) + \int_0^t \big(g(\xi(\tau), u_*(\tau)) - g(\eta_{x,u}(\tau), u(\tau))\big)\mathrm{d}\tau \\
&= (x_* - x) + \int_0^t \big(g(\xi(\tau), u_*(\tau)) - g(\eta_{x,u}(\tau), u_*(\tau))\big)\mathrm{d}\tau \\
&\quad + \int_0^t \sum_{i=1}^m (u_i(\tau) - u_{*i}(\tau))\big(g_i(\xi(\tau)) - g_i(\eta_{x,u}(\tau)\big)\,\mathrm{d}\tau \\
&\quad - \int_0^t \sum_{i=1}^m \big(u_i(\tau) - u_{*i}(\tau)\big) g_i(\xi(\tau))\mathrm{d}\tau\,.
\end{aligned}$$

The second summand here satisfies

$$\left\| \int_0^t \big(g(\xi(\tau), u_*(\tau)) - g(\eta_{x,u}(\tau), u_*(\tau))\big)\,\mathrm{d}\tau \right\| \le L \int_0^t \|\xi(\tau) - \eta_{x,u}(\tau)\|\,\mathrm{d}\tau\,,$$

where $L \ge 0$ is the global Lipschitz constant for g, for instance, given by the supremum of $\big\|\frac{\partial g}{\partial x}(x, u)\big\|$ over the compact range $\operatorname{supp}\rho \times U \subset \mathbb{R}^d \times \mathbb{R}^m$. The norm of the third summand is bounded above by

$$\begin{aligned}
&\tilde{L} \int_0^t \Big(\sum_{i=1}^m |u_i(\tau) - u_{*i}(\tau)|\Big) \|\xi(\tau) - \eta_{x,u}(\tau)\|\,\mathrm{d}\tau \\
&\le B\tilde{L} \int_0^t \|\xi(\tau) - \eta_{x,u}(\tau)\|\mathrm{d}\tau\,,
\end{aligned}$$

where $\tilde{L} \ge 0$ is the maximum of the global Lipschitz constants for $g_1, \dots, g_m$ and B is defined as

$$B := \sup_{u^1, u^2 \in U} \sum_{i=1}^m \big|u_i^1 - u_i^2\big|\,.$$

Here $B < \infty$ and $\operatorname{ess\,sup}_{\tau \in \mathbb{R}} \sum_{i=1}^m \|u_i(\tau) - u_{*i}(\tau)\| \le B$ since U is compact. Finally, the fourth summand tends to zero as $u \to u_*$, since the metric generates the weak* topology on $\mathcal{U}$, i.e., for every $\gamma > 0$, there is a $\delta > 0$ such $d(u, u_*) < \delta$ implies

$$\left\| \int_0^t \sum_{i=1}^m (u_i(\tau) - u_{*i}(\tau)) g_i(\xi(\tau))\,\mathrm{d}\tau \right\| < \gamma\,.$$

Combining these estimates gives

$$\|\xi(t) - \eta_{x,u}(t)\| \le (\delta + \gamma) + \big(L + \tilde{L}B\big) \int_0^t \|\xi(\tau) - \eta_{x,u}(\tau)\|\,\mathrm{d}\tau$$

for $\max\big\{\|x - x_*\|, d(u, u_*)\big\} < \delta$ and small enough δ. Since $\|\xi(\cdot) - \eta_{x,u}(\cdot)\|$ is continuous and nonnegative, the Gronwall inequality can be applied, and one obtains

$$\|\xi(t) - \eta_{x,u}(t)\| \le (\delta + \gamma) e^{t(L + \tilde{L}B)}\,.$$

Hence,

$$\|\varphi(t, x_*, u_*) - \psi(t, x, u)\| = \|\xi(t) - \eta_{x,u}(t)\| \le (\delta + \gamma) e^{(t^* + c)(L + \tilde{L}B)} =: C(\delta)$$

for all $t \in I$. Given $\varepsilon > 0$, one can choose δ small enough such that $C(\delta) < \varepsilon$ and $|t - t_*| < \delta$ imply that $t \in I$. Moreover, if ε is chosen small enough, then $\psi(t, x, u) \in V$ for all $t \in I$. Hence,

$$\psi(t, x, u) = x + \int_0^t g(\psi(\tau, x, u), u(\tau))\,\mathrm{d}\tau = x + \int_0^t f(\psi(\tau, x, u), u(\tau))\,\mathrm{d}\tau\,,$$

from which it follows that

$$\psi(t, x, u) = \varphi(t, x, u) \quad \text{for all } t \in I\,.$$

This finishes the proof of this theorem. □

Endnotes. The monograph of Liberzon [**157**] provides an overview of switching systems from the engineering perspective, see also Angeli [**5**], while Kloeden & Kozyakin [**127**] was motivated by issues in communication networks and scientific computing. The results on switching systems in this chapter are based mainly on the paper Kloeden [**121**], where a proof for Theorem 13.1 can be found, see also Boularas & Cheban [**24**]. See Yoshizawa [**239**] for Lyapunov functions and Molchanov & Pyatnitskiy [**168**] for stability criteria for inclusion difference and differential equation generated by control and switching systems. The section on affine control systems is based on the monograph of Colonius & Kliemann [**53**], the proof of Theorem 13.18 used here is due to Kawan [**106**].

CHAPTER 14

Random dynamical systems

A random dynamical system is very similar to a skew product flow except the driving system is a metric dynamical system (this terminology comes from ergodic theory) on a measure space and is measurable rather than continuous in the corresponding variable. This system represents the noise which drives the system, while a cocycle mapping represents the dynamics in the state space.

Let (X, d) be a complete metric space and $(\Omega, \mathcal{F}, \mathbb{P})$ be a probability space, i.e., with sample space Ω, a σ-algebra $\mathcal{F}$ of admissible events (measurable subsets of Ω) and a probability measure $\mathbb{P}$ on $\mathcal{F}$. Let $\mathbb{T} = \mathbb{R}$ or $\mathbb{Z}$, depending on whether a continuous time or discrete time system is being considered.

DEFINITION 14.1 (Random dynamical system). A *random dynamical system* (θ, φ) on $\Omega \times X$ consists of a metric dynamical system θ on Ω, i.e., a group family of measure preserving transformations $\theta_t : \Omega \to \Omega$, $t \in \mathbb{T}$, such that

(i) $\theta_0 = \text{id}$ and $\theta_t \circ \theta_s = \theta_{t+s}$ for all $t, s \in \mathbb{T}$,
(ii) the map $(t, \omega) \mapsto \theta_t(\omega)$ is measurable and invariant with respect to $\mathbb{P}$ in the sense that $\theta_t(\mathbb{P}) = \mathbb{P}$ for all $t \in \mathbb{T}$,

and a cocycle mapping $\varphi : \mathbb{T}_0^+ \times \Omega \times X \to X$ such that

(a) $\varphi(0, \omega, x_0) = x_0$ for all $x_0 \in X$ and $\omega \in \Omega$,
(b) $\varphi(s + t, \omega, x_0) = \varphi\big(s, \theta_t(\omega), \varphi(t, \omega, x_0)\big)$ for all $s, t \in \mathbb{T}_0^+$, $x_0 \in X$ and $\omega \in \Omega$,
(c) $(t, x_0) \mapsto \varphi(t, \omega, x_0)$ is continuous for each $\omega \in \Omega$,
(d) $\omega \mapsto \varphi(t, \omega, x_0)$ is $\mathcal{F}$-measurable for all $(t, x_0) \in \mathbb{T}^+ \times X$.

The notation $\theta_t(\mathbb{P}) = \mathbb{P}$ for the measure preserving property of θ_t with respect to $\mathbb{P}$ is just a compact way of writing

$$\mathbb{P}(\theta_t(A)) = \mathbb{P}(A) \quad \text{for all } t \in \mathbb{R} \text{ and } A \in \mathcal{F}.$$

A systematic treatment of the theory of random dynamical system theory is given in ARNOLD [**6**]. The reader is referred to ARNOLD [**6**] and KLOEDEN & PLATEN [**138**] for background material on stochastic processes and stochastic differential equations.

Ordinary differential equations with random coefficients or stochastic processes in their coefficients and stochastic differential equations with existence and uniqueness of global solutions generate continuous time random dynamical systems, while random difference equations generate discrete time random dynamical systems.

A simple example of random dynamical system is given by the random ordinary differential equation (RODE) is

$$\dot{x} = -x + W_t(\omega),$$

where W_t is a Wiener process. It is solved pathwise, i.e., for each $\omega \in \Omega$, as a deterministic ordinary differential equation and has the explicit solution

$$x(t) = x(t, \omega, x_0) := x_0 e^{-t} + e^{-t} \int_0^t e^s W_s(\omega) ds$$

with initial value x_0 at time $t = 0$ for the sample path $\omega \in \Omega$. These solutions define the cocycle mapping φ, specifically $\varphi(t, \omega, x_0) := x(t, \omega, x_0)$.

The driving system θ is not given directly in terms of the Wiener process W_t, but is defined in terms of shift operators θ_t on the canonical sample space $\Omega := C_0(\mathbb{R}, \mathbb{R})$ of continuous functions $\omega : \mathbb{R} \to \mathbb{R}$ with $\omega(0) = 0$, i.e., with $\theta_t(\omega(\cdot)) := \omega(t + \cdot) - \omega(\cdot)$ for all $t \in \mathbb{R}$. The σ-algebra of Borel subsets of $C_0(\mathbb{R}, \mathbb{R})$ is taken as the σ-algebra of events $\mathcal{F}$ and $\mathbb{P}$ is the corresponding Wiener measure. Note that no other topological properties of the space $C_0(\mathbb{R}, \mathbb{R})$ are used here apart from those defining the Borel sets (see ARNOLD **[6]** for more details).

A similar driving system is used in random dynamical systems generated by stochastic differential equations, some examples of which will be given later in the chapter.

1. Random attractors

A random dynamical system (θ, φ) on $\Omega \times X$ is not a skew product flow or an autonomous semi-dynamical system on $\Omega \times X$, since Ω (in the context considered here) is not a topological space. Nevertheless, skew product flows and random dynamical systems have many common properties, and concepts and results for one can be used with appropriate modifications for the other. The most significant modification concerns measurability and the nonautonomous sets under consideration are random sets.

DEFINITION 14.2. Let (X, d) be a complete and separable metric space (i.e., a Polish space).

(i) A family $\mathcal{D} = (D_\omega)_{\omega \in \Omega}$ of nonempty subsets of X is called a *random set* if the mapping $\omega \mapsto \mathrm{dist}(x, D_\omega)$ is $\mathcal{F}$-measurable for all $x \in X$.

(ii) A random set $\mathcal{D}$ is called a *random closed set* if D_ω is closed for each $\omega \in \Omega$ and a *random compact set* if D_ω is compact for each $\omega \in \Omega$.

(iii) A random set $\mathcal{D}$ is said to be *tempered* if there exists a $x_0 \in X$ such that

$$D_\omega \subset \{x \in X \,:\, d(x, x_0) \le r(\omega)\} \quad \text{for all } \omega \in \Omega,$$

where the random variable $r(\omega) > 0$ is tempered, i.e.,

$$\sup_{t \in \mathbb{R}} \left\{ r(\theta_t(\omega)) e^{-\gamma|t|} \right\} < \infty \quad \text{for all } \gamma > 0 \text{ and } \omega \in \Omega.$$

The collection of all tempered random sets in X will be denoted by $\mathfrak{D}$.

A random attractor of a random dynamical system is essentially a pullback attractor, which is a random set.

DEFINITION 14.3 (Random attractor). A random compact set $\mathcal{A} = (A_\omega)_{\omega\in\Omega}$ from $\mathfrak{D}$ is called a *random attractor* of a random dynamical system (θ, φ) on $\Omega \times X$ in $\mathcal{D}$ if $\mathcal{A}$ is a φ-invariant set, i.e.,

$$\varphi(t, \omega, A_\omega) = A_{\theta_t(\omega)} \quad \text{for all } t \geq 0 \text{ and } \omega \in \Omega,$$

and pathwise pullback attracting in $\mathfrak{D}$, i.e.,

$$\lim_{t\to\infty} \operatorname{dist}\big(\varphi(t, \theta_{-t}(\omega), D_{\theta_{-t}(\omega)}), A_\omega\big) = 0 \quad \text{for all } \omega \in \Omega \text{ and } \mathcal{D} \in \mathfrak{D}.$$

If the random attractor consists of singleton sets, i.e., $A_\omega = \{X^*(\omega)\}$ for some random variable X^* with $X^*(\omega) \in X$, then $\bar{X}_t(\omega) := X^*(\theta_t(\omega))$ is a stationary stochastic process on X.

THEOREM 14.4 (Existence of random attractors). *Let (θ, φ) be a random dynamical system on $\Omega \times X$ such that $\varphi(t, \omega, \cdot) : X \to X$ is a compact operator for each fixed $t > 0$ and $\omega \in \Omega$. If there exists a tempered random set $\mathcal{B} = (B_\omega)_{\omega\in\Omega}$ with closed and bounded component sets and a $T_{\mathcal{D},\omega} \geq 0$ such that*

$$\varphi\big(t, \theta_{-t}(\omega), D_{\theta_{-t}(\omega)}\big) \subset B_\omega \quad \textit{for all } t \geq T_{\mathcal{D},\omega} \tag{14.1}$$

for every tempered random set $\mathcal{D} = (D_\omega)_{\omega\in\Omega}$, then the random dynamical system (θ, φ) has a random pullback attractor $\mathcal{A} = (A_\omega)_{\omega\in\Omega}$ with the component sets defined for each $\omega \in \Omega$ by

$$A_\omega = \bigcap_{s>0} \overline{\bigcup_{t\geq s} \varphi(t, \theta_{-t}(\omega), B_{\theta_{-t}(\omega)})}\,. \tag{14.2}$$

The tempered random set $\mathcal{B} = (B_\omega)_{\omega\in\Omega}$ here is called a pullback absorbing random set.

The proof of Theorem 14.4 is essentially the same as its counterparts for deterministic skew product flows, Theorems 3.20 and 3.27 as well as Theorem 12.14 and Corollary 12.15. The only new feature is that of measurability, i.e., to show that $\mathcal{A} = (A_\omega)_{\omega\in\Omega}$ is a random set. This follows from the fact that the set-valued mappings $\omega \mapsto \varphi\big(t, \theta_{-t}(\omega), B_{\theta_{-t}(\omega)}\big)$ are measurable for each $t \in \mathbb{T}_0^+$ and that the union and intersection in (14.2) can be taken over a countable number of time instants (in the continuous time case).

EXERCISE 14.5. Show that the set-valued mapping $\omega \mapsto A_\omega$ defined by (14.2) is measurable and hence $\mathcal{A}$ is a random set.

It was shown in ARNOLD & SCHMALFUSS **[8]** that a random attractor is also a *forward* attractor in the weaker sense of convergence in probability, i.e.,

$$\lim_{t\to\infty} \int_\Omega \operatorname{dist}\big(\varphi(t, \omega, D_\omega), A_{\theta_t(\omega)}\big)\, \mathbb{P}(d\omega) = 0 \quad \text{for all } \mathcal{D} \in \mathfrak{D}.$$

This allows individual sample paths to have large deviations from the attractor, but for all to converge in this probabilistic sense.

2. The Ornstein–Uhlenbeck process

The Ornstein–Uhlenbeck process provides a simple example of a random attractor, which can be computed explicitly. It is the unique stochastic stationary solution a linear Itô stochastic differential equation (SDE) with additive noise and plays an important role in investigating random attractors in nonlinear SDE.

Specifically, consider the SDE with linear drift and additive noise,

$$\mathrm{d}X_t = -X_t \mathrm{d}t + \alpha \mathrm{d}W_t \,, \tag{14.3}$$

which has the explicit solution

$$X_t = X_{t_0} e^{-(t-t_0)} + \alpha e^{-t} \int_{t_0}^{t} e^s \, \mathrm{d}W_s \,.$$

Assuming that W_t is a two-sided Wiener process, i.e., defined for all $t \in \mathbb{R}$ and taking the pathwise pullback limit as $t_0 \to -\infty$ gives the Ornstein–Uhlenbeck process

$$O_t = \alpha e^{-t} \int_{-\infty}^{t} e^s \, \mathrm{d}W_s \,, \tag{14.4}$$

which is a solution of the SDE (14.3). It follows from Lemma 14.6 below that the Ornstein–Uhlenbeck process is the unique *stochastic stationary solution* of the SDE (14.3).

The following lemma is due to CARABALLO, KLOEDEN & SCHMALFUSS [**31**].

LEMMA 14.6. *There exists a θ-invariant subset $\Omega^* \in \mathcal{F}$ of $\Omega := C_0(\mathbb{R}, \mathbb{R})$ of full measure such that*

$$\lim_{t \to \pm\infty} \frac{1}{t} |\omega(t)| = 0 \quad \textit{for all } \omega \in \Omega^* \,,$$

and there exist a random variable O^* such that

$$O^*(\theta_t(\omega)) = O_t(\omega) \quad \textit{for all } \omega \in \Omega^* \textit{ and } t \in \mathbb{R} \,.$$

Moreover,

$$\lim_{t \to \pm\infty} \frac{1}{t} \int_0^t O^*(\theta_\tau(\omega)) \, \mathrm{d}\tau = 0 \quad \textit{for all } \omega \in \Omega^* \,.$$

For simplicity, and without loss of generality, the set Ω^* will be assumed to be the whole sample space Ω in the sequel.

REMARK 14.7. Here, and in the examples below, the cocycle mapping is defined by $\varphi(t, \omega, x_0) := X_t^{x_0}(\omega)$, where $X_t^{x_0}$ is the solution of the SDE with initial value x_0 at time $t = 0$, and θ_t is the shift operator on the canonical sample space mentioned in the introduction section to this chapter. Analogous results also hold for multi-dimensional SDE.

It is not difficult to show that the mapping φ defined by $\varphi(t, \omega, x_0) := X_t^{x_0}(\omega)$ satisfies the cocycle property (b) in Definition 14.1 for all ω except those in a null event $N_{s,t} \in \mathcal{F}$, i.e., with $\mathbb{P}(N_{s,t}) = 0$, for given $s, t \in \mathbb{T}_0^+$. This gives the *crude* cocycle property. However, Definition 14.1 requires the *perfect* cocycle property, for which the relationship (b) holds for all ω except those in a null event N which is independent of s, $t \in \mathbb{T}_0^+$. The original sample space Ω is then replaced by $\Omega^* := \Omega \setminus N$. It is much harder to show that the perfect cocycle property holds, see ARNOLD & SCHEUTZOW [**7**].

3. Random attractors for stochastic differential equations

Stochastic differential equations are, in fact, not differential equations at all, but rather integral equations and require the stochastic calculus for their treatment rather than the familiar deterministic calculus. The sample paths of their solutions, like those of the driving Wiener process, are not differentiable, so in order to use dissipative condition like the one-sided dissipative Lipschitz condition (14.5) below, these stochastic differential equations will be transformed into random ordinary differential equations (RODEs), for which the sample paths are differentiable are methods of deterministic calculus can be used. This kind of transformation goes back to DOSS [**72**] and SUSSMAN [**227, 228**].

The general form of a scalar stochastic differential equation is

$$\mathrm{d}X_t = f(t, X_t)\mathrm{d}t + g(t, X_t) \star \mathrm{d}W_t \,,$$

where f is called the drift coefficient and g the diffusion coefficient. In addition, W_t is a scalar Wiener process, which is assumed here to be two-sided, i.e., defined for all $t \in \mathbb{R}$. This is only a symbolic representation of the stochastic integral equation

$$X_t = X_{t_0} + \int_{t_0}^{t} f(s, X_s)\,\mathrm{d}s + \int_{t_0}^{t} g(s, X_s) \star \mathrm{d}W_s \,.$$

There are two types of stochastic differential equations, due to Itô and Stratonovich, which depend on how the stochastic integral is defined. The symbol $\star$ is omitted in the Itô case and is written $\circ$ in the Stratonovich case. Itô and Stratonovich SDE have the same solutions for additive noise, but not for multiplicative noise. See ARNOLD [**6**] and KLOEDEN & PLATEN [**138**] for definitions and more details.

In the two examples of random dynamical systems generated by scalar stochastic differential equations considered below, the drift coefficient $f : \mathbb{R} \to \mathbb{R}$ is assumed to be continuously differentiable and to satisfy the one-sided dissipative Lipschitz condition

$$\big\langle x_1 - x_2, f(x_1) - f(x_2)\big\rangle \le -L|x_1 - x_2|^2 \quad \text{for all } x_1, x_2 \in \mathbb{R}\,, \tag{14.5}$$

as well as the

Integrability condition. *There exists $m_0 > 0$ such that*

$$\int_{-\infty}^{t} e^{ms}|f(u(s))|^2\,\mathrm{d}s < \infty \tag{14.6}$$

holds for every $m \in (0, m_0]$ and any continuous function $u : \mathbb{R} \to \mathbb{R}^d$ with sub-exponential growth.

It can be assumed without loss of generality that $L \le m_0$.

It follows from this integrability condition and the one-sided dissipative Lipschitz condition that these examples have random attractors consisting of singleton subsets and is also pathwise attracting in the forward sense, i.e., the random attractor is, in fact, pathwise asymptotically stable stochastic stationary solution.

3.1. A nonlinear SDE with additive noise. Consider a nonlinear scalar SDE with additive noise

$$\mathrm{d}X_t = f(X_t)\mathrm{d}t + \alpha \mathrm{d}W_t \,, \tag{14.7}$$

where α is a positive constant and the drift coefficient $f : \mathbb{R} \to \mathbb{R}$ is continuously differentiable and satisfies the one-sided dissipative Lipschitz condition (14.5) as well as the integrability condition (14.6).

The SDE (14.7) is really the integral equation

$$X_t = X_0 + \int_0^t f(X_s)\,\mathrm{d}s + \alpha W_t\,.$$

Its solution paths are not differentiable, so in order to use the one-sided dissipative Lipschitz condition (14.5), consider the difference $X_t - O_t$, where O_t is the Ornstein–Uhlenbeck stationary process (14.4) satisfying the linear stochastic differential equation (14.3). This difference is pathwise differentiable since the paths are continuous and satisfy the integral equation

$$X_t - O_t = X_0 - O_0 + \int_0^t (f(X_s) + O_s)\,\mathrm{d}s$$

and are differentiable by the Fundamental Theorem of Calculus for this reason. This integral equation is thus equivalent to the pathwise random ordinary differential equation (RODE)

$$\dot{x} = f(x + O_t) + O_t\,,$$

where $x(t,\omega) := X_t(\omega) - O_t(\omega)$, or simply,

$$\frac{\mathrm{d}}{\mathrm{d}t}(X_t - O_t) = f(X_t) + O_t$$

for each $\omega \in \Omega$.

Now take the inner product with $X_t - O_t$ and apply the one-sided Lipschitz condition to this RODE to obtain

$$\begin{aligned}\frac{\mathrm{d}}{\mathrm{d}t}|X_t - O_t|^2 &= 2\big\langle X_t - O_t, f(X_t) - f(O_t)\big\rangle + 2\big\langle X_t - O_t, f(O_t) + O_t\big\rangle\\ &\le -2L|X_t - O_t|^2 + L|X_t - O_t|^2 + \frac{4}{L}|f(O_t) + O_t|^2\,.\end{aligned}$$

Hence,

$$|X_t - O_t|^2 \le |X_{t_0} - O_{t_0}|^2 e^{-L(t-t_0)} + \frac{4e^{-Lt}}{L}\int_{t_0}^t e^{Ls}|f(O_s) + O_s|^2\,\mathrm{d}s\,.$$

Pathwise pullback convergence (i.e., as $t_0 \to -\infty$) then gives pullback absorption (cf. Theorem 14.4)

$$|X_t - O_t|^2 \le R_X^2(\theta_t(\omega)) := 1 + \frac{4e^{-Lt}}{L}\int_{-\infty}^t e^{Ls}|f(O(\theta_s(\omega))) + O(\theta_s(\omega))|^2\,\mathrm{d}s$$

for all $t \ge T_{\mathcal{D}}$ for tempered families $\mathcal{D} = (D_\omega)_{\omega\in\Omega}$ of initial conditions. Note that the integrability condition (14.6) ensures the finiteness of $R_X^2(\theta_t(\omega))$.

Thus,

$$|X_t(\omega) - O_t(\omega)| \le R_X(\theta_t(\omega)) \quad \text{for all } t \ge T_{\mathcal{D}}\,,$$

or equivalently

$$|X_t(\omega)| \le |O_t(\omega)| + R_X(\theta_t(\omega)) \quad \text{for all } t \ge T_{\mathcal{D}}\,.$$

The family of compact balls B_ω centered on $O_0(\omega)$ with radius $R_X(\omega)$ is thus pullback absorbing, which means that this system has a random attractor $\mathcal{A} = (A_\omega)_{\omega\in\Omega}$. But the difference of any two solutions satisfies the differential inequality

$$\frac{\mathrm{d}}{\mathrm{d}t}\left|X_t^1 - X_t^2\right|^2 \le -2L\left|X_t^1 - X_t^2\right|^2,$$

which means all solutions converge pathwise to each other and thus the random attractor sets are singleton sets $A_\omega = \{X^*(\omega)\}$, i.e., the random attractor is formed by a stationary random process $\bar{X}_t(\omega) = X^*(\theta_t(\omega))$ which pathwise attracts all other solutions in the forward as well as the pullback sense.

EXERCISE 14.8. Adapt the proof of Theorem 3.21 to show that the above random attractor consists of singleton subsets.

3.2. A linear SDE with multiplicative noise. Consider the scalar Stratonovich SDE

$$\mathrm{d}X_t = (-X_t + 1)\mathrm{d}t + X_t \circ \mathrm{d}W_t\,.$$

This SDE can be transformed to the scalar RODE

$$\dot{x} = -x(1 + O_t(\omega)) + e^{-O_t(\omega)}\,, \tag{14.8}$$

with the transformation $x(t,\omega) = e^{-O_t(\omega)}\,X_t(\omega)$, where

$$O_t(\omega) := e^{-t}\int_{-\infty}^{t} e^s \mathrm{d}W_s(\omega)$$

is the stationary Ornstein–Uhlenbeck process and solution of the SDE

$$\mathrm{d}X_t = -X_t\mathrm{d}t + \mathrm{d}W_t\,.$$

The RODE (14.8) has the explicit solution

$$x(t,\omega) = e^{-(t-t_0)\left(1+\frac{1}{t-t_0}\int_{t_0}^{t} O_s(\omega)\,\mathrm{d}s\right)}x_0 + \int_{t_0}^{t} e^{-(t-\tau)-\int_\tau^t O_s(\omega)\,\mathrm{d}s - O_\tau(\omega)}\,\mathrm{d}\tau,$$

that is,

$$\begin{aligned}
&x(t,\omega)\\
&= e^{-(t-t_0)\left(1+\frac{1}{t-t_0}\int_{t_0}^{t} O_s(\omega)\,\mathrm{d}s\right)}x_0 + e^{-t-\int_0^t O_s(\omega)\,\mathrm{d}s}\int_{t_0}^{t} e^{\tau-\int_\tau^0 O_s(\omega)\,\mathrm{d}s - O_\tau(\omega)}\,\mathrm{d}\tau\\
&= e^{-(t-t_0)\left(1+\frac{1}{t-t_0}\int_{t_0}^{t} O_s(\omega)\,\mathrm{d}s\right)}x_0 + e^{-t-\int_0^t O_s(\omega)\,\mathrm{d}s}\int_{t_0}^{t} e^{\tau+\int_0^\tau O_s(\omega)\,\mathrm{d}s - O_\tau(\omega)}\,\mathrm{d}\tau\\
&= e^{-(t-t_0)\left(1+\frac{1}{t-t_0}\int_{t_0}^{t} O_s(\omega)\,\mathrm{d}s\right)}x_0 + e^{-t-\int_0^t O_s(\omega)\,\mathrm{d}s - O_0}\int_{t_0}^{t} e^{\tau-W_\tau(\omega)}\,\mathrm{d}\tau\\
&= e^{-(t-t_0)\left(1+\frac{1}{t-t_0}\int_{t_0}^{t} O_s(\omega)\,\mathrm{d}s\right)}x_0 + e^{-t-\int_{-\infty}^t O_s(\omega)\,\mathrm{d}s}\int_{t_0}^{t} e^{\tau-W_\tau(\omega)}\,\mathrm{d}\tau\,,
\end{aligned}$$

because

$$O_\tau(\omega) - O_0(\omega) = -\int_0^\tau O_s(\omega)\,\mathrm{d}s + W_\tau(\omega)$$

and

$$\int_0^t O_s(\omega)\,\mathrm{d}s + O_0 = \int_0^t O_s(\omega)\,\mathrm{d}s + \int_{-\infty}^0 O_s(\omega)\,\mathrm{d}s = \int_{-\infty}^t O_s(\omega)\,\mathrm{d}s\,.$$

The pullback limit as $t_0 \to -\infty$ gives the stationary stochastic solution which attracts all others pathwise,

$$\bar{X}(t,\omega) = e^{-t+\int_{-\infty}^{t} O_s(\omega)\,\mathrm{d}s} \int_{-\infty}^{t} e^{\tau - W_\tau(\omega)}\,\mathrm{d}\tau\,.$$

The integrals here and above exist due to the properties of the stationary Ornstein–Uhlenbeck process in Lemma 14.6.

In terms of the original SDE the limiting solution is

$$\bar{X}_t(\omega) = \bar{X}(t,\omega)e^{O_t(\omega)} = e^{-t+O_t(\omega)} \int_{-\infty}^{t} e^{\tau - W_\tau(\omega)}\,\mathrm{d}\tau\,,$$

that is

$$\bar{X}_t(\omega) = e^{-t+\int_{-\infty}^{t} O_s(\omega)\,\mathrm{d}s + O_t(\omega)} \int_{-\infty}^{t} e^{\tau - W_\tau(\omega)}\,\mathrm{d}\tau\,.$$

Hence,

$$\bar{X}_t(\omega) = e^{-t+W_t(\omega)} \int_{-\infty}^{t} e^{\tau - W_\tau(\omega)}\,\mathrm{d}\tau\,,$$

which does not involve the Ornstein–Uhlenbeck process.

Endnotes. The standard reference on random dynamical systems in the monograph of Ludwig Arnold [**6**]. See also Chueshov [**49**] for random dynamical systems with a monotone structure and Cong [**55**] for the linear theory. Random attractors were proposed independently at more or less the same time by Schmalfuß [**213, 214**] and Crauel & Flandoli [**61**], see also Crauel, Debussche & Flandoli [**59**]. Lyapunov functions characterizing random attractors were constructed by Arnold & Schmalfuß [**8**], who also showed that random attractors are forward attracting in the weaker sense of probability. The finite dimensionality of random attractors of infinite-dimensional random dynamical systems was considered by Debussche [**64**] and Langa [**153**] amongst many others. Kloeden, Keller & Schmalfuß [**123**] and Robinson [**202**] also consider the numerical approximation and perturbation of random dynamical systems. The equivalence of stochastic differential and random ordinary differential equations under transformation was first investigated by Doss [**72**] and Sussmann [**227**] for special cases in the 1970s and in the general case by Imkeller & Lederer [**97, 96**] and Imkeller & Schmalfuß [**98**] thirty years later. The examples of stochastic differential equations with a drift coefficient satisfying a one-sided dissipative Lipschitz condition are taken from Caraballo & Kloeden [**27**] and Caraballo, Kloeden & Neuenkirch [**30**]. The Multiplicative Ergodic Theorem of Oseledets is a very important result in the theory or random dynamical systems. For linear cocycles it provides an eigen-like decomposition consisting of deterministic Lyapunov exponents and corresponding eigen-like spaces, which are random, see Arnold [**6**].

CHAPTER 15

Synchronization

Consider two autonomous ordinary differential equations in $\mathbb{R}^d$,

$$\dot{x} = f(x) \quad \text{and} \quad \dot{y} = g(y),$$

which are sufficiently regular to ensure the forward existence and uniqueness of solutions and satisfy one-sided dissipative Lipschitz conditions

$$\langle x_1 - x_2, f(x_1) - f(x_2) \rangle \leq -L\|x_1 - x_2\|^2, \tag{15.1}$$

$$\langle y_1 - y_2, g(y_1) - g(y_2) \rangle \leq -L\|y_1 - y_2\|^2, \tag{15.2}$$

on $\mathbb{R}^d$ for some $L > 0$. They thus have unique equilibria $\bar{x}$ and $\bar{y}$, respectively, which are globally asymptotically stable (see Corollary 3.22).

Now consider the dissipatively coupled system

$$\dot{x} = f(x) + \nu(y - x), \qquad \dot{y} = g(y) + \nu(x - y) \tag{15.3}$$

with $\nu > 0$. It can be shown that this coupled system also satisfies a one-sided dissipative Lipschitz condition and thus also has a unique equilibrium $(\bar{x}^\nu, \bar{y}^\nu)$, which is globally asymptotically stable. Moreover, $(\bar{x}^\nu, \bar{y}^\nu) \to (\bar{z}, \bar{z})$ as $\nu \to \infty$, where $\bar{z}$ is the unique globally asymptotically stable equilibrium of the "averaged" system

$$\dot{z} = h(z) := \frac{1}{2}\big(f(z) + g(z)\big).$$

Note that the vector field h also satisfies a one-sided dissipative Lipschitz condition as above.

This phenomenon is known as *dissipative synchronization*. The coupling terms $\pm\nu(x - y)$ in (15.3) can be thought of as a feedback control, which the modeler has introduced to synchronize the systems. A similar result also holds for more general autonomous attractors than steady state solutions. This proofs follow as a special case of the situation for nonautonomous systems, which will be considered in the next section.

1. Deterministic nonautonomous systems

Consider two dissipative nonautonomous ordinary differential equations in $\mathbb{R}^d$, given by

$$\dot{x} = f(p, x), \quad \text{where } p \in P, \tag{15.4}$$

with driving system $\theta : \mathbb{R} \times P \to P$, and

$$\dot{y} = g(q, x), \quad \text{where } q \in Q, \tag{15.5}$$

with driving system $\psi : \mathbb{R} \times Q \to Q$.

Suppose that both systems are sufficiently regular to ensure the forward existence and uniqueness of solutions, so they generate nonautonomous dynamical systems on $P \times \mathbb{R}^d$ and $Q \times \mathbb{R}^d$, respectively. In particular, suppose that both satisfy a uniform dissipativity condition

$$\big\langle x, f(p,x)\big\rangle \leq K - L\|x\|^2 \quad \text{and} \quad \big\langle x, g(q,x)\big\rangle \leq K - L\|x\|^2 \tag{15.6}$$

for all $(p,q) \in P \times Q$. From these conditions follow the differential inequalities

$$\frac{1}{2}\frac{\mathrm{d}}{\mathrm{d}t}\|x(t)\|^2 \leq K - L\|x(t)\|^2 \quad \text{and} \quad \frac{1}{2}\frac{\mathrm{d}}{\mathrm{d}t}\|y(t)\|^2 \leq K - L\|y(t)\|^2$$

uniformly in $p \in P$ and $q \in Q$, respectively. Thus, as shown in Subsection 3.3 in Chapter 3, the closed ball

$$B_d\big[0, \sqrt{(K+1)/L}\big] := \big\{x \in \mathbb{R}^d : \|x\|^2 \leq (K+1)/L\big\}$$

is pullback absorbing and positively invariant in both cases, so by Theorem 3.20, both systems (15.4) and (15.5) have pullback attractors $\mathcal{A}^f$ and $\mathcal{A}^g$, respectively.

Consider now for some $\nu > 0$ the dissipatively coupled system

$$\dot{x} = f(p,x) + \nu(y - x), \qquad \dot{y} = g(q,x) + \nu(x - y) \tag{15.7}$$

with the product driving system $(\theta, \psi) : \mathbb{R} \times P \times Q \to P \times Q$. It follows from the uniform dissipativity condition (15.6) that

$$\begin{aligned}
\frac{\mathrm{d}}{\mathrm{d}t}\left(\|x\|^2(t) + \|y\|^2(t)\right) &= 2\big\langle x(t), \dot{x}(t)\big\rangle + 2\big\langle y(t), \dot{y}(t)\big\rangle \\
&= 2\big\langle x(t), f(\theta_t(p), x(t))\big\rangle + 2\big\langle x(t), \nu(y(t) - x(t)\big\rangle \\
&\quad + 2\big\langle y(t), g(\psi_t(q), y(t))\big\rangle + 2\big\langle x(t), \nu(x(t) - y(t)\big\rangle \\
&\leq 4K - 2L\left(\|x(t)\|^2 + \|y(t)\|^2\right),
\end{aligned}$$

from which it follows that the closed ball $B_{2d}[0, R_{K,L}]$ in $\mathbb{R}^{2d}$ with radius $R_{K,L} := \sqrt{(2K+1)/L}$ is a pullback absorbing and positively invariant for the coupled system (15.7), so by Theorem 3.20, the coupled system (15.7) has a pullback attractor in $\mathbb{R}^{2d}$ for each $\nu > 0$, which will be denoted by $\mathcal{A}^\nu$. In addition, abbreviate

$$\big(x^\nu(t), y^\nu(t)\big) = \big(x^\nu\big(t, (p,q), x_0, y_0\big), y^\nu\big(t, (p,q), x_0, y_0\big)\big)$$

for the solution of the coupled system (15.7) with initial parameter value (p,q) and initial state (x_0, y_0).

THEOREM 15.1. *For all* $(x_0, y_0) \in B_{2d}[0, R_{K,L}]$ *and* $(p,q) \in P \times Q$, *one has*

$$\lim_{\nu \to \infty} \big\|x^\nu(t) - y^\nu(t)\big\| = 0 \quad \textit{uniformly on bounded intervals } [T_1, T_2].$$

Define $\mathrm{Diag}(M) := \big\{(x,x) : x \in M\big\}$ for a set M. The next theorem follows from this and the fact that a pullback attractor consists of the entire trajectories.

THEOREM 15.2. *For all* $(p,q) \in P \times Q$, *one has*

$$\lim_{\nu \to \infty} \mathrm{dist}\left(A^\nu_{(p,q)}, \mathrm{Diag}(\mathbb{R}^d) \cap B_{2d}[0, R_{K,L}]\right) = 0\,.$$

In fact, one can say much more about the dynamics inside the pullback attractor $\mathcal{A}^\nu$ and what happens in the limit $\nu \to \infty$. Let $\big(x^\nu(t), y^\nu(t)\big)$ be an entire trajectory of the coupled system inside the pullback attractor $\mathcal{A}^\nu$, i.e., with

$$\big(x^\nu(t), y^\nu(t)\big) \in A^\nu_{(\theta_t(p),\psi_t(q))} \quad \text{for all } t \in \mathbb{R}$$

for some $p \in P$ and $q \in Q$.

THEOREM 15.3. *Given an entire trajectory $\big(x^\nu(t), y^\nu(t)\big)$ of the coupled system inside the pullback attractor $\mathcal{A}^\nu$. Then for any sequence $(\nu_n)_{n\in\mathbb{N}}$, there exists a subsequence $(\nu_{n_j})_{j\in\mathbb{N}}$ such that*

$$\lim_{j\to\infty} x^{\nu_{n_j}}(t) = z(t) \quad \textit{and} \quad \lim_{j\to\infty} y^{\nu_{n_j}}(t) = z(t)$$

uniformly on compact time subintervals in $\mathbb{R}$, where $z(t)$ is a solution of the nonautonomous differential equation

$$\dot{z} = h\big((p,q), z\big) := \frac{1}{2}\big(f(p,z) + g(q,z)\big) \tag{15.8}$$

on $\mathbb{R}^d$ with the product driving system $(\theta,\psi) : \mathbb{R} \times P \times Q \to P \times Q$.

It follows from the uniform dissipativity condition (15.6) that the averaged differential equation (15.8) also satisfies the uniform dissipativity condition (15.6) and hence that the closed ball $B_d\big[0, \sqrt{(K+1)/L}\big]$ is uniformly absorbing and positively invariant for the averaged system (15.8), which thus has a pullback attractor $\mathcal{A}^\infty$ in $\mathbb{R}^d$. Theorems 15.2 and 15.3 give the following statement.

COROLLARY 15.4. *One has*

$$\lim_{\nu\to\infty} \operatorname{dist}\left(A^\nu_{(p,q)}, \operatorname{Diag}\left(A^\infty_{(p,q)}\right)\right) = 0\,.$$

The proofs of Theorems 15.1, 15.2 and 15.3 and Corollary 15.4 can be found in KLOEDEN [**118**]. Simplified versions under a stronger one-sided dissipative Lipschitz conditions are given in the next subsection. Note that these results also hold for the autonomous systems in the introduction to the chapter (one simply has to discard the indices p and q above).

EXERCISE 15.5. Show that the averaged differential equation (15.8) satisfies the same uniform dissipative condition (15.6) as the decoupled systems (15.4) and (15.5).

1.1. Systems with one-sided dissipative Lipschitz conditions. In this subsection, the decoupled ordinary differential equations (15.4) and (15.5) are assumed to satisfy uniform one-sided dissipative Lipschitz conditions of the form

$$\big\langle x_1 - x_2, f(p,x_1) - f(p,x_2)\big\rangle \le -\tilde{L}\|x_1 - x_2\|^2 \quad \text{for all } p \in P\,,$$
$$\big\langle y_1 - y_2, g(q,y_1) - g(q,y_2)\big\rangle \le -\tilde{L}\|y_1 - y_2\|^2 \quad \text{for all } q \in Q\,,$$

on $\mathbb{R}^d$ for some $\tilde{L} > 0$. As in Subsection 3.3 in Chapter 3, it follows that both f and g satisfy the uniform dissipative condition (15.6) with constants

$$K = \frac{2}{\tilde{L}} \max\left\{\sup_{p\in P}\|f(p,0)\|, \sup_{q\in Q}\|g(q,0)\|\right\} \quad \text{and} \quad L = \frac{\tilde{L}}{2}\,.$$

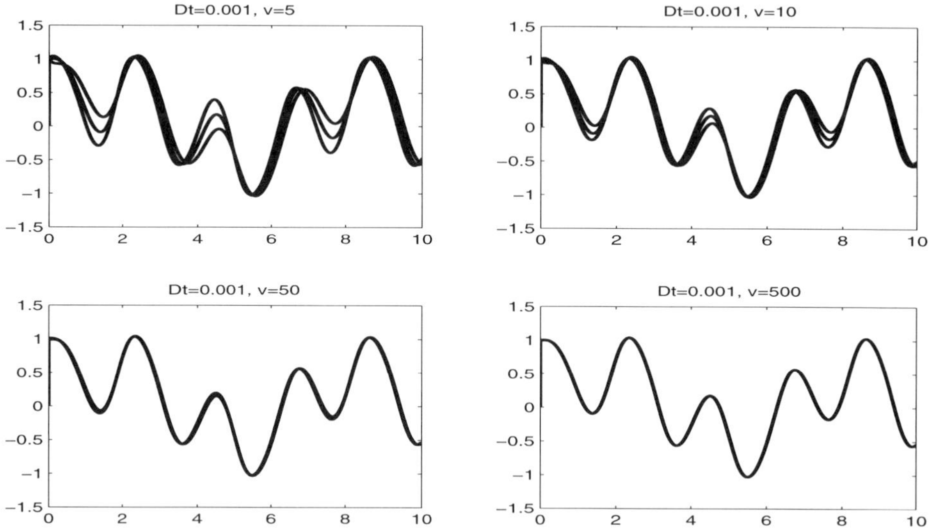

FIGURE 15.1. Trajectories of the coupled system $\dot{x} = -x + 2\sin t + \nu(y - x)$, $\dot{y} = -y - 2y^3 + 4\cos 3t + \nu(x - y)$ and the corresponding trajectories of the averaged system $\dot{z} = -z - z^3 + \sin t + 2\cos 3t$ for four values of ν, [**136**] copyright *Wiley-VCH Verlag GmbH & Co. KGaA*. Reproduced with permission.

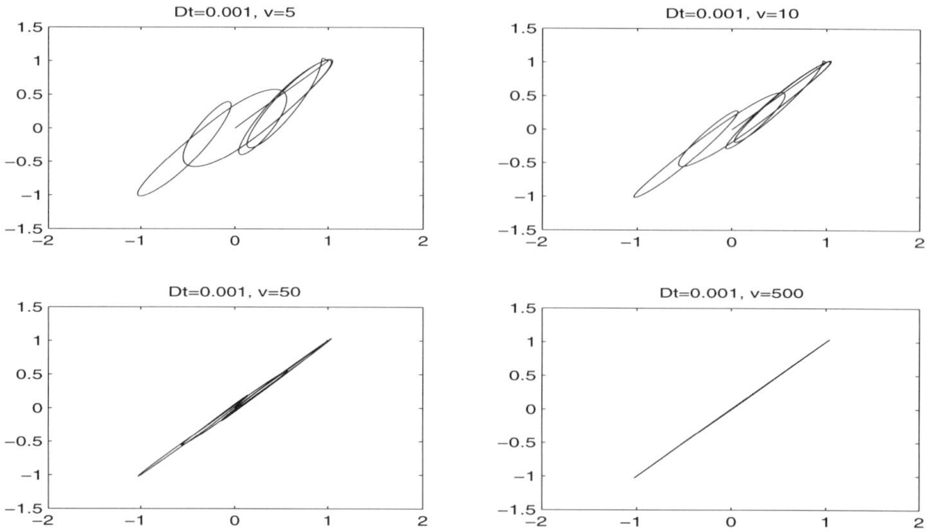

FIGURE 15.2. Trajectories of the coupled system in Figure 15.1 for four values of ν, [**136**] copyright *Wiley-VCH Verlag GmbH & Co. KGaA*. Reproduced with permission.

and that the closed ball $B_d\left[0, \sqrt{(K+1)/L}\right]$ is pullback absorbing and positively invariant for each of them. Thus, each of the decoupled systems has a pullback attractor with subsets in this ball, which, by Theorem 3.21, consist of singleton subsets.

Now both the dissipatively coupled differential equation (15.7) and the averaged differential equations (15.8) satisfy the uniform one-sided dissipative Lipschitz condition.

EXERCISE 15.6. Show that the dissipatively coupled differential equation (15.7) and the averaged differential equations (15.8) satisfy the uniform one-sided dissipative Lipschitz condition when each of the decoupled differential equations does.

The dissipatively coupled differential equation (15.7) and the averaged differential equations (15.8) thus also have pullback attractors $\mathcal{A}^\nu$ in $\mathbb{R}^{2d}$ and $\mathcal{A}^\infty$ in $\mathbb{R}^d$, respectively. Moreover, by Theorem 3.21, these consist of singleton component sets formed by entire solutions of the respective systems, which will be denoted by $a^\nu_{(\theta_t(p),\psi_t(q))}$ in $\mathbb{R}^{2d}$ and $a^\infty_{(\theta_t(p),\psi_t(q))}$ in $\mathbb{R}^d$, respectively.

LEMMA 15.7. *Write* $a^\nu_{(\theta_t(p),\psi_t(q))} = \big(x^\nu(t), y^\nu(t)\big)$ *for fixed* $(p,q) \in P \times Q$. *Then*

$$\lim_{\nu\to\infty} \|x^\nu(t) - y^\nu(t)\| = 0 \quad \textit{uniformly on bounded intervals } [T_1, T_2]\,. \tag{15.9}$$

PROOF. $t \mapsto \big(x^\nu(t), y^\nu(t)\big)$ is solution of the coupled system (15.7) with initial parameter value $(p,q) \in P \times Q$. For notational simplicity, the index ν will now be omitted until the end of the proof. From (15.7), it follows that the function $U(t) := x(t) - y(t)$ for $t \in \mathbb{R}$ satisfies the differential equation

$$\dot U(t) = -2\nu U(t) + f(\theta_t(p), x(t)) - g(\psi_t(q), y(t))\,,$$

which means that

$$U(t) = U(0)e^{-2\nu t} + e^{-2\nu t}\int_0^t e^{2\nu s}\big(f(\theta_s(p), x(s)) - g(\psi_s(q), y(s))\big)\,\mathrm{d}s\,,$$

and hence,

$$\begin{aligned}&\|U(t)\|\\ &\le \|U(0)\|e^{-2\nu t} + e^{-2\nu t}\int_0^t e^{2\nu s}\big(\|f(\theta_s(p), x(s))\| + \|g(\psi_s(q), y(s))\|\big)\,\mathrm{d}s\,.\end{aligned}$$

Now the component subsets of the pullback attractor are all contained in the ball $B_{2d}[0, R_{K,L}]$, where $R_{K,L} := \sqrt{(2K+1)/K}$, and so is the difference $x(t) - y(t) \in B_d[0, R_{K,L}]$ for all $t \ge 0$. Moreover, since $B_d[0, R_{K,L}]$, P and Q are compact, by the continuity of f and g, there exists a constant $M > 0$ such that

$$\|f(p,x)\| + \|g(q,y)\| \le M \quad \text{for all } (x,y) \in B_d[0, R_{K,L}]\,, p \in P \text{ and } q \in Q\,.$$

Thus,

$$\|U(t)\| \le \|U(0)\|e^{-2\nu t} + e^{-2\nu t}\int_0^t e^{2\nu s} M\,\mathrm{d}s\,,$$

from which it follows that

$$\|U(t)\| \le \|U(0)\|e^{-2\nu t} + \frac{M}{2\nu}\big(1 - e^{-2\nu t}\big)\,.$$

Thus, reinserting the index ν,

$$\big\|x^\nu(t) - y^\nu(t)\big\| = \|U^\nu(t)\| \to 0 \quad \text{as } \nu \to \infty$$

for all $t \in (0,T]$ with an arbitrary finite $T > 0$, and hence for any $t \in [T_1, T_2]$ for arbitrary finite $T_2 \geq T_1 > 0$. □

THEOREM 15.8. *Write* $a^{\nu}_{(\theta_t(p),\psi_t(q))} = \big(x^{\nu}(t), y^{\nu}(t)\big)$ *for fixed* $(p,q) \in P \times Q$. *Then*

$$\lim_{\nu\to\infty} x^{\nu}(t) = z(t) \quad \textit{and} \quad \lim_{\nu\to\infty} y^{\nu}(t) = z(t)$$

uniformly on compact subintervals $[T_1, T_2]$, *where* $z(t) := a^{\infty}_{(\theta_t(p),\psi_t(q))}$.

PROOF. Define

$$z^{\nu}(t) = \frac{1}{2}\big(x^{\nu}(t) + y^{\nu}(t)\big) \quad \text{for all } t \in \mathbb{R}.$$

Then

$$\begin{aligned} \dot{z}^{\nu}(t) &= \frac{1}{2}\big(f\left(\theta_t(p), x^{\nu}(t)\right) + g\left(\psi_t(q), x^{\nu}(t)\right)\big) \\ &= \frac{1}{2}\big(f\left(\theta_t(p), 2z^{\nu}(t) - y^{\nu}(t)\right) + g\left(\psi_t(q), 2z^{\nu}(t) - x^{\nu}(t)\right)\big) \end{aligned} \tag{15.10}$$

Thus,

$$\big\|\dot{z}^{\nu}(t)\big\| \leq \frac{1}{2}\big(\big\|f\big(\theta_t(p), x^{\nu}(t)\big)\big\| + \big\|g\big(\psi_t(q), y^{\nu}(t)\big)\big\|\big) \leq M,$$

where M is a bound of $\|f(p,x)\| + \|g(q,y)\|$ on the compact set $P \times Q \times B_{2d}[0, R_{K,L}]$. Thus, the sequence of functions z^{ν} is equi-continuous on any compact time interval and has a uniformly convergent subsequence on this interval. By a diagonal sequence argument this can be extended to uniform convergence on all time intervals of the form $[-T, T]$. Thus,

$$z^{\nu_n}(t) \to z(t) \quad \text{as } \nu_n \to \infty \quad \text{for all } t \in \mathbb{R},$$

where $z(t)$ is continuous. Now, by Lemma 15.7, one has

$$\begin{aligned} z^{\nu_n}(t) - y^{\nu_n}(t) &= \frac{1}{2}\big(x^{\nu_n}(t) - y^{\nu_n}(t)\big) \to 0, \\ z^{\nu_n}(t) - x^{\nu_n}(t) &= \frac{1}{2}\big(y^{\nu_n}(t) - x^{\nu_n}(t)\big) \to 0 \end{aligned}$$

as $\nu_n \to \infty$ for all $t \in \mathbb{R}$. Hence,

$$2z^{\nu_n}(t) - y^{\nu_n}(t) \to z(t) \quad \text{and} \quad 2z^{\nu_n}(t) - x^{\nu_n}(t) \to z(t),$$

as well as

$$x^{\nu_n}(t) \to z(t) \quad \text{and} \quad y^{\nu_n}(t) \to z(t)$$

as $\nu_n \to \infty$ for all $t \in \mathbb{R}$. The convergence here is, in fact, uniform on any compact interval in $\mathbb{R}$. Moreover, since the limit is the same for all such subsequences, convergence holds for all $\nu \to \infty$ (see Exercise 15.10).

Writing the differential equation (15.10) with ν in integral form, i.e., as

$$\begin{aligned} &z^{\nu}(t) \\ &= z^{\nu}(t_0) + \frac{1}{2}\int_{t_0}^{t} \big(f\big(\theta_s(p), 2z^{\nu}(s) - y^{\nu}(s)\big) + g\big(\psi_s(q), 2z^{\nu}(s) - x^{\nu}(s)\big)\big)\,\mathrm{d}s, \end{aligned}$$

by continuity it follows that the limit as $\nu \to \infty$ is

$$z(t) = z(t_0) + \frac{1}{2}\int_{t_0}^{t} \big(f(\theta_s(p), z(s)) + g(\psi_s(q), z(s))\big)\,\mathrm{d}s,$$

i.e., z is a solution of the nonautonomous differential equation (15.8), namely

$$\dot{z} = \frac{1}{2}\big(f(p,z) + g(q,z)\big)$$

with the product driving system $(\theta,\psi) : \mathbb{R} \times P \times Q \to P \times Q$. □

COROLLARY 15.9. *One has*

$$\lim_{\nu\to\infty} \operatorname{dist}\left(a^{\nu}_{(p,q)}, \operatorname{Diag}(\mathbb{R}^d) \cap B_{2d}[0, R_{K,L}]\right) = 0\,. \tag{15.11}$$

EXERCISE 15.10. Let $(x_n)_{n\in\mathbb{N}}$ be a sequence in a complete metric space (X,d) such that every subsequence $(x_{n_i})_{i\in\mathbb{N}}$ has a subsequence $\left(x_{n_{i_j}}\right)_{j\in\mathbb{N}}$ converging to a common limit x^*. Show that the sequence $(x_n)_{n\in\mathbb{N}}$ converges to x^*.

1.2. An example. Consider the scalar nonautonomous differential equations

$$\dot{x} = -x + \alpha(t) \quad \text{and} \quad \dot{y} = -y + \beta(t)\,, \tag{15.12}$$

where α and β are bounded continuous functions. Here the driving systems are defined by the shift operators defined by $\theta_t(\alpha(\cdot)) = \alpha(\cdot + t)$ and $\psi_t(\beta(\cdot)) = \beta(\cdot + t)$ for all $t \in \mathbb{R}$ and the base spaces P and Q are, respectively, the closed hulls of the functions α and β with an appropriate topology. However, the process notation will now be used since the derivation is more transparent in this context.

Note that both systems are strongly dissipative with

$$|x_1(t) - x_2(t)| \le e^{-t}|x_{0,1} - x_{0,2}| \quad \text{and} \quad |y_1(t) - y_2(t)| \le e^{-t}|y_{0,1} - y_{0,2}|$$

for any pair of initial values. Thus, both systems have singleton trajectory pullback attractors (which are also forward attracting) defined via

$$\bar{x}(t) = e^{-t}\int_{-\infty}^{t} e^{s}\alpha(s)\,\mathrm{d}s \quad \text{and} \quad \bar{y}(t) = e^{-t}\int_{-\infty}^{t} e^{s}\beta(s)\,\mathrm{d}s\,,$$

i.e., with component sets $A^f_{\alpha(\cdot)} = \{\bar{x}(0)\}$ and $A^g_{\beta(\cdot)} = \{\bar{y}(0)\}$.

The averaged system

$$\dot{z} = -z + \frac{1}{2}\big(\alpha(t) + \beta(t)\big)$$

is also strongly dissipative with a singleton trajectory pullback attractor (which is also forward attracting), given by

$$\bar{z}(t) = \frac{1}{2}e^{-t}\int_{-\infty}^{t} e^{s}\big(\alpha(s) + \beta(s)\big)\,\mathrm{d}s = \frac{1}{2}\big(\bar{x}(t) + \bar{y}(t)\big)\,,$$

i.e., the average of $\bar{x}$ and $\bar{y}$, which is a special situation due to the linearity of the equations.

The synchronized system

$$\dot{x} = -x + \alpha(t) + \nu(y - x)\,, \qquad \dot{y} = -y + \beta(t) + \nu(x - y)$$

has the general solution

$$\begin{pmatrix} x^{\nu}(t) \\ y^{\nu}(t) \end{pmatrix} = e^{A_\nu(t-t_0)}\begin{pmatrix} x_0 \\ y_0 \end{pmatrix} + \int_{t_0}^{t} e^{A_\nu(t-s)}\begin{pmatrix} \alpha(s) \\ \beta(s) \end{pmatrix}\mathrm{d}s\,,$$

where

$$A_\nu = \begin{pmatrix} -1-\nu & -\nu \\ -\nu & -1-\nu \end{pmatrix} \quad \text{and} \quad e^{A_\nu t} = \begin{pmatrix} e^{-t} + e^{-(1+\nu)t} & e^{-t} - e^{-(1+\nu)t} \\ e^{-t} - e^{-(1+\nu)t} & e^{-t} + e^{-(1+\nu)t} \end{pmatrix},$$

so

$$\begin{pmatrix} x^\nu(t) \\ y^\nu(t) \end{pmatrix} = \frac{1}{2}\begin{pmatrix} e^{-(t-t_0)}(x_0+y_0) + e^{-(1+\nu)(t-t_0)}(x_0-y_0) \\ e^{-(t-t_0)}(x_0+y_0) - e^{-(1+\nu)(t-t_0)}(x_0-y_0) \end{pmatrix}$$
$$+\frac{1}{2}\int_{t_0}^{t}\begin{pmatrix} e^{-(t-s)}(\alpha(s)+\beta(s)) + e^{-(1+\nu)(t-s)}(\alpha(s)-\beta(s)) \\ e^{-(t-s)}(\alpha(s)+\beta(s)) - e^{-(1+\nu)(t-s)}(\alpha(s)-\beta(s)) \end{pmatrix} ds\,.$$

Taking the pullback convergence limit $t_0 \to -\infty$ yields the singleton trajectory in the pullback attractor, given by

$$\begin{pmatrix} x^\nu(t) \\ y^\nu(t) \end{pmatrix} = \frac{1}{2}\int_{-\infty}^{t}\begin{pmatrix} e^{-(t-s)}(\alpha(s)+\beta(s)) + e^{-(1+\nu)(t-s)}(\alpha(s)-\beta(s)) \\ e^{-(t-s)}(\alpha(s)+\beta(s)) - e^{-(1+\nu)(t-s)}(\alpha(s)-\beta(s)) \end{pmatrix} ds\,.$$

Thus, taking $\nu \to \infty$ gives

$$\begin{pmatrix} x^\nu(t) \\ y^\nu(t) \end{pmatrix} \to \frac{1}{2}\int_{t_0}^{t} e^{-(t-s)}(\alpha(s)+\beta(s))\,ds \begin{pmatrix} 1 \\ 1 \end{pmatrix} = \bar{z}(t)\begin{pmatrix} 1 \\ 1 \end{pmatrix}.$$

Since the attractors here each consist of a single trajectory, they converge continuously as $\nu \to \infty$ rather than just upper semi-continuously.

EXERCISE 15.11. Rewrite the above solutions and pullback convergence in terms of the skew product notation.

2. Synchronization of systems with additive noise

The above synchronization effect is preserved in the presence of additive noise provided the equilibria are replaced by stationary random solutions. The analysis uses the theory of random dynamical systems from Chapter 14.

Consider two scalar Itô stochastic differential equations (SDEs) in $\mathbb{R}$, given by

$$dX_t = f(X_t)\,dt + \alpha\,dW_t^1 \quad \text{and} \quad dY_t = g(Y_t)\,dt + \beta\,dW_t^2\,, \tag{15.13}$$

where α and β are positive constants, W_t^1 and W_t^2 are independent two-sided scalar Wiener processes, and the drift coefficients f and g are continuously differentiable and satisfy the one-sided dissipative Lipschitz conditions (15.1) as well as the integrability condition (14.6) stated in Chapter 14.

It was seen in Section 3 of Chapter 14 that each of these stochastic systems has a pathwise asymptotically stable random attractor consisting of a single stationary random variable.

The system of coupled SDEs (15.13) to be synchronized reads as

$$dX_t = \big(f(X_t) + \nu(Y_t - X_t)\big)\,dt + \alpha\,dW_t^1\,, \tag{15.14}$$
$$dY_t = \big(g(Y_t) + \nu(X_t - Y_t)\big)\,dt + \beta\,dW_t^2\,. \tag{15.15}$$

It will shown that this system is dissipative and has a unique stationary solution $(\bar{X}_t^\nu, \bar{Y}_t^\nu)$, which is pathwise globally asymptotically stable with

$$(\bar{X}_t^\nu, \bar{Y}_t^\nu) \to (\bar{Z}_t^\infty, \bar{Z}_t^\infty) \quad \text{as } \nu \to \infty,$$

pathwise on finite time intervals $[T_1, T_2]$, where $\bar{Z}_t^\infty$ is the unique pathwise globally asymptotically stable stationary solution of the averaged SDE

$$\mathrm{d}Z_t = \frac{1}{2}\big(f(Z_t) + g(Z_t)\big)\,\mathrm{d}t + \frac{1}{2}\alpha\,\mathrm{d}W_t^1 + \frac{1}{2}\beta\,\mathrm{d}W_t^2\,. \tag{15.16}$$

EXAMPLE 15.12. For the SDEs with linear drift terms,

$$\mathrm{d}X_t = -X_t\,\mathrm{d}t + \alpha\,\mathrm{d}W_t^1 \quad \text{and} \quad \mathrm{d}Y_t = -Y_t\,\mathrm{d}t + \beta\,\mathrm{d}W_t^2\,, \tag{15.17}$$

the unique stationary solutions are

$$\bar{X}_t = \alpha e^{-t}\int_{-\infty}^t e^s\,\mathrm{d}W_s^1 \quad \text{and} \quad \bar{Y}_t = \beta e^{-t}\int_{-\infty}^t e^s\,\mathrm{d}W_s^2\,, \tag{15.18}$$

and are obtained by taking the pullback limit of the corresponding explicit solutions on a finite time interval. The averaged SDE, given by

$$\mathrm{d}Z_t = -Z_t\,\mathrm{d}t + \frac{1}{2}\alpha\,\mathrm{d}W_t^1 + \frac{1}{2}\beta\,\mathrm{d}W_t^2\,,$$

has the explicit stationary solution

$$\bar{Z}_t^\infty = \frac{1}{2}e^{-t}\left(\alpha\int_{-\infty}^t e^s\,\mathrm{d}W_s^1 + \beta\int_{-\infty}^t e^s\,\mathrm{d}W_s^2\right) = \frac{1}{2}\big(\bar{X}_t + \bar{Y}_t\big)\,.$$

This is the arithmetic average of the uncoupled stationary solutions, which is a special situation due to the linear structure of the SDEs.

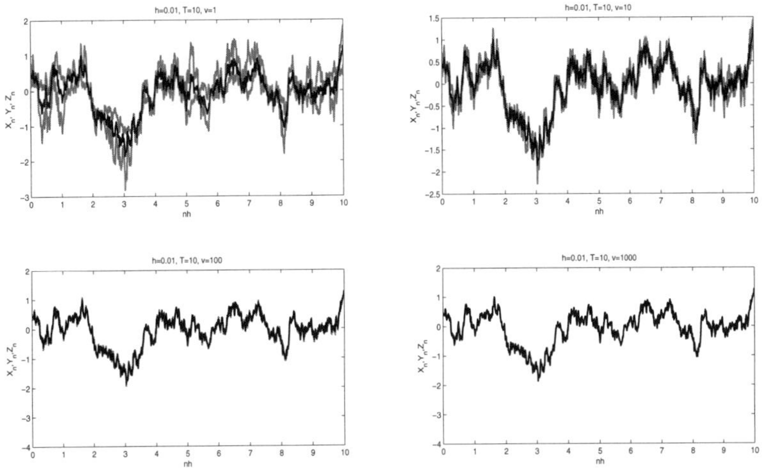

FIGURE 15.3. Trajectories of the coupled system $\mathrm{d}X_t = \big(-2X_t + \nu(Y_t - X_t)\big)\,\mathrm{d}t + \alpha\,\mathrm{d}W_t^1$, $\mathrm{d}Y_t = \big(-Y_t - Y_t^3 + \nu(X_t - Y_t)\big)\,\mathrm{d}t + \beta\,\mathrm{d}W_t^2$ and the corresponding trajectories of the averaged system $\mathrm{d}Z_t = -\frac{1}{2}\big(3Z_t + Z_t^3\big)\,\mathrm{d}t + \frac{1}{2}\,\mathrm{d}W_t^1 + \mathrm{d}W_t^2$ for four values of ν, [**136**] copyright *Wiley-VCH Verlag GmbH & Co. KGaA*. Reproduced with permission.

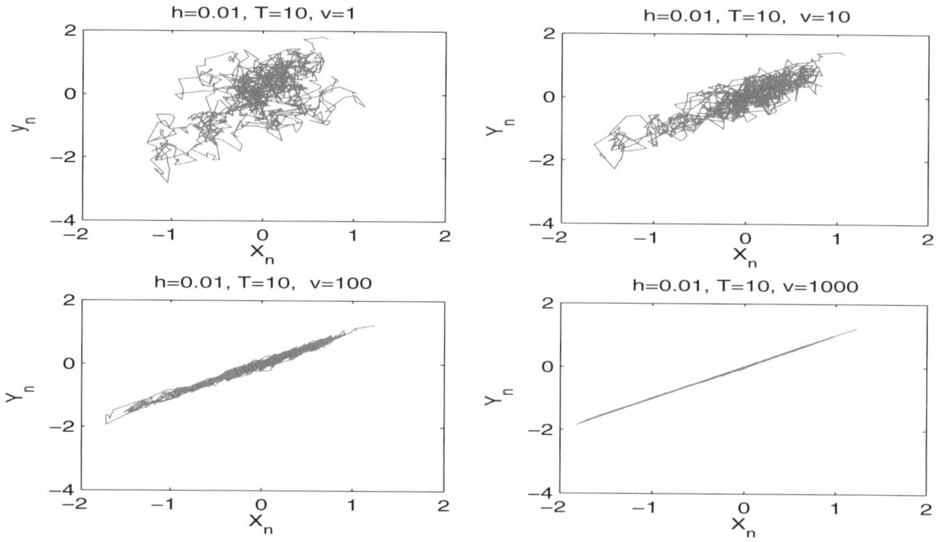

FIGURE 15.4. Trajectories of the coupled system in Figure 15.3 for four values of ν, [**136**] copyright *Wiley-VCH Verlag GmbH & Co. KGaA*. Reproduced with permission.

2.1. The coupled system with additive noise. The asymptotic behavior of the uncoupled systems with additive noise (15.13) was investigated in Chapter 14. In short, each has a random attractor consisting of singleton sets, i.e., stochastic stationary processes.

The coupled system (15.14)–(15.15) also satisfies a dissipative one-sided Lipschitz condition with the same constant L as the uncoupled systems (15.13). This is easier to see when the coupled system is written in vector form,

$$d\mathfrak{X}_t = \big(F(\mathfrak{X}_t) + \nu A\mathfrak{X}_t\big)\,dt + B\,d\mathfrak{W}_t$$

where

$$\mathfrak{X}_t = \begin{pmatrix} X_t \\ Y_t \end{pmatrix}, \qquad \mathfrak{W}_t = \begin{pmatrix} W_t^1 \\ W_t^2 \end{pmatrix}, \qquad X = \begin{pmatrix} x \\ y \end{pmatrix},$$
$$F(X) = \begin{pmatrix} f(x) \\ g(y) \end{pmatrix}, \qquad A = \begin{pmatrix} -1 & 1 \\ 1 & -1 \end{pmatrix}, \qquad B = \begin{pmatrix} \alpha & 0 \\ 0 & \beta \end{pmatrix}.$$

EXERCISE 15.13. Show that the mapping $G_\nu : \mathbb{R}^2 \to \mathbb{R}^2$, defined by $G_\nu(\mathfrak{X}) := F(\mathfrak{X}) + \nu B\mathfrak{X}$ satisfies a one-sided dissipative Lipschitz condition with constant $L+\nu$ (hence with a common lower bound L).

Similar estimates to those in Section 3 of Chapter 14 show the existence of a pullback absorbing family, so Theorem 14.4 can be applied to conclude that the coupled system has a random attractor $\mathcal{A}^\nu$ with fibers A_ω^ν, which are singleton sets formed by an ordered pair of stationary processes $\mathfrak{X}_t^\nu(\omega) = \big(\bar{X}_t(\omega), \bar{Y}_t^\nu(\omega)\big)^\top$.

2.2. The synchronization of stationary solutions as $\nu \to \infty$. The following lemma is taken from CARABALLO & KLOEDEN [**27**].

LEMMA 15.14. *Let W_t be a two-sided Wiener process (i.e., defined for all $t \in \mathbb{R}$). Then $\nu \int_{-\infty}^{t} e^{-\nu(t-s)}\, \mathrm{d}W_t(\omega)$ is pathwise uniformly bounded in $\nu > 0$ on finite time intervals $[T_1, T_2]$, and one obtains*

$$\int_{T_1}^{t} e^{-\nu(t-s)}\, \mathrm{d}W_t(\omega) \to 0 \quad \textit{as } \nu \to \infty$$

pathwise on finite time intervals $[T_1, T_2]$.

The next lemma is the first step in showing synchronization of the coupled solutions.

LEMMA 15.15. *One has $\bar{X}_t^{\nu}(\omega) - \bar{Y}_t^{\nu}(\omega) \to 0$ as $\nu \to \infty$ pathwise on finite time intervals $[T_1, T_2]$.*

PROOF. Subtracting the second from the first equation of (15.14)–(15.15) gives

$$\mathrm{d}(\bar{X}_t^{\nu} - \bar{Y}_t^{\nu}) = \left(-2\nu\left(\bar{X}_t^{\nu} - \bar{Y}_t^{\nu}\right) + f(\bar{X}_t^{\nu}) - g(\bar{Y}_t^{\nu})\right) \mathrm{d}t + \alpha\, \mathrm{d}W_t^1 - \beta\, \mathrm{d}W_t^2$$

or

$$\mathrm{d}\left(D_t^{\nu} e^{2\nu t}\right) = e^{2\nu t}\left(f(\bar{X}_t^{\nu}) - g(\bar{Y}_t^{\nu})\right) \mathrm{d}t + \alpha e^{2\nu t}\, \mathrm{d}W_t^1 - \beta e^{2\nu t}\, \mathrm{d}W_t^2,$$

where $D_t^{\nu} = \bar{X}_t^{\nu} - \bar{Y}_t^{\nu}$, so pathwise, one has

$$\begin{aligned} |D_t^{\nu}| \leq e^{-2\nu t}|D_{T_1}^{\nu}| &+ \int_{T_1}^{t} e^{-2\nu(t-s)}\left(|f(\bar{X}_s^{\nu})| + |g(\bar{Y}_s^{\nu})|\right) \mathrm{d}s \\ &+ |\alpha| \left| \int_{T_1}^{t} e^{-2\nu(t-s)}\, \mathrm{d}W_t^1 \right| + |\beta| \left| \int_{T_1}^{t} e^{-2\nu(t-s)}\, \mathrm{d}W_t^2 \right| \end{aligned} \tag{15.19}$$

By Lemma 15.14, one sees that

$$\begin{aligned} R_{\nu}(\theta_t(\omega)) := 1 &+ \int_{T_1}^{t} e^{-2\nu(t-s)}\left(|f(\bar{X}_s^{\nu}(\omega)| + |g(\bar{Y}_s^{\nu}(\omega)|\right) \mathrm{d}s \\ &+ |\alpha| \left| \int_{T_1}^{t} e^{-2\nu(t-s)}\, \mathrm{d}W_t^1(\omega) \right| + |\beta| \left| \int_{T_1}^{t} e^{-2\nu(t-s)}\, \mathrm{d}W_t^2(\omega) \right| \end{aligned}$$

is pathwise uniformly bounded on each bounded time interval $[T_1, T_2]$, so $|X_t^{\nu}(\omega)|$, $|Y_t^{\nu}(\omega)|$ and $|D_0^{\nu}(\omega)|$ are pathwise uniformly bounded on each bounded time interval $[T_1, T_2]$. Then, by Lemma 15.14, all integrals in (15.19) converge to zero as as $\nu \to \infty$ pathwise on the bounded time interval $[T_1, T_2]$. □

THEOREM 15.16. *One has $\left(\bar{X}_t^{\nu_n}, \bar{Y}_t^{\nu_n}\right) \to (Z_t^{\infty}, Z_t^{\infty})$ pathwise uniformly on bounded time intervals $[T_1, T_2]$ for any sequence $\nu_n \to \infty$, where Z_t^{∞} is the stationary stochastic solution of the averaged SDE*

$$\mathrm{d}Z_t = \frac{1}{2}\left(f(Z_t) + g(Z_t)\right) \mathrm{d}t + \frac{1}{2}\alpha \mathrm{d}W_t^1 + \frac{1}{2}\beta\, \mathrm{d}W_t^2. \tag{15.20}$$

PROOF. Define

$$Z_t^{\nu} := \frac{1}{2}\left(\bar{X}_t^{\nu} + \bar{Y}_t^{\nu}\right) \quad \text{for all } t \in \mathbb{R},$$

and observe that Z_t^{ν} satisfies the equation

$$\mathrm{d}Z_t^{\nu} = \frac{1}{2}\left(f(\bar{X}_t^{\nu}) + g(\bar{Y}_t^{\nu})\right) \mathrm{d}t + \frac{1}{2}\alpha\, \mathrm{d}W_t^1 + \frac{1}{2}\beta\, \mathrm{d}W_t^2.$$

Define

$$\bar{Z}_t := \frac{1}{2}(\bar{X}_t + \bar{Y}_t) \quad \text{for all } t \in \mathbb{R},$$

where $\bar{X}_t$ and $\bar{Y}_t$ are the Ornstein–Uhlenbeck processes satisfying the linear SDEs (15.17).

The difference $Z_t^\nu - \bar{Z}_t$ is pathwise continuous and thus by Lemma 15.14 (and the definitions of the respective absorbing sets) equi-bounded with respect to $\nu > 0$ on the bounded time interval $[T_1, T_2]$. Moreover, $Z_t^\nu - \bar{Z}_t$ is pathwise differentiable and satisfies the random differential expression

$$\frac{\mathrm{d}}{\mathrm{d}t}(Z_t^\nu - \bar{Z}_t) = \frac{1}{2}f(\bar{X}_t^\nu) + \frac{1}{2}g(\bar{Y}_t^\nu)\frac{1}{2}\bar{X}_t + \frac{1}{2}\bar{Y}_t .$$

Since

$$\begin{aligned}&\left|\frac{\mathrm{d}}{\mathrm{d}t}(Z_t^\nu(\omega) - \bar{Z}_t(\omega))\right| \\ &\leq \frac{1}{2}\left|f(\bar{X}_t^\nu(\omega))\right| + \frac{1}{2}\left|g(\bar{Y}_t^\nu(\omega))\right| + \frac{1}{2}\left|\bar{X}_t(\omega)\right| + \frac{1}{2}\left|\bar{Y}_t(\omega)\right| \\ &\leq M_{T_1,T_2}(\omega) < \infty\end{aligned}$$

by Lemma 15.14, one can use the Theorem of Arzelà–Ascoli to conclude that for any sequence $\nu_n \to \infty$, there is a (possibly) random subsequence $\nu_{n_j}(\omega) \to \infty$ such that

$$Z_t^{\nu_{n_j}}(\omega) - \bar{Z}_t(\omega) \to Z_t^\infty(\omega) - \bar{Z}_t(\omega) \quad \text{as } j \to \infty,$$

and thus that $Z_t^{\nu_{n_j}}(\omega) \to Z_t^\infty(\omega)$ as $j \to \infty$.

Now,

$$\begin{aligned}Z_t^{\nu_{n_j}}(\omega) - \bar{Y}_t^{\nu_{n_j}}(\omega) &= \frac{1}{2}\left(\bar{X}_t^{\nu_{n_j}}(\omega) - \bar{Y}_t^{\nu_{n_j}}(\omega)\right) \to 0, \\ Z_t^{\nu_{n_j}}(\omega) - \bar{X}_t^{\nu_{n_j}}(\omega) &= \frac{1}{2}\left(\bar{Y}_t^{\nu_{n_j}}(\omega) - \bar{X}_t^{\nu_{n_j}}(\omega)\right) \to 0\end{aligned}$$

as $\nu_{n_j} \to \infty$, so

$$\begin{aligned}\bar{X}_t^{\nu_{n_j}}(\omega) &= 2Z_t^{\nu_{n_j}}(\omega) - \bar{Y}_t^{\nu_{n_j}}(\omega) \to Z_t^\infty(\omega), \\ \bar{Y}_t^{\nu_{n_j}}(\omega) &= 2Z_t^{\nu_{n_j}}(\omega) - \bar{X}_t^{\nu_{n_j}}(\omega) \to Z_t^\infty(\omega)\end{aligned}$$

as $\nu_{n_j} \to \infty$. Moreover, with omitting ω for brevity,

$$\begin{aligned}Z_t^\nu - \bar{Z}_t = Z_t^\nu - \bar{Z}_t &+ \frac{1}{2}\int_{T_1}^t f(\bar{X}_s^\nu)\,\mathrm{d}s + \frac{1}{2}\int_{T_1}^t g(\bar{Y}_s^\nu)\,\mathrm{d}s \\ &+ \frac{1}{2}\int_{T_1}^t \bar{X}_s\,\mathrm{d}s + \frac{1}{2}\int_{T_1}^t \bar{Y}_s\,\mathrm{d}s,\end{aligned}$$

which converges pathwise to

$$\begin{aligned} Z_t^\infty &= Z_{T_1}^\infty + \frac{1}{2}\int_{T_1}^t f(Z_s^\infty)\,\mathrm{d}s + \frac{1}{2}\int_{T_1}^t g(Z_s^\infty)\,\mathrm{d}s \\ &\quad + \bar{Z}_t - \bar{Z}_{T_1} + \frac{1}{2}\int_{T_1}^t \bar{X}_s\,\mathrm{d}s + \frac{1}{2}\int_{T_1}^t \bar{Y}_s\,\mathrm{d}s \\ &= Z_{T_1}^\infty + \frac{1}{2}\int_{T_1}^t f(Z_s^\infty)\,\mathrm{d}s + \frac{1}{2}\int_{T_1}^t g(Z_s^\infty)\,\mathrm{d}s \\ &\quad + \frac{1}{2}\alpha\int_{T_1}^t \mathrm{d}W_s^1 + \frac{1}{2}\beta\int_{T_1}^t \mathrm{d}W_s^2\,, \end{aligned}$$

on the interval $[T_1, T_2]$, so Z_t^∞ is a solution of the SDE (15.20) for all $t \in \mathbb{R}$. The drift of this SDE satisfies the strongly dissipative one-sided Lipschitz condition (15.6), so it has a random attractor consisting of a singleton set formed by a stationary stochastic process which thus must be equal to Z_t^∞.

Finally, note that, pathwise, all possible subsequences here have the same limit, so by Exercise 15.10, every full sequence $Z_t^{\nu_n}$ in fact converges to Z_t^∞ as $\nu_n \to \infty$. □

COROLLARY 15.17. *$(\bar{X}_t^\nu, \bar{Y}_t^\nu) \to (Z_t^\infty, Z_t^\infty)$ as $\nu \to \infty$ pathwise on any finite time interval $[T_1, T_2]$.*

3. Synchronization of systems with linear noise

A form of synchronization can also occur in systems with linear noise, i.e., with linear functions as the diffusion coefficients. It is more convenient now to consider scalar Stratonovich stochastic differential equations, namely,

$$\mathrm{d}X_t = f(X_t)\,\mathrm{d}t + \alpha X_t \circ \mathrm{d}W_t^1\,, \qquad \mathrm{d}Y_t = g(Y_t)\,\mathrm{d}t + \beta Y_t \circ \mathrm{d}W_t^2\,, \tag{15.21}$$

where W_t^1 and W_t^2 are independent two-sided scalar Wiener processes, α and β are nonzero constants, and the drift coefficient functions f and g are assumed to be continuously differentiable and to satisfy the one-sided dissipative Lipschitz conditions (15.6).

The stochastic differential equations (15.21) can be transformed to the scalar random ordinary differential equations (RODEs)

$$\begin{aligned} \dot{x} &= F(x, O_t^1(\omega)) := e^{-O_t^1(\omega)} f\big(e^{O_t^1(\omega)}x\big) + O_t^1(\omega)x\,, \\ \dot{y} &= G(y, O_t^2(\omega)) := e^{-O_t^2(\omega)} g\big(e^{O_t^2(\omega)}y\big) + O_t^2(\omega)y\,, \end{aligned}$$

using the transformation

$$x(t,\omega) = e^{-O_t^1(\omega)} X_t(\omega) \quad \text{and} \quad y(t,\omega) = e^{-O_t^2(\omega)} Y_t(\omega)\,,$$

where

$$O_t^1 = \alpha e^{-t}\int_{-\infty}^t e^\tau\,\mathrm{d}W_\tau^1 \quad \text{and} \quad O_t^2 = \beta e^{-t}\int_{-\infty}^t e^\tau\,\mathrm{d}W_\tau^2 \quad \text{for all } t \in \mathbb{R}\,,$$

are stationary Ornstein–Uhlenbeck processes.

It follows from the theory of random dynamical systems that each of these RODEs has a pathwise asymptotically stable random attractor consisting of single stationary stochastic process, and hence so does each of the original Stratonovich SDEs

(15.21). Important for the proof is the fact that the vector fields F and G satisfy the one-sided dissipative Lipschitz condition in their first variables with the original constant L uniformly in their second variables. In particular, the function

$$\widetilde{F}(x,z) := e^{-z} f\left(e^{z} x\right) \tag{15.22}$$

in the first RODE in (15.21) satisfies a one-sided Lipschitz condition in its first variable uniformly in the second with the same constant as the original drift coefficient f.

EXERCISE 15.18. Show that the function $\widetilde{F}$ defined in (15.22) satisfies a one-sided Lipschitz condition in its first variable uniformly in the second with the same constant as in the one-sided Lipschitz condition of the function f.

Thus, any of two solutions of the first RODE satisfy pathwise the differential inequality

$$\frac{\mathrm{d}}{\mathrm{d}t}|x_1(t) - x_2(t)|^2 \le \left(-2L + 2O_t^1\right)|x_1(t) - x_2(t)|^2, \tag{15.23}$$

and hence, by the Ornstein–Uhlenbeck limit in Lemma 14.6, one gets

$$|x_1(t) - x_2(t)|^2 \le e^{-2t\left(L - \frac{1}{t}\int_0^t O_\tau^1(\omega)\,\mathrm{d}\tau\right)}\,|x_1(0) - x_2(0)|^2 \to 0 \quad \text{as } t \to \infty,$$

which means all solutions converge pathwise to each other. A similar situation holds for the second RODE.

EXERCISE 15.19. Prove the inequality (15.23).

Consider now the vector field of coupled RODEs

$$\dot{x} = F\left(x, O_t^1(\omega)\right) + \nu(y - x), \quad \dot{y} = G\left(y, O_t^2(\omega)\right) + \nu(x - y), \tag{15.24}$$

which also satisfies a one-sided dissipative Lipschitz condition. Hence, it also has a pathwise asymptotically stable random attractor consisting of a single stationary stochastic process $(\bar{x}_\nu(\omega), \bar{y}_\nu(\omega))$ and that

$$\lim_{\nu\to\infty}\left(\bar{x}_\nu(\omega), \bar{y}_\nu(\omega)\right) = \left(\bar{z}(\omega), \bar{z}(\omega)\right),$$

where $\bar{z}(\omega)$ is the pathwise asymptotically stable solution of the averaged RODE

$$\dot{z} = \frac{1}{2}\left(F\left(z, O_t^1\right) + G\left(z, O_t^2\right)\right), \tag{15.25}$$

that is,

$$\dot{z} = \frac{1}{2}\Big(e^{-O_t^1(\omega)} f\left(e^{O_t^1(\omega)} z\right) + e^{-O_t^2(\omega)} g\left(e^{O_t^2(\omega)} z\right) + \left(O_t^1(\omega) + O_t^2(\omega)\right) z\Big),$$

which is equivalent to the Stratonovich SDE

$$\mathrm{d}Z_t = \frac{1}{2}\left(e^{-\eta_t} f(e^{\eta_t} Z_t) + e^{\eta_t} g(e^{-\eta_t} Z_t)\right)\mathrm{d}t + \frac{1}{2}\alpha Z_t \circ \mathrm{d}W_t^1 + \frac{1}{2}\beta Z_t \circ \mathrm{d}W_t^2,$$

where $\eta_t := \frac{1}{2}\left(O_t^1 - O_t^2\right)$.

In terms of the original system of SDEs (15.21), the coupled RODE (15.24) has the form

$$\begin{aligned}
\mathrm{d}X_t &= \left(f(X_t) + \nu\left(e^{2\eta_t} Y_t - X_t\right)\right)\mathrm{d}t + \alpha X_t \circ \mathrm{d}W_t^1, \\
\mathrm{d}Y_t &= \left(g(Y_t) + \nu\left(e^{-2\eta_t} X_t - Y_t\right)\right)\mathrm{d}t + \beta Y_t \circ \mathrm{d}W_t^2.
\end{aligned}$$

This system has a unique stationary stochastic solution $\left(\bar{X}^\nu \circ \theta_t, \bar{Y}^\nu \circ \theta_t\right)$, which is pathwise globally asymptotically stable with

$$\left(\bar{X}^\nu(\theta_t(\omega)), \bar{Y}^\nu(\theta_t(\omega))\right) \to \left(\bar{z}(\theta_t(\omega))e^{-O_t^1(\omega)}, \bar{z}(\theta_t(\omega))e^{-O_t^2(\omega)}\right) \quad \text{as } \nu \to \infty,$$

pathwise on finite time intervals $[T_1, T_2]$.

REMARK 15.20. When $\eta_t \equiv 0$, i.e., when $\alpha W_t^1 \equiv \beta W_t^2$, then synchronization occurs as in the additive noise case. However, for independent noise processes the synchronization is modulo a factor which takes into account the fact that the noise is different in each of the original systems.

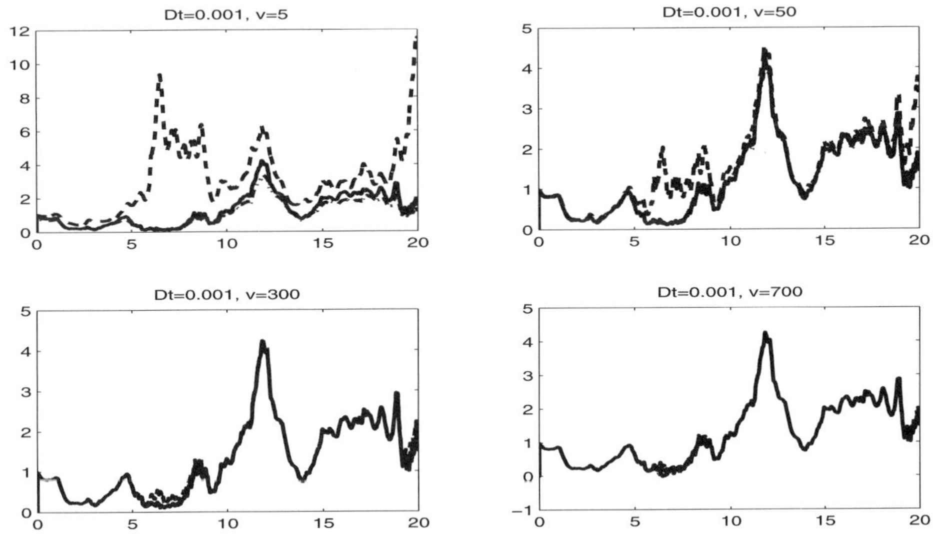

FIGURE 15.5. Trajectories of the coupled system (15.24) with $f(x) = 2 - x$ and $g(y) = -y - 2y^3$ for four values of ν, [**136**] copyright *Wiley-VCH Verlag GmbH & Co. KGaA*. Reproduced with permission.

Endnotes. Strogatz [**225**] is a readable popularization of synchronization, while the monograph by Pikovsky, Rosenblum & Kurths [**179**] is at a more serious scientific level. Dissipative synchronization for autonomous deterministic systems was investigated by Carvalho, Rodrigues & Dłotko [**36**] and Rodrigues [**203**], as well as Kloeden [**120**] for discrete time systems and van Geene [**231**] for lattice systems. Afraimovich & Rodrigues [**2**] also considered nonautonomous deterministic systems. The results in this chapter on nonautonomous deterministic systems using pullback attractors are taken from Kloeden [**118**], where more general non-singleton fiber sets are also considered. The effects of additive noise on synchronization were investigated in Caraballo & Kloeden [**27**] and of multiplicative noise in Caraballo, Kloeden & Neuenkirch [**30**]. The effects of discretization on the synchronization of noisy systems was considered in Kloeden, Neuenkirch & Pavani [**135**]. The figures in this chapter were taken from the overview article Kloeden & Pavani [**136**].

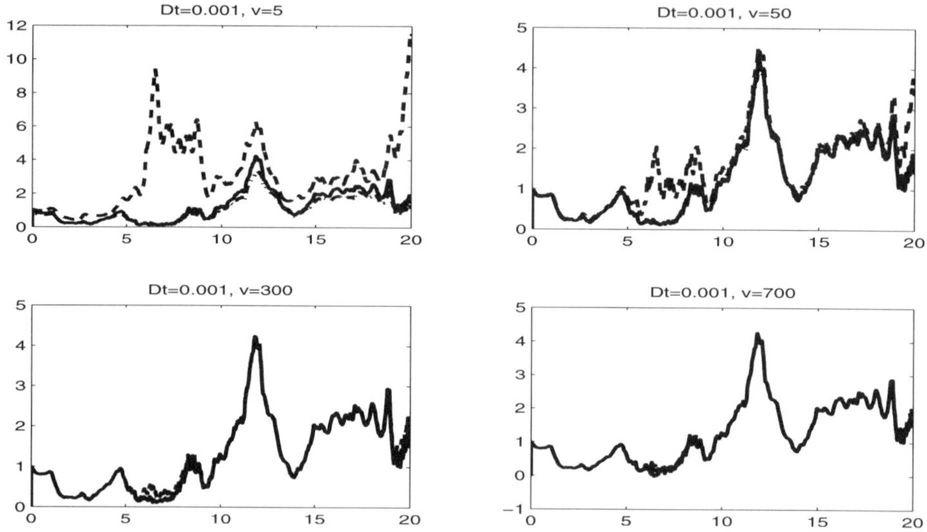

FIGURE 15.6. Trajectories of the coupled system compared with those of the averaged system (15.25) in Figure 15.5 for four values of ν, [**136**] copyright *Wiley-VCH Verlag GmbH & Co. KGaA.* Reproduced with permission.

Appendix

Notation. Denote by $\mathbb{Z}$ and $\mathbb{R}$ the sets of all integers and reals, respectively, and define $\overline{\mathbb{R}} := \mathbb{R} \cup \{-\infty, \infty\}$. Given an arbitrary set $A \subset \mathbb{R}$ and $\kappa \in \mathbb{R}$, set $A^+ := A \cap (0, \infty)$, $A^+_\kappa := A \cap [\kappa, \infty)$, $A^- := A \cap (-\infty, 0)$ and $A^-_\kappa := A \cap (-\infty, \kappa]$. Moreover, define $\mathbb{N} := \mathbb{Z}^+$, and $\mathbb{R}^{M \times N}$ is the set of all real $M \times N$ matrices. $\mathbb{1}$ stands for the unit matrix. The *range* of a matrix $B \in \mathbb{R}^{M \times N}$ is defined by $\mathfrak{R}(B) := \{y \in \mathbb{R}^M : y = Bx \text{ for some } x \in \mathbb{R}^N\}$, and the *null space* is given by $\mathfrak{N}(B) := \{x \in \mathbb{R}^N : Bx = 0\}$.

Given a differentiable function $g : X \subset \mathbb{R}^N \to \mathbb{R}^M$, write $Dg : X \to \mathbb{R}^{M \times N}$ for its derivative and $D_i g : X \to \mathbb{R}^M$ for its partial derivative with respect to the i-th variable, $i \in \{1, \dots, N\}$. Higher order derivatives $D^n g$ or $D_i^n g$ are defined inductively.

Given a metric space (X, d) and $\varepsilon > 0$, write $B_\varepsilon(x_0) = \{x \in X : d(x, x_0) < \varepsilon\}$ for the ε-neighborhood of a point $x_0 \in X$ and $B_\varepsilon(A) = \bigcup_{x \in A} B_\varepsilon(x)$ for the ε-neighborhood of a set $A \subset X$. Sometimes, the notations $B[x_0, \varepsilon]$ and $B[A, \varepsilon]$ will be used for the neighborhoods. The set of all inner points of a nonempty set $A \subset X$ is denoted by $\operatorname{int} A$; write $\overline{A}$ for the closure of A and ∂A for the boundary of A. Define the *distance* of a point $x \in X$ to a nonempty set $A \subset X$ by $\operatorname{dist}(x, A) := \inf_{y \in A} d(x, y)$ and the *Hausdorff semi-distance* of two nonempty sets $A, B \subset X$ by

$$\operatorname{dist}(A, B) := \sup_{x \in A} \operatorname{dist}(x, B).$$

In addition, if both A and B are empty, set $d(A, B) := 0$. The *Hausdorff distance* of A and B is defined by

$$h(A, B) := \max\{\operatorname{dist}(A, B), \operatorname{dist}(B, A)\}.$$

The *diameter* of a nonempty set $A \subset X$ is given by $\operatorname{diam}(A) := \sup\{d(x, y) : x, y \in A\}$.

Projective space. For $d \in \mathbb{N}$, the Euclidian space $\mathbb{R}^d$ is equipped with the Euclidian norm $\|\cdot\|$, which is induced by the scalar product $\langle \cdot, \cdot \rangle$, defined by $\langle x, y \rangle := \sum_{i=1}^d x_i, y_i$, where $x = (x_1, \dots, x_d), y = (y_1, \dots, y_d) \in \mathbb{R}^d$. To introduce the real projective space $\mathbb{P}^{d-1}$ of the $\mathbb{R}^d$, one says that two elements $x, y \in \mathbb{R}^d \setminus \{0\}$ are equivalent if there exists a $c \in \mathbb{R}$ such that $x = cy$. The equivalence class of $x \in \mathbb{R}^d$ is denoted by $\mathbb{P}x$, and the set of all equivalent classes is called the *projective space* $\mathbb{P}^{d-1}$. Equipped with the metric $d_{\mathbb{P}} : \mathbb{P}^{d-1} \times \mathbb{P}^{d-1} \to [0, \sqrt{2}]$, given by

$$d_{\mathbb{P}}(\mathbb{P}v, \mathbb{P}w) = \min\left\{ \left\| \frac{v}{\|v\|} - \frac{w}{\|w\|} \right\|, \left\| \frac{v}{\|v\|} + \frac{w}{\|w\|} \right\| \right\} \quad \text{for all } v, w \in \mathbb{R}^d,$$

the projective space is a compact metric space. For any $v \in \mathbb{P}^{d-1}$, define $\mathbb{P}^{-1}v := \{x \in \mathbb{R}^d : \mathbb{P}x = v\} \cup \{0\}$. The following lemma is a direct consequence of COLONIUS & KLIEMANN [**53**, Lemma B.1.17., p. 538].

LEMMA A.1. *For all $\varepsilon > 0$, there exists a $\delta \in (0,1)$ such that for all nonzero $v, w \in \mathbb{R}^N$ with*

$$\frac{\langle v, w\rangle^2}{\|v\|^2\|w\|^2} \geq 1 - \delta \,,$$

one has

$$d_{\mathbb{P}}(\mathbb{P}v, \mathbb{P}w) \leq \varepsilon \,.$$

Almost periodicity. Only the basic definitions will be reviewed here. For the theory of almost periodic functions, see CORDUNEANU [**58**], FINK [**77**] and KRASNOSEL'SKIĬ, BURD & KOLESOV [**150**].

A function $f : \mathbb{R} \to \mathbb{R}^N$ is called *(Bohr) almost periodic* if the set

$$\mathcal{T}(f, \varepsilon) := \big\{\tau \in \mathbb{R} : \big\|f(t) - f(\tau + t)\big\| < \varepsilon \text{ for all } t \in \mathbb{R}\big\}$$

is relatively dense in $\mathbb{R}$ for all $\varepsilon > 0$. Note that a set $L \subset \mathbb{R}$ is *relatively dense* if and only there exists a $T > 0$ such that $[t, t+T] \cap L \neq \emptyset$ for all $t \in \mathbb{R}$. An almost periodic function is necessarily uniformly continuous on $\mathbb{R}$ (see CORDUNEANU [**58**, Theorem 1.3] or FINK [**77**, Corollary 1.15]).

An almost periodic differential equation $\dot{x} = F(t, x)$, where $F : \mathbb{R} \times \mathbb{R}^N \to \mathbb{R}^N$ is continuous, is called *(Bohr) almost periodic* if for all compact sets $K \subset \mathbb{R}^N$ and $\varepsilon > 0$, the set

$$\mathcal{T}(F, \varepsilon, K) := \big\{\tau \in \mathbb{R} : \|F(t, x) - F(\tau + t, x)\| < \varepsilon \text{ for all } t \in \mathbb{R} \text{ and } x \in K\big\}$$

is relatively dense in $\mathbb{R}$. The right hand side F of an almost periodic differential equations is necessarily uniformly continuous on sets of the form $\mathbb{R} \times K$, where $K \subset \mathbb{R}^N$ is compact (see, e.g., SELL [**217**]).

Bibliography

[1] R.H. Abraham, J.E. Marsden, and T. Ratiu. *Manifolds, Tensor Analysis, and Applications.* Springer, New York, 1988.

[2] V. Afraimovich and H.M. Rodrigues. Uniform dissipativeness and synchronization on nonautonomous equations. In *International Conference on Differential Equations (Lisboa, 1995)*, pages 3–17. World Scientific, River Edge, New Jersey, 1998.

[3] R.P. Agarwal. *Difference Equations and Inequalities.* Marcel Dekker Inc., New York, 1992.

[4] E. Akin. *The General Topology of Dynamical Systems.* Number 1 in Graduate Studies in Mathematics. American Mathematical Society, Providence, Rhode Island, 1993.

[5] D. Angeli. A Lyapunov approach to the incremental stability properties. *IEEE Transactions on Automatic Control*, 47(3):410–421, 2002.

[6] L. Arnold. *Random Dynamical Systems.* Springer, Berlin, Heidelberg, New York, 1998.

[7] L. Arnold and M. Scheutzow. Perfect cocycles through stochastic differential equations. *Probability Theory and Related Fields*, 101(1):65–88, 1995.

[8] L. Arnold and B. Schmalfuß. Lyapunov's second method for random dynamical systems. *Journal of Differential Equations*, 177(1):235–265, 2001.

[9] J.P. Aubin and H. Frankowska. *Set-Valued Analysis*, volume 2 of *Systems and Control: Foundations and Applications.* Birkhäuser, Boston, 1990.

[10] B. Aulbach. A reduction principle for nonautonomous differential equations. *Archiv der Mathematik*, 39:217–232, 1982.

[11] B. Aulbach, M. Rasmussen, and S. Siegmund. Approximation of attractors of nonautonomous dynamical systems. *Discrete and Continuous Dynamical Systems B*, 5(2):215–238, 2005.

[12] B. Aulbach and S. Siegmund. The dichotomy spectrum for noninvertible systems of linear difference equations. *Journal of Difference Equations and Applications*, 7(6):895–913, 2001.

[13] B. Aulbach and T. Wanner. Integral manifolds for Carathéodory type differential equations in Banach spaces. In B. Aulbach and F. Colonius, editors, *Six Lectures on Dynamical Systems.* World Scientific, Singapore, 1996.

[14] B. Aulbach and T. Wanner. The Hartman-Grobman theorem for Carathéodory-type differential equations in Banach spaces. *Nonlinear Analysis. Theory, Methods & Applications*, 40(1–8):91–104, 2000.

[15] B. Aulbach and T. Wanner. Topological simplification of nonautonomous difference equations. *Journal of Difference Equations and Applications*, 12(3–4):283–296, 2006.

[16] A.V. Babin and M.I. Vishik. Maximal attractors of semigroups corresponding to evolution equation. *Mathematics of the USSR-Sbornik*, 54(2):387–408, 1986.

[17] L. Barreira and C. Valls. *Stability of Nonautonomous Differential Equations*, volume 1926 of *Springer Lecture Notes in Mathematics.* Springer, Berlin, 2008.

[18] P.W. Bates and C.K.R.T. Jones. Invariant manifolds for semilinear partial differential equations. In U. Kirchgraber and H.O. Walther, editors, *Dynamics Reported*, volume 2, pages 89–169. Wiley & Sons, B.G. Teubner, Stuttgart, 1989.

[19] P.W. Bates, K. Lu, and C. Zeng. Existence and persistence of invariant manifolds for semiflows in Banach space. *Memoirs of the American Mathematical Society*, 135(645), 1998.

[20] A. Ben-Artzi and I. Gohberg. Dichotomies of perturbed time varying systems and the power method. *Indiana University Mathematics Journal*, 42(3):699–720, 1993.

[21] A. Ben-Artzi, I. Gohberg, and M.A. Kaashoek. Invertibility and dichotomy of differential operators on a half-line. *Journal of Dynamics and Differential Equations*, 5(1):1–36, 1993.

[22] W.-J. Beyn and W. Kleß. Numerical Taylor expansions of invariant manifolds in large dynamical systems. *Numerische Mathematik*, 80(1):1–37, 1998.

[23] N.P. Bhatia and G.P. Szegö. *Stability Theory of Dynamical Systems*. Springer, Berlin, Heidelberg, New York, 1970.

[24] D. Boularas and D. Cheban. Asymptotic stability of switching systems. *EJDJ*, 2010(21):1–18, 2010.

[25] H. Brézis. *Analyse fonctionnelle*. Masson, Paris, 1983.

[26] M. Brin and G. Stuck. *Introduction to Dynamical Systems*. Cambridge University Press, Cambridge, 2002.

[27] T. Caraballo and P.E. Kloeden. The persistence of synchronization under environmental noise. *Proceedings of the Royal Society of London. Series A. Mathematical, Physical and Engineering Sciences*, 461(2059):2257–2267, 2005.

[28] T. Caraballo, P.E. Kloeden, and J.A. Langa. Atractores globales para sistemas diferenciales no autonomos. *Cubo Matematica Educacional*, 5(2):305–329, 2003.

[29] T. Caraballo, P.E. Kloeden, and P. Marín-Rubio. Global and pullback attractors of set-valued skew product flows. *Annali di Matematica Pura ed Applicata, Series IV*, 185:S23–S45, 2006.

[30] T. Caraballo, P.E. Kloeden, and A. Neuenkirch. Synchronization of systems with multiplicative noise. *Stochastics and Dynamics*, 8(1):139–154, 2008.

[31] T. Caraballo, P.E. Kloeden, and B. Schmalfuß. Exponentially stable stationary solutions for stochastic evolution equations and their perturbation. *Applied Mathematics and Optimization*, 50(3):183–207, 2004.

[32] T. Caraballo and J.A. Langa. On the upper semicontinuity of cocycle attractors for non-autonomous and random dynamical systems. *Dynamics of Continuous, Discrete and Impulsive Systems A*, 10(4):491–513, 2003.

[33] T. Caraballo, J.A. Langa, and J.C. Robinson. A stochastic pitchfork bifurcation in a reaction-diffusion equation. *Proceedings of the Royal Society of London. Series A. Mathematical, Physical and Engineering Sciences*, 457(2013):2041–2061, 2001.

[34] J. Carr. *Applications of Centre Manifold Theory*, volume 35 of *Applied Mathematical Sciences*. Springer, Berlin, Heidelberg, New York, 1981.

[35] A.N. Carvalho, J.A. Langa, and J.C. Robinson. Attractors of Infinite Dimensional Nonautonomous Dynamical Systems. to be published by Springer.

[36] A.N. Carvalho, H.M. Rodrigues, and T. Dłotko. Upper semicontinuity of attractors and synchronization. *Journal of Mathematical Analysis and Applications*, 220(1):13–41, 1998.

[37] L. Cesari. *Asymptotic Behavior and Stability Problems in Ordinary Differential Equations*. Springer, Berlin, 1963.

[38] D.N. Cheban. *Global Attractors of Non-Autonomous Dissipative Dynamical Systems*, volume 1 of *Interdisciplinary Mathematical Sciences*. World Scientific, Hackensack, New Jersey, 2004.

[39] D.N. Cheban and D.S. Fakhih. *The Global Attractors of Dynamical Systems without Uniqueness*. Sigma, Kishinev, 1994.

[40] D.N. Cheban, P.E. Kloeden, and B. Schmalfuß. Pullback attractors in dissipative nonautonomous differential equations under discretization. *Journal of Dynamics and Differential Equations*, 13(1):185–213, 2001.

[41] D.N. Cheban, P.E. Kloeden, and B. Schmalfuß. The relationship between pullback, forward and global attractors of nonautonomous dynamical systems. *Nonlinear Dynamics and Systems Theory*, 2(2):125–144, 2002.

[42] V.V. Chepyzhov and A.Y. Goritsky. Global integral manifolds with exponential tracking for nonautonomous equations. *Russian Journal of Mathematical Physics*, 5(1):9–28 (1998), 1997.

[43] V.V. Chepyzhov and M.I. Vishik. *Attractors for Equations of Mathematical Physics*, volume 49 of *Colloquium Publications*. American Mathematical Society, Providence, Rhode Island, 2002.

[44] C. Chicone and Y. Latushkin. Center manifolds for infinite-dimensional nonautonomous differential equations. *Journal of Differential Equations*, 141(2):356–399, 1997.

[45] C. Chicone and Y. Latushkin. *Evolution Semigroups in Dynamical Systems and Differential Equations*, volume 70 of *Mathematical Surveys and Monographs*. American Mathematical Society, Providence, Rhode Island, 1999.

[46] C. Chicone and R.C. Swanson. Spectral theory for linearizations of dynamical systems. *Journal of Differential Equations*, 40(2):155–167, 1981.

[47] Shui-Nee Chow, Cheng Zhi Li, and Duo Wang. *Normal Forms and Bifurcation of Planar Vector Fields.* Cambridge University Press, 1994.

[48] Shui-Nee Chow and Kening Lu. Invariant manifolds and foliations for quasiperiodic systems. *Journal of Differential Equations*, 117(1):1–27, 1995.

[49] I. Chueshov. *Monotone Random Systems – Theory and Applications*, volume 1779 of *Springer Lecture Notes in Mathematics.* Springer, Berlin, Heidelberg, New York, 2002.

[50] C.V. Coffman and J.J. Schäffer. Dichotomies for linear difference equations. *Mathematische Annalen*, 172:139–166, 1967.

[51] F. Colonius and W. Kliemann. The Lyapunov spectrum of families of time-varying matrices. *Transactions of the American Mathematical Society*, 348(11):4389–4408, 1996.

[52] F. Colonius and W. Kliemann. The Morse spectrum of linear flows on vector bundles. *Transactions of the American Mathematical Society*, 348(11):4355–4388, 1996.

[53] F. Colonius and W. Kliemann. *The Dynamics of Control.* Birkhäuser, Boston, 2000.

[54] F. Colonius, P.E. Kloeden, and M. Rasmussen. Morse spectrum for nonautonomous differential equations. *Stochastics and Dynamics*, 8(3):351–363, 2008.

[55] N.D. Cong. *Topological Dynamics of Random Dynamical Systems.* Oxford Mathematical Monographs. The Clarendon Press Oxford University Press, New York, 1997.

[56] C.C. Conley. *Isolated Invariant Sets and the Morse Index.* Number 38 in Regional Conference Series in Mathematics. American Mathematical Society, Providence, Rhode Island, 1978.

[57] W.A. Coppel. *Dichotomies in Stability Theory*, volume 629 of *Springer Lecture Notes in Mathematics.* Springer, Berlin, Heidelberg, New York, 1978.

[58] C. Corduneanu. *Almost Periodic Functions.* Number 22 in Interscience Tracts in Pure and Applied Mathematics. Interscience Publishers, New York, 1968.

[59] H. Crauel, A. Debussche, and F. Flandoli. Random attractors. *Journal of Dynamics and Differential Equations*, 9(2):307–341, 1997.

[60] H. Crauel, L. H. Duc, and S. Siegmund. Towards a Morse Theory for Random Dynamical Systems. *Stochastics and Dynamics*, 4(3):277–296, 2004.

[61] H. Crauel and F. Flandoli. Attractors for random dynamical systems. *Probability Theory and Related Fields*, 100(3):365–393, 1994.

[62] C.M. Dafermos. An invariance principle for compact processes. *Journal of Differential Equations*, 9:239–252, 1971.

[63] J.L. Daleckiĭ and M.G. Kreĭn. *Stability of Solutions of Differential Equations in Banach Spaces*, volume 43 of *Translations of Mathematical Monographs.* American Mathematical Society, Providence, Rhode Island, 1974.

[64] A. Debussche. On the finite dimensionality of random attractors. *Stochastic Analysis and Applications*, 15(4):473–491, 1997.

[65] K. Deimling. *Ordinary Differential Equations in Banach Spaces*, volume 596 of *Springer Lecture Notes in Mathematics.* Springer, Berlin, 1977.

[66] K. Deimling. *Nonlinear Functional Analysis.* Springer, Berlin, 1985.

[67] Li Desheng and P.E. Kloeden. Equi-attraction and the continuous dependence of attractors on parameters. *Glasgow Mathematical Journal*, 46(1):131–141, 2004.

[68] Li Desheng and P.E. Kloeden. Equi-attraction and the continuous dependence of pullback attractors on parameters. *Stochastics and Dynamics*, 4(3):373–384, 2004.

[69] Li Desheng and P.E. Kloeden. Equi-attraction and continuous dependence of strong attractors of set-valued dynamical systems on parameters. *Set-Valued Analysis*, 13(4):405–416, 2005.

[70] P. Diamond and P.E. Kloeden. Spatial discretization of mappings. *Computers & Mathematics with Applications*, 25(6):85–94, 1993.

[71] P. Diamond, P.E. Kloeden, and V.S. Kozyakin. Semi-hyperbolicity and bi-shadowing in nonautonomous difference equations with Lipschitz mappings. *Journal of Difference Equations and Applications*, 14(10–11):1165–1173, 2008.

[72] H. Doss. Liens entre équations différentielles stochastiques et ordinaires. *Annales de l'Institut Henri Poincaré. Section B.*, 13(2):99–125, 1977.

[73] T. Eirola and J. von Pfaler. Numerical Taylor expansions for invariant manifolds. *Numerische Mathematik*, 99(1):25–46, 2005.

[74] R. Fabbri, R. A. Johnson, and F. Mantellini. A nonautonomous saddle-node bifurcation pattern. *Stochastics and Dynamics*, 4(3):335–350, 2004.

[75] R. Fabbri and R.A. Johnson. On a saddle-node bifurcation in a problem of quasi-periodic harmonic forcing. In F. Dumortier, H. Broer, J. Mawhin, A. Vanderbauwhede, and V. Lunel, editors, *EQUADIFF 2003. Proceedings of the International Conference on Differential Equations, Hasselt, Belgium*, pages 839–847, 2005.

[76] K. Falconer. *Fractal Geometry.* John Wiley & Sons Ltd., Chichester, 1990.

[77] A.M. Fink. *Almost Periodic Differential Equations*, volume 377 of *Springer Lecture Notes in Mathematics.* Springer, Berlin, Heidelberg, 1974.

[78] F. Flandoli and B. Schmalfuß. Random attractors for the 3-D stochastic Navier-Stokes equation with mulitiplicative white noise. *Stochastics and Stochastics Reports*, 59(1–2):21–45, 1996.

[79] C. Foias, O. Manley, and R. Temam. Modelling of the interaction of small and large eddies in two-dimensional turbulent flows. *RAIRO Modélisation Mathématique et Analyse Numérique*, 22(1):93–118, 1988.

[80] C. Foias, G.R. Sell, and R. Temam. Inertial manifolds for nonlinear evolutionary equations. *Journal of Differential Equations*, 73(2):309–353, 1988.

[81] C. Foiaş and R. Temam. Some analytic and geometric properties of the solutions of the evolution Navier-Stokes equations. *Journal de Mathématiques Pures et Appliquées. Neuvième Série*, 58(3):339–368, 1979.

[82] P. Giesl. *Construction of Global Lyapunov Functions Using Radial Basis Functions*, volume 1904 of *Springer Lecture Notes in Mathematics.* Springer, Berlin, Heidelberg, New York, 2007.

[83] P. Glendinning. Global attractors of pinched skew products. *Dynamical Systems*, 17(3):287–294, 2002.

[84] L. Grüne and P.E. Kloeden. Discretization, inflation and perturbation of attractors. In B. Fiedler, editor, *Ergodic Theory, Analysis, and Efficient Simulation of Dynamical Systems*, pages 399–416. Springer, Berlin, 2001.

[85] L. Grüne, P.E. Kloeden, S. Siegmund, and F.R. Wirth. Lyapunov's second method for nonautonomous differential equations. *Discrete and Continuous Dynamical Systems*, 18(2–3):375–403, 2007.

[86] J. Guckenheimer and P. Holmes. *Nonlinear Oscillation, Dynamical Systems, and Bifurcations of Vector Fields*, volume 42 of *Applied Mathematical Sciences.* Springer, New York, 1983.

[87] J. Guckenheimer, G. Oster, and A. Ipaktchi. The dynamics of density dependent population models. *Journal of Mathematical Biology*, 4(2):101–147, 1977.

[88] J. Hadamard. Sur l'itération et les solutions asymptotiques des équations différentielles. *Bulletin de la Société Mathématique de France*, 29:224–228, 1901.

[89] W. Hahn. *Stability of Motion.* Springer, Berlin, 1967.

[90] J.K. Hale. *Asymptotic Behavior of Dissipative Systems*, volume 25 of *Mathematical Surveys and Monographs.* American Mathematical Society, Providence, Rhode Island, 1988.

[91] Y. Hamaya. Bifurcation of almost periodic solutions in difference equations. *Journal of Difference Equations and Applications*, 10(3):257–297, 2004.

[92] J. Härterich, B. Sandstede, and A. Scheel. Exponential dichotomies for linear non-autonomous functional differential equations of mixed type. *Indiana University Mathematics Journal*, 51(5):1081–1109, 2002.

[93] D. Henry. *Geometric Theory of Semilinear Parabolic Equations*, volume 840 of *Springer Lecture Notes in Mathematics.* Springer, Berlin, Heidelberg, New York, 1981.

[94] M.W. Hirsch, C.C. Pugh, and M. Shub. *Invariant Manifolds*, volume 583 of *Springer Lecture Notes in Mathematics.* Springer, Berlin, Heidelberg, New York, 1977.

[95] M.W. Hirsch, S. Smale, and R.L. Devaney. *Differential Equations, Dynamical Systems, and an Introduction to Chaos*, volume 60 of *Pure and Applied Mathematics.* Elsevier/Academic Press, Amsterdam, second edition, 2004.

[96] P. Imkeller and C. Lederer. On the cohomology of flows of stochastic and random differential equations. *Probability Theory and Related Fields*, 120(2):209–235, 2001.

[97] P. Imkeller and C. Lederer. The cohomology of stochastic and random differential equations, and local linearization of stochastic flows. *Stochastics and Dynamics*, 2(2):131–159, 2002.

[98] P. Imkeller and B. Schmalfuß. The conjugacy of stochastic and random differential equations and the existence of global attractors. *Journal of Dynamics and Differential Equations*, 13(2):215–249, 2001.

[99] R.A. Johnson and P.E. Kloeden. Nonautonomous attractors of skew-product flows with digitized driving systems. *Electronic Journal of Differential Equations*, (58):1–16, 2001.

[100] R.A. Johnson, P.E. Kloeden, and R. Pavani. Two-step transition in nonautonomous bifurcations: An explanation. *Stochastics and Dynamics*, 2(1):67–92, 2002.

[101] R.A. Johnson and F. Mantellini. A nonautonomous transcritical bifurcation problem with an application to quasi-periodic bubbles. *Discrete and Continuous Dynamical Systems*, 9(1):209–224, 2003.

[102] R.A. Johnson, K.J. Palmer, and G.R. Sell. Ergodic properties of linear dynamical systems. *SIAM Journal on Mathematical Analysis*, 18(1):1–33, 1987.

[103] D.A. Jones and A.M. Stuart. Attractive invariant manifolds under approximation. Inertial manifolds. *Journal of Differential Equations*, 123(2):588–637, 1995.

[104] J. Kato, A.A. Martynyuk, and A.A. Shestakov. *Stability of Motion of Nonautonomous Systems (Method of Limiting Equations)*, volume 3 of *Stability and Control: Theory, Methods and Applications*. Gordon & Breach Publishers, Philadelphia, 1996.

[105] A. Katok and B. Hasselblatt. *Introduction to the Modern Theory of Dynamical Systems*, volume 54 of *Encyclopedia of Mathematics and its Applications*. Cambridge University Press, Cambridge, 1995.

[106] C. Kawan. Invariance Entropy for Control Systems. PhD Thesis, University of Augsburg, 2009.

[107] A. Kelley. Stability of the center-stable manifold. *Journal of Mathematical Analysis and Applications*, 18:336–344, 1967.

[108] A. Kelley. The stable, center-stable, center, center-unstable, unstable manifolds. *Journal of Differential Equations*, 3:546–570, 1967.

[109] U. Kirchgraber and K.J. Palmer. *Geometry in the Neigborhood of Invariant Manifolds of Maps and Flows and Linearization*, volume 233 of *Pitman Research Notes in Mathematical Series*. Longman, Burnt Mill, 1990.

[110] P.E. Kloeden. General control systems without backwards extension. In *Differential Games and Control Theory (Proc. NSF-CBMS Regional Res. Conf., Univ. Rhode Island, Kingston, 1973)*, volume 10 of *Lecture Notes in Pure and Applied Mathematics*, pages 49–58. Dekker, New York, 1974.

[111] P.E. Kloeden. General control systems. In *Mathematical Control Theory (Proc. Conf., Australian Nat. Univ., Canberra, 1977)*, volume 680 of *Lecture Notes in Mathematics*, pages 119–137. Springer, Berlin, 1978.

[112] P.E. Kloeden. On Sharkovsky's cycle coexistence ordering. *Bulletin of the Australian Mathematical Society*, 20(2):171–177, 1979.

[113] P.E. Kloeden. Chaotic difference equations in $\mathbb{R}^n$. *Australian Mathematical Society. Journal. Series A*, 31(2):217–225, 1981.

[114] P.E. Kloeden. Lyapunov Functions for Cocycle Attractors in Nonautonomous Difference Equations. *Izvetsiya Akad Nauk Rep Moldovia Mathematika*, 26:32–42, 1998.

[115] P.E. Kloeden. A Lyapunov function for pullback attractors of nonautonomous differential equations. In *Proceedings of the Conference on Nonlinear Differential Equations (Coral Gables, 1999)*, Electronic Journal of Differential Equations Conference 05, pages 91–102, San Marcos, Texas, 2000. Southwest Texas State University.

[116] P.E. Kloeden. Pullback attractors in nonautonomous difference equations. *Journal of Difference Equations and Applications*, 6(1):91–102, 2000.

[117] P.E. Kloeden. Pullback attractors of nonautonomous semidynamical systems. *Stochastics and Dynamics*, 3(1):101–112, 2003.

[118] P.E. Kloeden. Synchronization of nonautonomous dynamical systems. *Electronic Journal of Differential Equations*, (39):10 pp., 2003.

[119] P.E. Kloeden. Pitchfork and transcritical bifurcations in systems with homogeneous nonlinearities and an almost periodic time coefficient. *Communications on Pure and Applied Analysis*, 3(2):161–173, 2004.

[120] P.E. Kloeden. Synchronization of discrete time dynamical systems. *Journal of Difference Equations and Applications*, 10(13-15):1133–1138, 2004.

[121] P.E. Kloeden. Nonautonomous attractors of switching systems. *Dynamical Systems*, 21(2):209–230, 2006.

[122] P.E. Kloeden. Upper semicontinuous dependence of pullback attractors on time scales. *Journal of Difference Equations and Applications*, 12(3–4):357–368, 2006.

[123] P.E. Kloeden, H. Keller, and B. Schmalfuß. Towards a theory of random numerical dynamics. In H. Crauel and M. Gundlach, editors, *Stochastic Dynamics*. Springer, Berlin, Heidelberg, New York, 1999.

[124] P.E. Kloeden and V.S. Kozyakin. The inflation of attractors and their discretization: the autonomous case. *Nonlinear Analysis. Theory, Methods & Applications*, 40(1–8):333–343, 2000.

[125] P.E. Kloeden and V.S. Kozyakin. The inflation of nonautonomous systems and their pullback attractors. *Transactions of the Russian Academy of Natural Sciences, Series MMMIU*, 4:144–169, 2000. 1–2.

[126] P.E. Kloeden and V.S. Kozyakin. The perturbation of attractors of skew-product flows with a shadowing driving system. *Discrete and Continuous Dynamical Systems*, 7(4):883–893, 2001.

[127] P.E. Kloeden and V.S. Kozyakin. Single parameter dissipativity and attractors in discrete time asynchronous systems. *Journal of Difference Equations and Applications*, 7(6):873–894, 2001.

[128] P.E. Kloeden and V.S. Kozyakin. Uniform nonautonomous attractors under discretization. *Discrete and Continuous Dynamical Systems*, 10(1-2):423–433, 2004.

[129] P.E. Kloeden and J.A. Langa. Flattening, squeezing and the existence of random attractors. *Proceedings of the Royal Society of London. Series A. Mathematical, Physical and Engineering Sciences*, 463(2077):163–181, 2007.

[130] P.E. Kloeden and Zhong Li. Li–Yorke chaos in higher dimensions: a review. *Journal of Difference Equations and Applications*, 12(3–4):247–269, 2006.

[131] P.E. Kloeden and J. Lorenz. Stable attracting sets in dynamical systems and in their one-step discretizations. *SIAM Journal on Numerical Analysis*, 23(5):986–995, 1986.

[132] P.E. Kloeden and P. Marín-Rubio. Negatively invariant sets and entire solutions. to appear in: Journal of Dynamics and Differential Equations.

[133] P.E. Kloeden and P. Marín-Rubio. Negatively invariant sets and entire trajectories of set-valued dynamical systems. to appear in: Set-Valued and Variational Analysis.

[134] P.E. Kloeden and P. Marín-Rubio. Weak pullback attractors of non-autonomous difference inclusions. *Journal of Difference Equations and Applications*, 9(5):489–502, 2003.

[135] P.E. Kloeden, A. Neuenkirch, and R. Pavani. Synchronization of noisy dissipative systems under discretization. *Journal of Difference Equations and Applications*, 15(8–9):785–801, 2009.

[136] P.E. Kloeden and R. Pavani. Dissipative synchronization of nonautonomous and random systems. *GAMM-Mitteilungen*, 32(1):80–92, 2009.

[137] P.E. Kloeden and S.I. Piskarev. Discrete convergence and the equivalence of equi-attraction and the continuous convergence of attractors. *International Journal of Dynamical Systems and Differential Equations*, 1(1):38–43, 2007.

[138] P.E. Kloeden and E. Platen. *Numerical Solution of Stochastic Differential Equations*, volume 23 of *Applications of Mathematics*. Springer, Berlin, 1992.

[139] P.E. Kloeden and H.M. Rodrigues. Dynamics of a class of ODEs more general than almost periodic. *Nonlinear Analysis. Theory, Methods & Applications*, 74:2695–2719, 2011.

[140] P.E. Kloeden and B. Schmalfuß. Lyapunov functions and attractors under variable time-step discretization. *Discrete and Continuous Dynamical Systems*, 2(2):163–172, 1996.

[141] P.E. Kloeden and B. Schmalfuß. Cocycle attractors of variable time-step discretizations of Lorenzian systems. *Journal of Difference Equations and Applications*, 3(2):125–145, 1997.

[142] P.E. Kloeden and B. Schmalfuß. Nonautonomous systems, cocycle attractors and variable time-step discretization. *Numerical Algorithms*, 14(1–3):141–152, 1997.

[143] P.E. Kloeden and B. Schmalfuß. Asymptotic behaviour of nonautonomous difference inclusions. *Systems & Control Letters*, 33(4):275–280, 1998.

[144] P.E. Kloeden and S. Siegmund. Bifurcations and continuous transitions of attractors in autonomous and nonautonomous systems. *International Journal of Bifurcation and Chaos*, 15(3):743–762, 2005.

[145] P.E. Kloeden and D.J. Stonier. Cocycle attractors in nonautonomously perturbed differential equations. *Dynamics of Continuous, Discrete and Impulsive Systems*, 4(2):211–226, 1998.

[146] P.E. Kloeden and J. Valero. Attractors of weakly asymptotically compact set-valued dynamical systems. *Set-Valued Analysis*, 13(4):381–404, 2005.

[147] H.W. Knobloch and F. Kappel. *Gewöhnliche Differentialgleichungen.* B.G. Teubner, Stuttgart, 1974.

[148] N. Koksch and S. Siegmund. Pullback attracting inertial manifolds for nonautonomous dynamical systems. *Journal of Dynamics and Differential Equations*, 14(4):889–941, 2002.

[149] A.M. Krasnosel'skiĭ and J. Mawhin. Remark on some type of bifurcation at infinity. *Académie Royale de Belgique. Bulletin de la Classe des Sciences. 6e Série*, 3(12):293–297, 1992.

[150] M.A. Krasnosel'skiĭ, V.Sh. Burd, and Yu.S. Kolesov. *Nonlinear Almost Periodic Oscillations.* Halsted Press, New York-Toronto, Ontario, 1973.

[151] O. Ladyzhenskaya. *Attractors for Semigroups and Evolution Equations.* Cambridge University Press, Cambridge, 1991.

[152] J.S.W. Lamb, M. Rasmussen, and C.S. Rodrigues. Topological bifurcations of minimal invariant sets for set-valued dynamical systems. submitted.

[153] J.A. Langa. Finite-dimensional limiting dynamics of random dynamical systems. *Dynamical Systems*, 18(1):57–68, 2003.

[154] J.A. Langa, J.C. Robinson, and A. Suárez. Stability, instability and bifurcation phenomena in non-autonomous differential equations. *Nonlinearity*, 15(3):887–903, 2002.

[155] J.A. Langa, J.C. Robinson, and A. Suárez. Bifurcations in non-autonomous scalar equations. *Journal of Differential Equations*, 221(1):1–35, 2006.

[156] Y. Latushkin and A. Pogan. The dichotomy theorem for evolution bi-families. *Journal of Differential Equations*, 245(8):2267–2306, 2008.

[157] D. Liberzon. *Switching in Systems and Control.* Birkhäuser, Boston, 2003.

[158] Zhenxin Liu. The random case of Conley's theorem. III. Random semiflow case and Morse decomposition. *Nonlinearity*, 20(12):2773–2791, 2007.

[159] E.N. Lorenz. Deterministic nonperiodic flow. *Journal of the Atmospheric Sciences*, 20:130–141, 1963.

[160] A.M. Lyapunov. *The General Problem of the Stability of Motion.* Mathematical Society of Kharkov, Kharkov, 1892. (in Russian).

[161] A.M. Lyapunov. Problème générale de la stabilité de mouvement. *Annales de la Faculte des Sciences de Toulouse*, 9:203–474, 1907. (in French).

[162] A.M. Lyapunov. *Stability of Motion*, volume 30 of *Mathematics in Science and Engineering.* Academic Press, New York, London, 1966. Translated from Russian by F. Abramovici and M. Shimshoni.

[163] Q. Ma, S. Wang, and C. Zhong. Necessary and sufficient conditions for the existence of global attractors for semigroups and applications. *Indiana University Mathematics Journal*, 51(6):1541–1559, 2002.

[164] J. Mallet-Paret. Morse decompositions for delay-differential equations. *Journal of Differential Equations*, 72(2):270–315, 1988.

[165] J.L. Massera and J.J. Schäffer. *Linear Differential Equations and Function Spaces.* Academic Press, New York, London, 1966.

[166] J.N. Mather. Characterization of Anosov diffeomorphisms. *Nederl. Akad. Wetensch. Proc. Ser. A 71 = Indag. Math.*, 30:479–483, 1968.

[167] V.S. Melnik and J. Valero. On attractors of multivalued semi-flows and differential inclusions. *Set-Valued Analysis*, 6(1):83–111, 1998.

[168] A.P. Molchanov and Ye.S. Pyatnitskiy. Criteria of asymptotic stability of differential and difference inclusions encountered in control theory. *Systems & Control Letters*, 13(1):59–64, 1989.

[169] C. Núñez and R. Obaya. A non-autonomous bifurcation theory for deterministic scalar differential equations. *Discrete and Continuous Dynamical Systems B*, 9(3–4):701–730, 2008.

[170] G. Ochs. Weak random attractors. Report Nr. 449, Institut für Dynamische Systeme, Universität Bremen, 1999.

[171] K. Palmer. Exponential dichotomies and transversal homoclinic points. *Journal of Differential Equations*, 55:225–256, 1984.

[172] K. Palmer. On the stability of the center manifold. *Zeitschrift für Angewandte Mathematik und Physik*, 38(2):273–278, 1987.

[173] K. Palmer. *Shadowing in Dynamical Systems*, volume 501 of *Mathematics and its Applications.* Kluwer Academic Publishers, Dordrecht, 2000. Theory and applications.

[174] K. Palmer and S. Siegmund. Generalized attractor-repeller pairs, diagonalizability and integral separation. *Advanced Nonlinear Studies*, 4:189–207, 2004.

[175] Kenneth J. Palmer. Exponential dichotomies and Fredholm operators. *Proceedings of the American Mathematical Society*, 104(1):149–156, 1988.

[176] K.J. Palmer. A perturbation theorem for exponential dichotomies. *Proceedings of the Royal Society of Edinburgh. Section A*, 106(1–2):25–37, 1987.

[177] O. Perron. Über Stabilität und asymptotisches Verhalten der Integrale von Differentialgleichungssystemen. *Mathematische Zeitschrift*, 29:129–160, 1928.

[178] O. Perron. Die Stabilitätsfrage bei Differentialgleichungen. *Mathematische Zeitschrift*, 32:703–728, 1930.

[179] A. Pikovsky, M. Rosenblum, and J. Kurths. *Synchronization*, volume 12 of *Cambridge Nonlinear Science Series*. Cambridge University Press, Cambridge, 2001.

[180] V.A. Pliss. Principle reduction in the theory of the stability of motion. *Izv. Akad. Nauk SSSR, Mat. Ser.*, 28:1297–1323, 1964. (in Russian).

[181] V.A. Pliss and G.R. Sell. Robustness of exponential dichotomies in infinite-dimensional dynamical systems. *Journal of Dynamics and Differential Equations*, 11(3):471–513, 1999.

[182] H. Poincaré. Mémoire sur les courbes définie par une équation différentielle IV. *Journal de Mathématiques pures et appliquées*, 2:151–217, 1886. (in French).

[183] C. Pötzsche. Nonautonomous continuation and bifurcation of bounded solutions II: A shovel bifurcation pattern.

[184] C. Pötzsche. Exponential dichotomies of linear dynamic equations on measure chains under slowly varying coefficients. *Journal of Mathematical Analysis and Applications*, 289(1):317–335, 2004.

[185] C. Pötzsche. Attractive invariant fiber bundles. *Applicable Analysis*, 86(6):687–722, 2007.

[186] C. Pötzsche. Discrete inertial manifolds. *Mathematische Nachrichten*, 281(6):847–878, 2008.

[187] C. Pötzsche. Topological decoupling, linearization and perturbation on inhomogeneous time scales. *Journal of Differential Equations*, 245(5):1210–1242, 2008.

[188] C. Pötzsche. *Geometric Theory of Nonautonomous Discrete Dynamical Systems and Discretizations*, volume 1907 of *Springer Lecture Notes in Mathematics*. Springer, Berlin, Heidelberg, New York, 2010.

[189] C. Pötzsche. Nonautonomous continuation and bifurcation of bounded solutions I: A Lyapunov–Schmidt approach. *Discrete and Continuous Dynamical Systems B*, 14(2):739–776, 2010.

[190] C. Pötzsche. Nonautonomous continuation of bounded solutions. *Communications on Pure and Applied Analysis*, 10(3):937–961, 2011.

[191] C. Pötzsche and M. Rasmussen. Taylor approximation of invariant fiber bundles of nonautonomous difference equations. *Nonlinear Analysis. Theory, Methods & Applications*, 60(7):1303–1330, 2005.

[192] C. Pötzsche and M. Rasmussen. Taylor approximation of integral manifolds. *Journal of Dynamics and Differential Equations*, 18(2):427–460, 2006.

[193] M. Rasmussen. Towards a bifurcation theory for nonautonomous difference equations. *Journal of Difference Equations and Applications*, 12(3–4):297–312, 2006.

[194] M. Rasmussen. *Attractivity and Bifurcation for Nonautonomous Dynamical Systems*, volume 1907 of *Springer Lecture Notes in Mathematics*. Springer, Berlin, Heidelberg, New York, 2007.

[195] M. Rasmussen. Morse decompositions of nonautonomous dynamical systems. *Transactions of the American Mathematical Society*, 359(10):5091–5115, 2007.

[196] M. Rasmussen. Nonautonomous bifurcation patterns for one-dimensional differential equations. *Journal of Differential Equations*, 234(1):267–288, 2007.

[197] M. Rasmussen. All-time Morse decompositions of linear nonautonomous dynamical systems. *Proceedings of the American Mathematical Society*, 136(3):1045–1055, 2008.

[198] M. Rasmussen. Bifurcations of asymptotically autonomous differential equations. *Set-Valued Analysis*, 16(7–8):821–849, 2008.

[199] M. Rasmussen. Dichotomy spectra and Morse decompositions of linear nonautonomous differential equations. *Journal of Differential Equations*, 246(6):2242–2263, 2009.

[200] C. Robinson. *Dynamical Systems. Stability, Symbolic Dynamics and Chaos*. CRC Press, Boca Raton, 2 edition, 1999.

[201] J.C. Robinson. *Infinite-Dimensional Dynamical Systems.* Cambridge Texts in Applied Mathematics. Cambridge University Press, Cambridge, 2001.

[202] J.C. Robinson. Stability of random attractors under perturbation and approximation. *Journal of Differential Equations*, 186(2):652–669, 2002.

[203] H.M. Rodrigues. Abstract methods for synchronization and applications. *Applicable Analysis*, 62(3-4):263–296, 1996.

[204] E.O. Roxin. On generalized dynamical systems defined by contingent equations. *Journal of Differential Equations*, 1:188–205, 1965.

[205] E.O. Roxin. Stability in general control systems. *Journal of Differential Equations*, 1:115–150, 1965.

[206] K.P. Rybakowski. *The Homotopy Index and Partial Differential Equations.* Springer, Berlin, Heidelberg, New York, 1987.

[207] K.P. Rybakowski. Formulas for higher-order finite expansions of composite maps. In *The mathematical heritage of C.F. Gauss*, pages 652–669. World Scientific, River Edge, New Jersey, 1991.

[208] R.J. Sacker and G.R. Sell. Existence of dichotomies and invariant splittings for linear differential systems I. *Journal of Differential Equations*, 15:429–458, 1974.

[209] R.J. Sacker and G.R. Sell. A spectral theory for linear differential systems. *Journal of Differential Equations*, 27:320–358, 1978.

[210] K. Sakamoto. Estimates on the strength of exponential dichotomies and application to integral manifolds. *Journal of Differential Equations*, 107(2):259–279, 1994.

[211] D. Salamon and E. Zehnder. Flows on vector bundles and hyperbolic sets. *Transactions of the American Mathematical Society*, 306(2):623–649, 1988.

[212] B. Sandstede. Stability of travelling waves. In *Handbook of Dynamical Systems, Vol. 2*, pages 983–1055. North-Holland, Amsterdam.

[213] B. Schmalfuß. Backward cocycles and attractors of stochastic differential equations. In V. Reitmann, T. Riedrich, and N. Koksch, editors, *International Seminar on Applied Mathematics — Nonlinear Dynamics: Attractor Approximation and Global Behaviour*, pages 185–192. Technische Universität Dresden, 1992.

[214] B. Schmalfuß. The random attractor of the stochastic Lorenz system. *Zeitschrift für Angewandte Mathematik und Physik*, 48(6):951–975, 1997.

[215] J.F. Selgrade. Isolated invariant sets for flows on vector bundles. *Transactions of the American Mathematical Society*, 203:359–390, 1975.

[216] G.R. Sell. Stability theory and Lyapunov's second method. *Archive for Rational Mechanics and Analysis*, 14:108–126, 1963.

[217] G.R. Sell. Nonautonomous differential equations and dynamical systems. *Transactions of the American Mathematical Society*, 127:241–283, 1967.

[218] G.R. Sell. *Topological Dynamics and Ordinary Differential Equations.* Van Nostrand Reinhold Mathematical Studies, London, 1971.

[219] G.R. Sell. The structure of a flow in the vicinity of an almost periodic motion. *Journal of Differential Equations*, 27:359–393, 1978.

[220] G.R. Sell and Y. You. *Dynamics of Evolutionary Equations*, volume 143 of *Applied Mathematical Sciences.* Springer, New York, 2002.

[221] W. Shen and Y. Yi. *Almost Automorphic and Almost Periodic Dynamics in Skew-Product Semiflows.* Number 647 in Memoirs of the AMS. American Mathematical Society, Providence, Rhode Island, 1998.

[222] M. Shub. *Global Stability of Dynamical Systems.* Springer, Berlin, Heidelberg, New York, 1987.

[223] K.S. Sibirsky. *Introduction to Topological Dynamics.* Noordhoff International Publishing, Leiden, 1975.

[224] P. Storck. Diskretisierung von Bifurkationen partieller logistischer Differentialgleichungen. Diploma Thesis, J. W. Goethe Universität Frankfurt am Main, 2009. (in German).

[225] S. Strogatz. *Sync: How order emerges from chaos in the universe, nature, and daily life.* Hyperion Books, New York, 2003.

[226] A.M. Stuart and A.R. Humphries. *Dynamical Systems and Numerical Analysis*, volume 2 of *Cambridge Monographs on Applied and Computational Mathematics.* Cambridge University Press, Cambridge, 1996.

[227] H. Sussmann. An interpretation of stochastic differential equations as ordinary differential equations which depend on the sample point. *Bulletin of the American Mathematical Society*, 83(2):296–298, 1977.

[228] H. Sussmann. On the gap between deterministic and stochastic ordinary differential equations. *The Annals of Probability*, 6(1):19–41, 1978.

[229] G.P. Szegő and G. Treccani. *Semigruppi di transformazioni multivoche*, volume 101 of *Springer Lecture Notes in Mathematics*. Springer, Berlin, 1969.

[230] R. Temam. *Infinite-Dimensional Dynamical Systems in Mechanics and Physics*, volume 68 of *Applied Mathematical Sciences*. Springer, New York, 1988.

[231] S. van Geene. Synchronisation von Reaktions-Diffusions-Gleichungen. Diploma Thesis, J. W. Goethe Universität Frankfurt am Main, 2009. (in German).

[232] A. Vanderbauwhede. Center manifolds, normal forms and elementary bifurcations. In U. Kirchgraber and H.O. Walther, editors, *Dynamics Reported*, volume 2, pages 89–169. Wiley & Sons, B.G. Teubner, Stuttgart, 1989.

[233] M.I. Vishik. *Asymptotic Behaviour of Solutions of Evolutionary Equations*. Cambridge University Press, Cambridge, 1992.

[234] Y. Wang, C.K. Zhong, and S. Zhou. Pullback attractors of nonautonomous dynamical systems. *Discrete and Continuous Dynamical Systems*, 16(3):587–614, 2006.

[235] S. Wiggins. *Introduction to Applied Nonlinear Dynamical Systems and Chaos*, volume 2 of *Texts in Applied Mathematics*. Springer, New York, 1990.

[236] S. Wiggins. *Normally Hyperbolic Invariant Manifolds in Dynamical Systems*, volume 105 of *Applied Mathematical Sciences*. Springer, New York, 1994.

[237] Wang Yejuan, Li Desheng, and P.E. Kloeden. On the asymptotical behavior of nonautonomous dynamical systems. *Nonlinear Analysis. Theory, Methods & Applications*, 59(1–2):35–53, 2004.

[238] Y. Yi. A generalized integral manifold theorem. *Journal of Differential Equations*, 102(1):153–187, 1993.

[239] T. Yoshizawa. *Stability Theory by Liapunov's Second Method*. Number 9 in Publications of the Mathematical Society of Japan. The Mathematical Society of Japan, Tokyo, 1966.

Index